WOMEN MINERS IN DEVELOPING COUNTRIES

Voices in Development Management

Series Editor:
Margaret Grieco
Napier University

The Voices in Development Management series provides a forum in which grass roots organisations and development practitioners can voice their views and present their perspectives alongside the conventional development experts. One of the goals of the series is to provide a forum for explicit debates between various voices in development. Many important and hitherto neglected development issues such as gender and transport or the micro credit needs of low income communities are now given the public and professional attention they richly deserve.

Also in the series

Accountability of the International Monetary Fund
Edited by Barry Carin and Angela Wood
ISBN 0 7546 4523 1

Social Exclusion and the Remaking of Social Networks
Robert Strathdee
ISBN 0 7546 3815 4

Design and Determination
The Role of Information Technology in Redressing Regional Inequities in the Development Process
Stephen E. Little
ISBN 0 7546 1099 3

The Digital Challenge: Information Technology in the Development Context
Edited by S. Krishna and Shirin Madon
ISBN 0 7546 3445 0

Women Miners in Developing Countries

Pit Women and Others

Edited by

KUNTALA LAHIRI-DUTT
The Australian National University, Australia

and

MARTHA MACINTYRE
University of Melbourne, Australia

Routledge
Taylor & Francis Group
LONDON AND NEW YORK

First published 2006 by Ashgate Publishing

2 Park Square, Milton Park, Abingdon, Oxon OX14 4RN
711 Third Avenue, New York, NY 10017, USA

Routledge is an imprint of the Taylor & Francis Group, an informa business

First issued in paperback 2016

British Library Cataloguing in Publication Data
Women miners in developing countries : pit women and
others. - (Voices in development management)
1.Women miners - Developing countries - Social conditions
2.Mineral industries - Developing countries
I.Lahiri-Dutt, Kuntala, 1956- II.Macintyre, Martha
331.4'822'091724

Library of Congress Cataloging-in-Publication Data
Women miners in developing countries : pit women and others / edited by Kuntala Lahiri-Dutt and Martha Macintyre.
p. cm. -- (Voices in development management)
Includes bibliographical references and index.
ISBN 0-7546-4650-5
1. Women miners--Developing countries. 2. Sex role in the work environment--Developing countries. 3. Ethnicity--Developing countries. 4. Women--Economic conditions--Developing countries. 5. Women's rights--Developing countries. I. Lahiri-Dutt, Kuntala, 1956- II. Macintyre, Martha. III. Series.

HD6073.M61D446 2006
331.4'822091724--dc22

2005031655

ISBN 13: 978-0-7546-4650-1 (hbk)
ISBN 13: 978-1-138-26440-3 (pbk)

Contents

List of Figures

List of Tables

Preface and Acknowledgements

> In many ways mining and social research are parallel activities. Both intervene in and disturb a landscape by probing and digging for a rich lode of ore or layer of stratum that has hitherto lain covered, or unknown, perhaps until now unvalued. Like bauxite, women's knowledge has lain potentially accessible, part of the surficial landscape, for centuries untouched by the technical intervention of dynamite and front-end loader, tape-recorder and questionnaire, simply because society has had no use for it. Today, bauxite is mined by aluminium can makers and women's knowledge is mined by feminist researchers.
>
> J. K. Gibson-Graham (1994) in 'Stuffed if I know!: Reflections on Post-Modern Feminist Social Research', *Gender, Place and Culture*, vol. 1, no. 2, pp. 206-7

The analogies between mining and social research, including feminist research, can appear to be striking, especially in relation to 'unearthing' the hidden stories of women. However, as asserted by Gibson-Graham, the similarities end there as we begin to recognize that the approaches to researching women's stories can be as diverse as practices of mining itself. We acknowledge this diversity and by choosing this interdisciplinary platform of 'Pit women and others' to explore the theme of women miners in developing countries, render our commitment to 'engendering' the development mining can, and should, bring.

Women Miners in Developing Countries uses the gendered constructions of mines and mining to both celebrate women miners and grounds then in the contemporary debates on gender and mining in the developing countries. The conventional hegemonic notions of mining as large-scale and masculine tend to hide important questions relating to gender and development that the authors of this book explore.

Two major transformation processes are on the horizon of the contemporary mining landscape. The first is that organized mining is expanding beyond the conventional geographies of national borders into the mineral rich tracts of developing countries. This large scale formal mining industry is at the same time under pressure to improve its environmental, social and community commitments in these countries. Each of these commitments needs to incorporate a gendered outlook, if their responsibility is to yield meaningful and equitable developmental results. The second change is the the increased attention on the wide diversity of livelihoods and conditions within the fold of informal and subsistence practices loosely called artisanal and small mining (ASM) – the other extreme of 'mining'. These artisanal, small and sometimes unauthorized forms of mining have a large number of women as part of its labour force. The involvement of women in these forms of mining can be seen as part of the global process of informalization of women's work throughout the poorer world.

These changes have important gendered consequences for the developing countries. Mining as a globalized operation is known to subordinate women across the regions of operation through social upheavals and environmental ills. Whilst women bear more than their fair share of the negative impacts, there is no reason why they do not directly partake of the positive economic gains that participation can bring to them and their families. The participation of women in mining, according to us, is not only as part of 'the community', in community 'consultations', but as important economic actors in their own accord. Accepting the hegemonic representations of mining as masculine and large has the potential to marginalize the important ways women make a living in the mine pits – traditional and non-traditional, informal and formal, small- and large-scale mining by communities and companies all over the developing world.

Our standpoint then, is to make way for a more inclusive approach that empowers a woman and sees her as an economic and political agent – much more than a secretary, a miner's wife or a prostitute – in mining induced changes in developing countries. It is indeed reassuring that both the civil society and international mining establishment have initiated efforts towards this direction, and our genuine hope is this book will make a major contribution to, and enhance, the 'engendering' process of mining in the world.

Women Miners in Developing Countries would not have materialized without the active contribution of several people. First of all, we would thank Dr David Williams, Conjoint Fellow, The University of Newcastle and Honorary Fellow, Commonwealth Scientific and Industrial Research Organization, Australia, for devoting his time and expertise to the technical matters of editing. Thanks also to Dr Colin Filer, Convenor, Resource Management in Asia Pacific Program for his continued support of the project, and to Professor Katherine Gibson and Dr Kathy Robinson of the Research School of Pacific and Asian Studies for academic nourishment. The book began as an idea nourished by some of us – Gill Burke and Sachiko Sone especially deserve a warm thank you for their encouragements in fostering this vision into its present physical form as a book. We are pleased to acknowledge the financial contributions of National Institute for Asia and the Pacific and National Institute of Social Sciences and Law, The Australian National University for supporting the international workshop on women miners in Asia Pacific countries. We thank the Executive officers Ms Christine Debono and Mr Richard Thompson of these institutes. However, as appreciated by the participants of the workshop, without Africa and Latin American experiences, our understanding of gender and mining would have remained incomplete. Consequently, the book has expanded far beyond the workshop and its regional focus to reflect a diversity of approaches and experiences, some of which are truly 'bottom up' views from feminist activists living and fighting for women mineworkers. Finally, we would like to thank the individual contributors, dedicated activists and experts in their fields, for sharing their experiences with us. This book would not have been possible without the support of any of the people mentioned above.

Notes on Contributors

Lindsay Barnes

Lindsay Barnes is a researcher and activist, living and working in a remote and rural part of Jharkhand colliery tracts. She came to India from the UK in the early 1980s for a two year Masters degree, and has stayed on since then. She was introduced to the women of the Jharia coalmines through activists of the left wing trade union, the Bihar Colliery *Kamgar* Union. Intrigued by the women of the coalmines, she returned many times to learn more about their lives and struggles. She went on to complete her doctorate from Jawaharlal Nehru University, New Delhi, on *Women, Work and Struggle in the Bhowra Colliery*. Lindsay is now involved in organizing village-based women's groups, and in setting up a grass-roots, rural, women's federation, the Mahila Mandal Samiti. She has also helped train health activists providing women-centred, low cost health care. Lindsay also takes up women's health-related research, and is involved in state and national level advocacy efforts to improve women's health care. She is currently involved in research on reproductive health services, especially during childbirth as Jharkhand has one of the highest maternal mortality rates in India.

Gill Burke

Gill Burke has worked as a consultant minerals economist and development advisor for the past twenty years. After her doctorate on *The Cornish Miner and the Cornish Mining Industry 1870–1923* (London, 1982), Gill taught at the University of Malayasia, Kuala Lumpur. With the collapse of the International Tin Council in 1985 she diversified her research into gold and other metallic minerals. This led in turn to the development of her passion – small scale/illegal/artisanal/ wildcat miners and mining. The bulk of Gill's work has been in Southeast Asia for UNIDO, UNDP, ESCAP, and she did the initial research for the ILO Report on small-scale mining. She was a member of the Expert Group on the World Non-Ferrous Metals Industry preparing for the UNIDO First Consultative Meeting in Budapest, 1987. She has enjoyed a collaborative relationship for many years with the Raw Materials Group, Stockholm, most lately (2001) as part of the team headed by the French Geological Survey conducting an assessment of mining in PNG for the European Union. Gill was an Associate Research Fellow in the Research School of Pacific and Asian Studies at ANU from 1991–1997.

Evelyn J. Caballero

Evelyn Caballero applies Ecological Anthropology concepts and methodologies including participatory rural appraisal (PRA) to natural resource management disaggregated by gender in community development programs. After completing her PhD from the Department of Anthropology, University of Hawaii in 1992, she worked as the Senior Anthropologist and Consultant for Newmont Philippines, Inc. from 1995 to 1996. She has been a member of Indigenous Peoples Advisory Group of The World Bank, Head of Office of the Presidential Adviser on Indigenous People Affairs Secretariat, Philippines, and has worked with indigenous communities involved with coastal, forestry and mining resources. At Ateneo de Manila University, Evelyn has conducted a project funded by Toyota Foundation on a cultural analysis of the Indigenous Peoples Rights Act (IPRA) in the Philippines. She has recently been the key actor in organizing an Asia-Pacific Learning Event on community and state interests in small scale mining in Manila, held in 2005.

Minerva Chaloping-March

Minerva Chaloping-March is completing her PhD from the School of Anthropology, Geography and Environmental Studies, the University of Melbourne. Her experience in mining derives primarily from her growing up in a mining campsite community in the Mankayan mining district of the Cordillera region, Northern Philippines where she witnessed the growth and changes within these communities. Her Masters thesis on gold panning as a complementary livelihood to subsistence agriculture in Itogon and her Master of Philosophy thesis from the University of Cambridge examining ethnic consciousness in the Cordilleras provide a key issue in the conceptualization of regional autonomy and identity. In the Philippines, Minerva worked with the Environmental Management Bureau as a member of the Environmental Impact Assessment Review Committee in 1998–99. She spearheaded the review of 'social acceptability' of mining projects as a requirement for environmental compliance and rendered consulting services for Rio Tinto for almost four years. She has also provided advice in the area of social development and community affairs.

Geoff Crispin

Geoff Crispin is an independent consultant who has been working with indigenous people both in Australia and the developing world since the 1970s. His work on small scale miners in Papua New Guinea has been well-recognized in Australia and abroad. Consequently, he has been involved in several projects examining income generation processes in small scale mining projects. His work on the community and environmental issues in the mining of industrial minerals as well as mining of gold has attracted major developmental and training programs to these areas. He is a member of the Technical Advisory Group for Communities and Small Scale Mining (CASM), an international think-tank in the area.

Jeannette Graulau

Jeannette Graulau is a doctorate in Political Science from Syracuse University, NY, and after a brief teaching stint in Political Science at the University of Puerto Rico, joined the faculty at the City University of New York, USA. She has written extensively about peasant mining in Brazilian Amazonia. Her research interests are gender and globalization, globalization of raw material markets, urbanization of tropical mining frontiers and women's labour in extractive economies of the developing countries of Latin America. Jeannette's current research project analyzes the proliferation of corporatist mining operations in Third World Countries and their impacts on the sustainability of natural resource bases.

Barbara E. Hinton

Barbara E. Hinton holds a Masters in Sociology specializing in Community Development. As an independent consultant and applied researcher for mining companies in Canada and Brazil, Barbara has worked on the socio-economic impacts of artisanal and large scale mining on aboriginal and non-aboriginal communities. In addition to her work in mining, Barbara has extensive experience with not-for-profit organizations, where she has worked with the developmentally disabled, abused children, and inner city women and children. In addition to being an avid contemporary artist, Barbara is also a support worker for women in crisis.

Jennifer J. Hinton

Jennifer J. Hinton is completing her doctoral degree in the Department of Mining Engineering at the University of British Columbia. Jennifer's research has mainly focused on the social measures to mitigate risks in artisanal and small-scale gold mining communities. Recently, she has worked as an artisanal and small-scale mining consultant for the United Nations Industrial Development Programme (UNIDO) and Brazilian Centre for Mineral Technology (CETEM), and as an environmental and social consultant to a number of mining companies. She is also working with the World Bank and Uganda Department of Mines researching the gender-differentiated impacts and benefits of artisanal mining in Uganda. In addition to a number of projects related to large scale mining, she has conducted extensive studies around the world on the roles of women in artisanal mining. Jennifer is a Killam Predoctoral Fellow and a Fellow in the interdisciplinary CIHR/MSFHR Bridge Program, a ground-breaking initiative created to strengthen linkages between engineering, public health and public policy.

Els Van Hoecke

Els Van Hoecke studied psychology to work in JOC, a young workers movement that formed the basis of her future advocacy work with men and women workers and migrants in Belgium, her mother country, in France and later in her adopted country, Bolivia. With wide experience working with NGOs, she began to work in Bolivia in the 1970s. In 1979, she founded the NGO Cepromin – Centre for Mining Promotion, where she worked with and for women mineworkers. In 1987, she set up SENTEC, an institute providing training in mechanics in the city of Oruro for the children of miners. She was instrumental in organizing the first Conference of the International Network Women and Mining or RIMM in 1997. This Network coordinates and strengthens the actions of the mining sector in different countries focusing on a life with dignity for all women related to mining. In 2000, she helped to organize the second International Conference 'Women and Mining' in Bolivia. She was also the legal representative of RIMM till 2004.

Amarjit Kaur

Amarjit Kaur is Professor of Economic History at the University of New England in Armidale, Australia. Her main research contributions have been in the study of Malaysian and Southeast Asian economic development and social change, with a focus on labour issues (women, child and migrant workers) and the environmental impacts of development. She is currently involved in three major international research projects: the Economic History of Southeast Asia (ECHOSEA) project; the Changing Labour Relations in Asia project; and the Border Controls and Labour Migration in Southeast Asia project. Her most recent publications include *Wage Labour in Southeast Asia since 1840: Globalization, the International Division of Labour and Labour Transformations* (Palgrave/Macmillan, 2004), and an edited book, *Women Workers in Industrialising Asia: Costed, not Valued* (Palgrave/Macmillan, 2004).

Kuntala Lahiri-Dutt

Kuntala Lahiri-Dutt has researched gender and mining since 1993, and much of her work has been done on women in the eastern Indian collieries and in the small quarries in India and elsewhere in South Asia. These research projects were funded by the University Grants Commission and the Ministry of Environment and Forests, Government of India. Since 2004, she has researched gender issues in the large Indonesian coal mines. She has advised international agencies on issues related to community perception and participation in mining regions. Kuntala has been active in the recent initiatives taken up by both the non-governmental organizations and international agencies operating at different levels on gender mainstreaming in mining. She has written widely on gender and mining in developing countries, and has authored a major report on gender concerns in South Asian artisanal and small scale mines prepared on behalf of the Communities and Small Mines (CASM) group. In 2002, Kuntala coordinated an Oral Testimony Project among the mining displaced indigenous communities of Jharkhand, India for the Panos Institute, UK. Kuntala is currently a research fellow at the Research School of Pacific and Asian Studies, The Australian National University.

Ingrid Macdonald

Ingrid Macdonald has been the Mining Ombudsman at Oxfam Community Aid Abroad since February 2000 – a platform for a formal complaints mechanism for the Australian mining industry, established by Oxfam Community Aid Abroad. Ingrid has dealt with cases in Papua New Guinea, the Philippines, Fiji, Indonesia and Peru where communities have complaints in respect of the activities of Australian mining companies. She also convened the Oxfam Community Aid Abroad workshop 'Tunnel Vision: Women Mining and Communities' in 2002 which brought together speakers from the Asia-Pacific and Indigenous Australia to discuss the impacts of mining on women living in local communities. Ingrid has been involved in the international, regional and local workshops and processes, from the World Bank Extractive Industries Review to the Papua New Guinea 'Women and Mining: Voices in the Wilderness' workshop in 2004. She is the Pacific coordinator for the International Women and Mining Network (www.oxfam.org.au).

Martha Macintyre

Martha Macintyre is a Senior Lecturer in Medical Anthropology at The University of Melbourne. Martha completed her undergraduate studies in History and English Literature at the University of Melbourne, and then went on to Cambridge University where she studied Social Anthropology. She gained her PhD from the Research School of Pacific Studies at the Australian National University for a thesis on the people of Tubetube in Papua New Guinea. She has held research positions at Monash University and the Australian National University and taught in Anthropology and Women's Studies at La Trobe University. Martha's research interests include historical ethnography of the Pacific region, anthropology of gender, immigrant communities in Australia, development studies, and medical anthropology. Over the last twenty-five year's research in Papua New Guinea, Martha has published numerous articles on women and gender. One of her major publications is *Human Rights and Gender Politics: Asia Pacific Perspectives* (jointly edited with Anne-Marie Hilsdon, Vera Mackie and Maila Stivens) published by Routledge, London. She has been involved as a consultant on two major mining projects and has been monitoring the social and economic impacts of the Lihir goldmine since 1995.

Shashank S. Sinha

Shashank S. Sinha has taught history at PGDAV and Deshbandhu College in the University of Delhi. With a keen interest in development, Shashank has been involved with government policy reports for the National Commission for Women, India (on the first ever study cum survey of child prostitution in Delhi) and the Ministry of Tribal Affairs, Government of India (on minor forest produce management strategies in Jharkhand, Orissa and Chattisgarh). Shashank's book *Restless Mothers and Turbulent Daughters: Situating Tribes in Gender Studies* (Stree, Calcutta) is scheduled for publication in 2006. He has written articles and presented papers on gender, tribes, witchcraft, and identity formations and has been involved with reputed media organizations on documentary films and other infotainment programs, and has edited a series of sports books for Penguin India. He is based in New Delhi and his current research interests include gender, social and environmental history and the social history of medicine.

W. Donald Smith

W. Donald Smith taught at the University of Illinois and University of Alabama after receiving his MA at Columbia University and PhD at the University of Washington. Donald's research has drawn out the little known history of Korean and women workers in the prewar Japanese coal mines. Donald has also written about Japanese wartime labour mobilization, and Korean ethnic education under the US Occupation of Japan. After a stint in journalism, he currently works for the US government in East Asia, distributing his time between the US, Japan and Thailand.

Sachiko Sone

Sachiko Sone is a lecturer in the Japanese program in Asian Studies, University of Western Australia. Her academic interest is cross-cultural studies. Her recent articles include 'Exploitation or Expectation? Child Labour in Japan's Coalmines before World War II', (*Critical Asian Studies*, formerly the *Bulletin of Concerned Asian Scholars*) and 'The Life and Circumstances of the Japanese Prostitutes in Asia before World War II', in Ann-Marie Metcalfe and Carolyn Brewer (eds) *Researching the Fragments: Histories of Women in the Asian Context*, New Day Publisher, Quezon City, 2000, pp. 103-140.

Marcello M. Veiga

Marcello M. Veiga is an Associate Professor in the Department of Mining Engineering at the University of British Columbia, and has worked in artisanal mining over the past two decades. As a special consultant for UNIDO, Marcello has worked extensively on environmental, social and economic issues related to artisanal gold mining throughout Latin American, Africa and Southeast Asia. His expertise also extends to fields including mineral processing, environmental geochemistry and the social impacts of mining. In addition to research, he has consulted for numerous companies throughout South America and Canada, including CETEM, CVRD, IDRC, the Peruvian Ministry of Mines and Brazilian Goldfields, Ltd.

Linqing Yao

Linqing Yao is an Associate Professor in the Medium Management School, the Beijing Broadcast Institute, China. Before joining this institution, she was first a Lecturer and then an Associate Professor in the Department of Management, the China University of Mining and Technology in Beijing. She is trained in labour economics from the Capital University of Economics and Business. Linqing has done several research projects on the mining labour force in China and has published on the mining workforce in academic and professional journals. Her past research had examined the labour situation in the Chinese mining industries. She is also part of the Communities and Small Mines, China network.

This book is dedicated to all those women who have found a living in the mines particularly those in the developing countries: the palliris, the kamins, the dulangs and many others. We hope that the contributions from the different authors in this book will increase their visibility in mining, make a case for involving them in mine-related development and assist in empowering them.

Introduction

Where Life is in the Pits (and Elsewhere) and Gendered[1]

Kuntala Lahiri-Dutt and Martha Macintyre

The 'pit'?

> *Pit*: 1. a hole, shaft or cavity in the ground; 2. mine; 3. a. a scooped out place used for burning something as charcoal, b. an area often sunken or depressed below the adjacent floor area; 4. hell – used with the 's' - a place or situation of futility, misery or degradation.
>
> *Encyclopedia Britannica*, 2002

Mining literature yields much more than the sense of depth implied in the dictionary definition. Social scientists – economic and labour historians, human geographers, anthropologists and development experts – do see more than just depth; they see people busy in various jobs in the mine pits. However, people in the 'pit' are almost always seen as being 'pit men', as though pits are, and always have been, inhabited by men. Words associated with pit men are also masculine, such as 'labouring men', creating a 'proletarian solidarity'[2], 'a man from the picks', and above all 'the working class'. The plunge down the shaft and the darkness at the end led trade union leaders (Dange, 1945) to use the expression 'death pits of our land'. This is rendered heroic, as writers laud 'the nonchalance with which our diggers daily face death itself' (Coleman, 1945). Carter Goodrich (1925) defined the miner as 'an isolated piece worker on a rough sort of work, who

---

[1] By gender, we imply an asymmetrical social relationship between women and men based on perceived social differences, and an ideology regarding their roles, rights, and values as workers, owners, citizens, and parents. We understand it as a process and a structure and use it to draw attention to the gendered nature of mine production.

[2] This solidarity is expressed not only in strong traditions of unionism, but a 'marraship' or mateship among co-workers, identification with a particular seam, and cavilling or grouping.

sees his boss less often than once a day'[3]. People in the pits of that era were seen as men undertaking dangerous, dirty, risky and hazardous work and were characterized by masculinity suitable for independent heroes (see Harrison, 1978). Robinson (1996: 137) comments:

> ...mining provides us with powerful visual and metaphorical images; it is gargantuan, dangerous, heroic and mysterious, involving destruction and penetration of the earth's surface ... Images of mining as human endeavour incorporates the imperatives of physical strength, endurance and filth, all characteristics of masculinised work.

In the pits, according to Connell (1987: 85) the physicality is 'one of the main ways in which the power of men becomes "naturalized", i.e. seen as part of the order of nature'. Thus, pit life[4] was perceived as a uniquely male world where the sharing of risks contributed to the formation of a particular form of male solidarity, endowing their manual labour with attributes of masculinity (Garside, 1971).

Women in pit communities, therefore, were seen as the quintessential 'other' of pit men, and are commonly seen as miners' wives, staying at home and supporting their working class men, who endanger their lives by going down the mines to earn the bread for their families (Carr, 2003). Gibson-Graham (1994: 207) dissects this position:

> ...miners' wives are still constituted within a discourse that brings to the foreground the functional overlap of capitalist exploitation and patriarchal oppression. Working class first and women second, miners' wives are subsumed within the fictional identity 'working class' and relegated to the status of Other within this totalizing conception.

Women in mining communities are not seen as active participants in the economy; they are represented as working class women shouldering the burden of domestic responsibilities in deplorable living conditions, and the dangerous, dirty and irregular work of their husbands, fathers and brothers[5]. Once this perception becomes established and entrenched, processes of change, especially those imposed by the State, the agencies and companies from above begin to use gender-neutral lenses in dealing with 'the community'.

When this perception of women as only miners' wives, as part of 'the community', is held by the heads of community development experts, the consequences become potentially dangerous. For instance, this fact is immediately

---

[3] 'One of the redeeming features of pit work, and one that miners will fight to maintain, is that of independent job control...' (Douglass, 1972: 2). Thus we see the pits becoming also the 'frontier of control' (Goodrich, 1920) that miners resist (Parkinson, 1920: 20).

[4] We even have 'Pit yacker' (Hitchin, 1962) but that's not the subject of our interest here.

[5] Rothermund's (1994) report on the relationship between the degree of political awareness and levels of income amongst working women of the coalfields of Dhanbad, India, showed little positive relationship with the concentration of economic development.

apparent with a casual reading of development planning instruments that are currently being offered to the mining industry. In its 125 pages or so of the Community Development Toolkit of the International Council on Mining and Metals/World Bank (2005), gender does not merit great mention. The Toolkit does not assume the community to be an undifferentiated mass: there are checklists of possible stakeholders and instructions on their identification (pp. 18-20), but takes for granted that the benefits would automatically reach women. Even in locations where women have not been part of the workforce, Hipwell et al. (2002: 11) suggests that the impacts of mining operations are gender differentiated. As mining companies move in developing countries[6] in the contemporary context offering only a few jobs and many ills, changing gender roles and robbing women of their rights of participating in determining the benefits, one must search for ways of making the developments more inclusive and beneficial for both women and men, so that pits are not seen as a masculine domain of work and women are placed at home.

Pit women

However, once we are in the pit, and our eyes are accustomed to the darkness, we begin to see them as gendered places. We start to see figures of women working alongside men. These are our 'pit women' who are at the centre of this book. These near invisible pit women – 'pit brow lasses' (John, 1980) or *kamin*s or *dulang*s or *palliri*s – are crucial to building a gendered understanding of mining. Whatever the name, women's work in the mines has remained obscure and hidden, forgotten and devalued. Thus mine pits have come to be the playing fields of men, the work either excluding women altogether or attributing a lower place to women. Yet, it would be a folly to construct a homogeneous and universal category of 'women in mining' or 'mining women', partly because the countries we are dealing with have experienced mining in different ways and also because of the multitude of identities of women in developing countries[7].

Mining has played an important role in the modernization process in the developed world; although mining has a long artisanal history, in developing countries recently it is taking increasingly diverse and complicated forms. Mining is now a truly global enterprise with ramifications from the local to national to

---

[6] The Third World, less developed countries or the global South.

[7] Koczberski (1998: 404) notes her reservations about treating women as a separate socio-economic category: 'by compartmentalising women the (gender analysis) framework renders irrelevant the linkages between women's various work roles, kinship relationships and the social, cultural and poltical systems of which they are a part. Women's work activities, their access to resources and their needs are not isolated fragments, rather they are intricately embedded within communities in which they live. To ignore these linkages predispose simplistic and unrealistic analyses and constructs images far removed from reality'.

global. In several parts of the developing world, women have worked in the past and continue to work as miners, either independently, as part of a family unit of labour, or as employees of others. As more and more mining 'breaks ground' in the developing countries (MMSD, 2002), these roles become important, raising some questions. Do we see women as victims of the ill-effects of mining on the community and suffering without agency, at best fit for soft servicing jobs as 'secretaries and office clerks', or do we see them as equal and productive partners in an empowering process of development[8]? There is not yet enough literature that illuminates how gendered identities and inequalities are constructed and sustained in mines in developing countries. This book intends to begin filling that gap, with a view to engendering the mining related development processes, and put forth an interdisciplinary approach in examining women in the mine pits and quarries of the past and the present.

Mining and gendered constructions

Poor working conditions in the pits of the past gave rise to a strong sense of occupational identity, which often extended to entire mining communities, but nevertheless rendered women and their work in the mines invisible. The heavy manual character of the work, the dirt and risk tend to be emphasized and tend to make the male miner the typical labourer, and to protect their interests over those of women workers (see Eveline, 1995; Metcalfe, 1987). The frontier nature of mining, according to Burke (1995), has been seen as creating a situation in which 'the ideal cultural goal was a hard-nosed style of doing things in management, supervisors and men' which is 'unashamedly sexist' (Williams, 1981: 50). The masculine work culture arouses masculinist analysis such as this (Allen, 1981: 4):

> ...mining evokes popular images of hard unrefined men, distinct and separate from other workers, hewing in mysterious dungeons of coal: dirty, strange men, in some ways frightening and for this reason repellent, yet attractive because they are masculine and sensuous...

The miner occupies a special place in western popular culture, rising almost to the status of an icon. Orwell's (1937) words resound of the cruelty and hellish romance of the coal miner's occupation, which builds an aura of sentiment: 'all of us owe the comparative decency of our lives to poor drudges underground, blackened to the eyes with their throats full of coal dust, driving their shovels forward with arms and belly muscles of steel'. According to Campbell (1984):

> Miners are men's love object. They bring together all the necessary elements of romance... It is the nature of the work that produces a tendency among men to see it as essential and elemental, all those images of men down in the abdomen of earth, raiding

[8] See Lahiri-Dutt, 2000b; 2001.

> its womb for the fuel that makes the world go round. The intestinal metaphors foster the cult of this work as dark and dangerous, an exotic oppression … it constructs the miner as earth-man and earth-man is true man. And it completed the equation between some idea of elemental work and essential masculinity. This romance is duly mirrored in working class politics – miners are the Clark Gables, the Reds of class struggle.

The masculinity that emerges from the images of the miner as a man is revealed both in the industry and in the politics of socialism in which miners have historically played a central part. John (1980) believed that the root of this masculinism lies in the circumstances surrounding the exclusion of women from underground work in the name of protecting them from the risky and hard mining jobs. In some British collieries in the eighteenth century, women actually stood or doubled, often in knee-deep water, in deep and thin shafts. Rev. Eddy (1854: 296-7) gave a vivid description:

> …females submit to labour in places which no man, or even lad could be got to labour in; they work in bad roads, up to their knees in water, in a position nearly double; they are below till the last hour of pregnancy; their limbs and ankles swell, and they are prematurely brought to the grave, or what is worse, a lingering existence.

A sixteen-year-old girl working as windlass woman is quoted by Rev. Eddy as saying: 'We wind up 800 loads (a day). Men do not like the winding. It is too hard work for them'. Yet, the masculinity of mining continued to be represented. For us, such representations are important because they initiated protective interventions around women's work in the mines. For example, a self-congratulatory public outcry in Britain resulted in the exclusion of women from the mines. Similar legislation followed in several European countries who exported this legal protectionism into their colonies. As is well known, protectionism is notorious for causing double hardship for women. For example, in the case of Britain, after the submission of petitions that women may be restored to the 'privilege' of working in the mines that they might not starve, they were permitted to work in the mines again. The result was that working conditions for women were not improved and they re-entered the mines subdued and at the mercy of the owners. More complicated examples can be cited from the contemporary situation in developing countries. Menstruation leave, for example, provided by the Indonesian state as a gender right to all working women (but not commonly utilized), has recently posed intricately complex issues and dilemmas for women workers in the mine pits in Indonesian collieries (Lahiri-Dutt and Robinson, 2006).

If mines are seen as masculine work areas, then the use of machines has differential impacts on women and men's work. The increased use of technology in Indian mines has been largely responsible for the exclusion of women mineworkers (Lahiri-Dutt, 2000). Exclusion of women from the mines was supported sometimes by superstitions, but also by the protective laws that restrict women from working in the mines. Such laws are not uncommon in other Asian countries as well[9]. However, in some of the large mines owned by multinational companies, technological advances have transformed jobs that were formerly considered to be 'heavy' for women and enable women to work in areas such as plant operation. Indeed, in a Thai mine, women were preferred as vehicle drivers because of their better safety record. In Papua New Guinea women work in geological exploration, plant operations and other fields that were in the past male domains (see Macintyre, this volume). The hypocrisy of patriarchal social values and fixed notions about a woman's place reinforce the delineation and distinction of male and female roles in mines. In prohibiting their participation in wage labour, they effectively confine women to domestic and marginal economic roles.

Mining, as a range of practices, has gendered orientations that in this book we put under a lens – scrutinizing intersections of space, place, race, ethnicity and class. Looking through this lens, we also reflect upon the gendered nature of mines as a special place, attributing a specific place to women in its economic organization. Through this lens, the mine-pits emerge as gendered places and the 'miners' become visible as gendered subjects defying conventional geographies and histories of space and industrial work. Traditionally, the history of exclusion of women from the pits has been represented as a masculine victory, protecting the gentler sex from the rigours of dangerous and heavy work. The differential positions of women and men in the spheres of industrial production reflect the social relations of gender and are perpetuated by gender ideologies that infantilize women; economic differences among women result from the inequalities of class and ethnicity, structured by the mode of production. The social organization of mining imposes unequal economic and social relationships on women and men, helping to subordinate the position of women both within the industry and outside it, in the community, directly as well as indirectly.

Pit women as social subjects

In the large literature on the social and cultural attributes of mining communities, pit men are represented as a certain kind of subject – independent, free-willed and rational. How do we then place pit women? Metcalfe (1988) examined the drunkenness, gambling, assault and rioting stereotypes of male miners. The coal miner is seen as a 'pit horse', a 'big hewer' and these physical characteristics lead to the glorification of his masculinity. Moore, in his 1974 book, *Pit Men, Preachers and Politics*, tried to negate the macho image and show that pit men were actually god-fearing religious people. He stated that 'miners were traditionally conservative and resisted traditional proletarian social imagery' (p. 17). The early work of Kerr and Siegel (1954) suggested that the isolation of miners from the wider community (both in their work and their lives outside the

[9] Examples that come to mind are Japan, China, India and Indonesia – all major coal/mineral producers now or in the past.

mine) made them a particularly cohesive group who tended to be characterized as the 'other' of the 'normal' society.

One theorization of mining and gender has come from the early conceptualizations of mineral wealth being equated with the 'natural endowment', the basis of wealth accumulation. Minerals are, however, found hidden from humans, needing searching for them and prying open the 'womb' or the belly of the earth in a manner of violating the mother earth, penetrating her secrets, raping her of her prized wealth. This powerful organismic metaphoric view, put together with the fact of mining being associated with the expansion of imperialism and capitalism, challenges the 'moral framework of organic unity, of an animated natural world, and allowed values to be turned upside down', so that the commercial exploitation became equivalent to rape of the earth (see Robinson, 1996: 138-9; Burke, this volume, for elaboration of this view). Within this cultural framework, women are positioned outside productive labour and those who engage in mining work are perceived as masculinized and 'denatured'.

There have been other theorizations of gender in mining. In her assessment of the place of women in mines in Zambia during the colonial period, Parpart (1983: 1) used an economic analysis of class (although women as the exploited class rings throughout), affirming the gender inequalities:

> Women spent most of their day ensuring the daily reproduction of labour. As with their husbands, capital-labour relations defined much of their lives. Thus on the surface, the African women in copper mining areas seemed truly working class wives, while willing to engage in class struggle against capital, these women also struggled against their male 'protectors'.

This is not unusual, as Gibson noted (1992: 29):

> ...unlike the miner, whose class position defines for him a seemingly clear role in capitalist class struggles, the miner's wife's class position appears to involve a problematic relationship to working class politics.

She argues that this ambivalence robs women of their agency and fails to see their productive roles (p. 30):

> The construction of women as unknowable or unpredictable political subjects is a politically debilitating result of theoretically locating women only in relationship to capitalist class processes (which in effect places them as spectators or, at best, reserve players in a game in which their husbands are actively involved).

Mining all over the world is an industry with strong working class traditions. This male miner, as a more or less powerful or conscious agent of the working class, is also a psychologically, emotionally and politically powerful agent of patriarchy. Feminist geographers explain this solidarity as arising out of a sexual division of labour grounded in 'place', that is, in a spatial division of labour. They speculated (Massey and McDowell, 1984: 129):

> ...danger and drudgery: male solidarity and female oppression – this sums up life ... Here the separation of men and women's lives was virtually total: men were the breadwinners, women the domestic labourers, though hardly the 'angels of the house' that featured so large in the middle class Victorian's idealization of women.

This separation, they noted, was caused by the notion that women's place is at 'home': 'For miners' wives almost without exception, and for many of their daughters, unpaid work in the home was the only and time-consuming option'. Many scholars saw women as being unproductive and isolated, unable to resist domestic oppression; they saw women staying at home caring and cleaning for husbands and sons who often worked different shifts and came home dirty. Using Mitchell's (1975) expression about the wives of radical labour movement, Gibson (1992) described this phenomenon as women turning into domesticated 'hewers of cake and drawers of tea'. Men on the other hand, were seen as gathering together in union halls or local pubs and sharing a sense of community (Library of Congress, 2001).

However, seeing women miners as social subjects in developing countries renders these views problematic, although both form important backgrounds. For developing countries, one must look at mining in its entirety, both as large and small, as oppressive and emancipating, as highly capitalized as well as a livelihood activity for women. In other words, one must look at the complexity of mining and gendered roles therein. Large scale mining does challenge the 'normative values' of indigenous and poor peoples and the figure of the earth as a nurturing mother is often presented as a contrary image to that of mining. Robinson quotes a mine manager (1996: 141):

> I asked the Canadian personnel manager at PT INCO why the workforce was all male (except for a small number of women employed in service activities). He looked surprised at my question – as an INCO man, the 'naturalness' of the male workforce was unremarkable. 'It's government legislation,' he said. Indeed, Indonesian labour legislation (inherited from the Dutch) does prohibit the employment of women in mines. This serves to underscore my point about the discursive framing of mining as a masculine activity, as man's work ... Mining is so naturally 'masculine', its gender effects are invisible.

From the perspective of development, seeing women as equal and productive partners in mining in developing countries is a positive and affirmative step. This is achieved in the book by an engendering of the mines of developing countries, through looking at the intersections of gender with class, race, caste and ethnicity, and by reasserting the multiple identities of these pit women and their counterparts on the surfaces of quarries and elsewhere.

And others

The 'other' usually plays a role in our conceptualizations of the self. However, in this volume, we talk about 'and others' and not 'the other', pointing to the fact that as mining as a work area is diverse, so is its workforce. Just as both women and men have worked in mining jobs, mining too has not always been concentrated only in pits, and is not necessarily a large scale operation. Throughout most of human history, mining has been a rather small, scattered, low production and low technology affair. Early humans were able to obtain certain minerals through what we now call 'artisanal' or 'small scale mining'. It was combined with agricultural and pastoral subsistence activities and the human endeavours and technological inventions successively giving rise to bronze and iron 'ages'. Many of the mining jobs, especially in mineral-rich tracts of the developing world, still work this way, within the context of similar subsistence-based, small and informal economic activities, providing a livelihood often for communities (Labonne and Gilman, 1999). According to Jennings (1999), approximately 15-20 per cent of the world's non-fuel mineral production comes from such mines, employing around 13 million people. The range of mining practices are lumped together as 'artisanal and small scale mining' (ASM) and are often seen as in dire need of 'regulation' and formalization', and as disruptive of the purity of the environment of their areas of operation. The scale of environmental damage caused by artisanal mining varies – but is usually commensurate with the scale of the activity itself. The health risks and dangers often reflecting the poverty of those who quarry, pan and smelt using primitive technologies. As the studies of artisanal mining, particularly the work by Hinton illustrates, people do expose themselves to dangerous chemicals and safety standards are low – but the reasons for this are not those that apply in large scale mining. Low levels of education mean that the dangers of chemical substances (especially mercury which is used to extract gold) are often not appreciated; lack of capital prevents artisanal pit miners from erecting and maintaining safe worksites; poverty drives people to labour without attention to long-term consequences. Engineers differentiate between the various processes of tunnelling, digging, quarrying, washing, panning and mining, with accompanied differences in labour, technology and areal spread, often related to the occurrence of the mineral itself. As a result, we are initiated into a learning that sees a dichotomy in mining practices – large and small – although there is a continuum between these two extremes as evidenced by recent reports (for example, see Hentschel et al., 2002).

Poverty-driven and generating low levels of capital, ASM can be traditional, such as in the case of the Philippines or Bolivia, or it can be a cash-income, need-induced recent phenomenon as in many African countries. It can be seasonal, or encouraged by economic recession as in the case of Brazil. Regardless of the specific nature, women play important roles in such mining practices, whether in Africa, Latin America, or in the Asia-Pacific region. Of these enormous numbers of miners, as many as 30-50 per cent are women in certain countries and this percentage is actually growing (Noestaller et al., 2004; Hinton et al., 2003; Heemskerke, 2000)[10]. Jennings noted (1999: 30):

> ...the impact of structural adjustment programs, low commodity prices, or drought on private and public sector employment, trading, farming and inflation has led many people, especially women who relied on subsistence agriculture, to seek new, alternative or additional paid employment for a better quality of life, more usually, just to survive.

We see a close relationship between poverty and women's need to maintain subsistence for families; as Hentschel et al., (2002: 22) observe: 'increasing numbers of women are entering (the) ASM sector as an alternative to subsistence agriculture, and it has been found that women are more likely to spend their income on family maintenance compared to men who may be tempted to spend their income on prostitution, gambling and alcohol. Empowering women in these communities could lead to substantial alleviation of poverty.' Yet, neither are the women owners of mines in most cases, nor are they able to access credit from external sources to invest in their small mines due to their inability to provide collateral for bank loans.

Our understanding of mining as a gendered and contemporary 'work' in this segment of mining in countries remains limited. The relations between the lack of ownership of land by women and their consequent subordination in economies that flourish on the resources offered by the land, also need further investigation. In the absence of legal rights over land, continued work burdens at home and protective laws that disempower those involved, engendering the ASM sector might be an uphill task. For example, several features such as the entry of machines limiting women's work in the mines generally remain valid in ASM too; but our understanding of women's roles in the supply of food, drink, tools and equipment, and as traders of gold and gemstones, remains incomplete (although some recent work such as that by Amutabi and Lutta-Mekhebi, 2001, provides a wealth of information). Also incomplete is our knowledge of the control of women over the incomes earned through their participation in ASM (see Lahiri-Dutt, 2003). Women's work remains invisible, as is evident in a comprehensive report on the Ninja gold miners of Mongolia (MBDA, 2003: 206-7):

> A sharp division of labour exists between male and female ninjas ... Men undertaking most of the arduous tasks such as digging, sledge hammering and hoisting, whereas women predominate in assisting from the surface, preparing bags of ore for transport, panning and in provision of *ger*-based services (cooking, washing, trading, etc) ... the division of labour is rational, based on physical strength and stamina...

[10] Noestaller et al. (2004: 15) note: 'Significant numbers of women and children are involved in ASM activities, mainly in mineral processing, as well as in peripheral service activities such as providing food and water. Estimates on women participation range from 25 per cent (Tanzania, 1996) to 50 per cent (Ghana, 2001; Zimbabwe, 2001)'.

But then goes on to observe:

> Women placer ninjas commonly work underground in Zaaamar *Soum* and Shariin Gol *Soum* as tunnellers when the placer is thin, as women's bodies are more flexible, smaller and more compact, rendering them quicker at tunnelling and more maneuverable at bagging and turning ... The main job for women placer ninjas is as panners. This is a tedious backbreaking work, especially difficult while panning for long hours using cold water.

Clearly, there is a great need to begin comprehending the entire range of practices within the broad field of 'mining and quarrying' as a gendered field and throwing light on the range of work that women artisanal miners perform would enable us to see these 'others' of pit women working on quarries in developing countries. Our definitions of 'mines', 'mine pits', and 'mining' are changing. They are indeed no longer centred upon Europe and North America, and the hegemonic notions of masculinity are no longer useful. Yet, new approaches are yet to emerge. In this volume, the authors are attempting to recover gendered histories but are also trying to re-position women in the mines and engendering the vision of the present situation within the mines. Pit women and others, in the developing countries, past or present, encourage us to examine the specificities of contexts and explore the possibilities of drawing temporal and spatial parallels of gender as socially embedded practice in the mines. The essays in this volume reveal the need for further research into particular cultural, political and economic contexts for the gendered nature of work in mines and critical analysis of the way the work in the mines is represented. A gendered approach entails a reexamination of what is commonly meant by labour militancy and uncovering the rights and resistances of women miners, and the consideration of the gendered results of global capitalism and/or technology on the mining industry in all its variants.

Mining in developing countries

The ways in which women have experienced and are experiencing mining and related socio-economic changes in the developing countries may be different from the ways in which women have related to mining in other parts of the world. Many of the countries that are experiencing varying types of economic change were colonies of European countries in the past. They are now facing the challenges of providing better standards of living to increasingly large populations with more subtle but equally hegemonic forms of foreign economic domination in the form of globalization. Exploitation of minerals plays a significant role in their participation in the process whereby they are linked to the economies of Europe, the United States and other industrialized capitalist nations (see Connell and Howitt, 1991). Mineral extraction plays a significant role in developing economies, giving rise to numerous questions the answers to which are yet to be settled upon. What are the socio-cultural impacts of mining in developing countries? How does an indigenous society rooted in landed traditions react when capital-intensive resource extraction descends upon it? How do the social and economic processes triggered by mining change gender roles? Experiences from some Pacific countries can give contextual answers to these questions, but the variations and specificities are only now becoming the focus of research (see Ballard and Banks, 2003).

Mining in some developing countries was part of the colonial experience, developing as enclaves at the periphery serving the primary metropolitan enclaves, while in others it fuelled industrialism, imperialism or global capitalism. Companies' staff from other nations and workers from the host countries carried gender ideologies that were transcribed on the mining communities – often in ways that were novel, drawing on both and generating new roles for women as workers and as wives, mothers and managers of transformed household economies. While the roles of women in the mining industry have been underplayed, there is little doubt that the intensity of the activity of women in developing countries is of a higher level than that in large-scale mining, and is a subject that needs to be explored. The fact that women took and still take part in a non-traditional occupation such as mining challenges us to review the stereotypes and representatives of 'third world women'. Were the experiences of women miners in these countries similar to those of Europe and North America during the industrial revolution and nineteenth century when the mining industry was expanding there? If not, can we move from there towards an understanding of gendered mining cultures of the developing countries? The project assumes greater importance given that in many of these countries, mining is seen as the harbinger of modernity, bringing 'sustainable development', part of the neo-liberal developmentalism and global capital circulations (for an example of some recent work, see Ericsson and Noras, 2005; Roonwal et al., 2005). These are neither singular nor gender-neutral processes, although the mining establishment would prefer to believe that. Capital-intensive mining is rapidly changing, choosing to use more area extensive mechanized processes making many older mines seem relatively smaller.

Multinational mining companies in countries such as the Philippines, Indonesia and Papua New Guinea are increasingly employing women in administrative jobs that are conventionally seen as feminine but also in non-conventional jobs such as those in technical areas and as equipment operators. The long history of gender discrimination continues to influence employment patterns, but, as minerals are increasingly mined by machines, and women in the developing world gain access to education, so their participation in previously male-dominated areas grows. Again, seeing the mines as a large, capital-intensive and exploitative industry is only one representation of the whole gamut of processes that make up mining, and processes where a large number of the world's poorest are concentrated in return of a low level subsistence living. As we are aware, this kind of work – heavily manual and unsafe digging, panning, transporting and processing – forms that grey area of informal, artisanal and small scale mining. But mining in these circumstances provides livelihoods not only for families but entire communities in developing countries (Hinton, 2003). We need to remember, whilst we celebrate the past contributions and recover the histories, that mining continues to be a basis of the livelihood of millions of women and men in their daily struggles for survival.

Queens of coal and gold

However, the question as to why coal and gold feature so prominently in this book needs to be addressed. Coal has been the backbone of modern development, truly the king among all the minerals, and women mineworkers, largely unseen and unsung, were the 'queens' of coal whose histories and contributions are celebrated in this book. Gold is another valuable mineral, forming the basis of prosperity, with almost as much symbolic value in certain Asian cultures as coal in modern Europe. It is but natural that these two minerals would appear in several studies in this book, offering different contexts. About coal, Coleman (1945: 16) says: 'Here is the mineral upon which industrial civilization, East and West, ultimately rests'[11]. Women miners in these coal mines, whether in Britain, in America, or in the Japanese, Chinese and Indian collieries, and in the gold mines of Papua New Guinea, the Philippines, Brazil and Bolivia, thus provide that invisible connecting thread to reinterpret modern labour and post-colonial histories and economies.

The search for mineral wealth inspired and drove imperial expansion. The Spanish sought gold and silver in South America, the Colonial British brought their knowledge of systems of mining and control of mineworkers to Africa, Asia and Australia. In some countries, women were seen by them as more useful labour, easier to put to work at a lower wage. In India, the early coal mines employed women of indigenous families as part of the 'family labour unit'. Women mineworkers thus played important roles in colonial imperialism and empire-building. In India, the exclusion of women from the collieries took place in a completely different circumstance, primarily through the introduction of shaft-sinking technology, but began as a similar measure of 'protection' for women from the harsh working conditions of these mines (Lahiri-Dutt, 2001). In several countries, the 'protective' legal frameworks are still in place, although some other countries such as South Africa have revoked such laws. Although women's labour in many other forms of work remains predominantly manual, hard, time-consuming, tedious and dangerous, it is mining that remains shrouded and protected as a male bastion.

In America, Moore (1996) showed that whereas early women miners used to hand load coal, and work manually, the affirmative action policies and the biggest coal boom in a half-century in the 1970s brought several thousands of women into the high-wage jobs in the coal mines. Old superstitions that the presence of a woman underground could bring disaster were demolished and women miners became role models for American women pioneering in non-traditional fields.

[11] It's strange, but true, that coal was called 'cole' till the end of 19th century in the Appalachian areas of the United States.

Women emerged as activists in the United Mine Workers of America and their non-profit organization, the Coal Employment Project. In Canada, too, where 94 per cent of coal is produced from large surface operations, women drive specialized earthmoving vehicles, and work on mining machines (Espley, Francis and Castonguay, 2004).

Gold was first mined 6,000 years ago, since which time humans have been obsessed with this beautiful and valuable mineral, leading to 'Gold Rushes' and upheavals of society. It is said that objects made of gold survived better than others, making gold artefacts a source of power and a symbol of wealth. That it came to be so loved in so many cultures testifies to the mysterious ways that humans confer value on resources, but it is through the attribution of value to this substance that gold has served as the most important medium of exchange and money standard throughout history. The term 'gold' continues to conjure images of the good and the powerful. In modern times, gold has been used in industry, computers, electronic circuits and in aero-space industry. Gold for women, however, is not only as an object of desire for jewellery but mining of the ore has been and still is a means of subsistence. From the forest-clad hills and mountains of Papua New Guinea to the Philippines, from the African continent to Latin America whether it is Bolivia or Brazil, a large number of women today are making a livelihood in gold mining. Many of these women work at low income levels, and live risky and precarious lives. At the same time, large-scale gold mining is a huge and intensely capitalized business, usually granting only marginal or token roles to women. Yet technological advances transform the labour process so that male 'brawn and muscle' are no longer required to extract ore – heavy machines with computerized operating systems are far more efficient. Improvements in access to education ensure that women can gain qualifications that equip them to work in all areas of mining operations. In these changing circumstances the ideological nature of female exclusion from industrial mining is exposed as discriminatory. Can gold today be turned into a mineral resource that empowers women and brings prosperity to their communities? Efforts from all sectors are underway, as we noted before, to engender mining. We hope that our book will positively contribute to the process.

The chapters in the book

Placing chapters into sections or parts provides a means of ordering the contributions, but because these come from a wide range of experts with varied experiences of mining and from different methodological backgrounds, these groupings have a measure of subjectivity and are not without overlap – indeed other groupings could have been selected.

Chapters in the first part reconstruct the gendered histories of the mine pits through examples derived from a range of countries. The first of the four chapters begins with an historical overview of women in Asian mining. Gill Burke looks at women miners in Asia at a general and contextual level during the modern period. She suggests that in the majority of instances women have always participated in small-scale, artisanal mining and, indeed, continue to do so in many countries today. In larger scale operations however, women's participation apparently occurred only at the early stages of industrialized mining. She examines the reasons for this, and the subsequent development of the masculinist model for mine workers. Although the main focus of the chapter is historical, it links with contemporary practices in developing countries. Sachiko Sone follows the Japanese pit women in the course of technological, political and economic changes that affected the coal industry. Within the patriarchal structure of the Japanese mines, women miners were placed even lower than their male counterparts. In the world under the ground surface, however, these women defied the prevailing cultural norms and laboured alongside men as equals. Though women were legally prohibited from working underground in the late 1920s, Sachiko shows that the demise of the female underground miner in Chikuhō had as much – or more – to do with family labour practices, technology and geology, as it did with legal sanctions. Amarjit Kaur describes the first half of the last century and analyzes race and gender concerns in the tin mines of Malaya. Her focus on Chinese *dulang* workers or panners provides not only an authentic history but also an important case study of gender at work: the structure of the industry, the occupational categories involved, the division of labour, and spells out why and how gender mattered in colonial Malaya. By reserving a particular job category – a method of recovery in this case – mining industry attributed women a place in its social and economic hierarchy. Amarjit argues that the job enabled and empowered women and provided them economic independence and security that the colonial state failed to provide. Shashank S. Sinha elaborates this theme, revealing the continuities and disjunctures associated with changing economic and political contexts in the Chotanagpur collieries in colonial India.

In the second part, we analyze these 'gendered' identities in the mine pits by examining how other aspects of identity such as race/ethnicity and caste can produce extremely complex situations. W. Donald Smith shows how the 'double minority' of Korean women in prewar Japanese coal mines was shaped by gender and ethnicity. Mining was a traditional occupation for women, who comprised primarily the overwhelming majority of coal haulers until the ban on underground female labour. Among Koreans, however, male coal haulers outnumbered women and women were placed in various low-paid jobs on the surface. Evelyn Caballero throws light on a very old and traditional culture of small scale miners in the Philippines whose subsistence base has predominantly been mining with some shifting cultivation and a few other occupations. The roles of women in these processes are described carefully and in fine detail. Martha Macintyre in her chapter points out how traditional gender role expectations can conflict with the coming of the mines, bringing a whole new set of gender discourses. Mining today encompasses a range of technologies that mean women's participation in the industry covers the whole range of jobs. As the chapter by Macintyre demonstrates, women workers in large-scale mining are beginning to challenge the image of the miner as they penetrate the bastions of masculinism and take advantage of the changing technologies that enable them to work in technical fields. Kuntala Lahiri-Dutt shows that this may not be the case everywhere; women in the Indian collieries, the *kamin*s working on the field, do not construct their work practices favourably. If this is seen as 'identity work', then the women's active attempt to construct identity through various representations of reality can be seen as a resistance to us forming their identity as against that of the researcher. Clearly, it is not possible to give a single and objective description of life in the mines. Many alternative narratives are constructed on mining as a work for women with available discursive material.

The globalization of large scale mining has not resulted in the diminution of ASM as the chapters by Els Van Hoecke, Jennifer J. Hinton, Barbara E. Hinton and Marcello E. Veiga, Geoff Crispin and Jeanette Graulau show. The third part, on the place of gender in mining economies, also includes Linqing Yao's detailed study of data on women's participation in the coal mining industry in China. She shows that in spite of the apparent absence of women in official mining figures, they continue to work in the mines in China, albeit in far fewer numbers than those in ASM in Africa reported by Hinton et al., (2003) They show that many women in Africa work part-time at informal mining operations or occupy 'ancillary roles', yet their participation is viewed as being less relevant. Women are often associated with transporting and processing materials as opposed to digging, and hence are not always identified as 'miners'. Els Van Hoecke uses her activist knowledge of the ground in writing about Bolivian women in the mines, a very different perspective from that presented in Nash's (1979) classic ethnography. Minerva Chaloping-March describes the nature and extent of women's work in formal and informal mining in the Philippines, and traces the various factors that explain why their participation in the sector expands or contracts. Geoff Crispin describes his experiences in finding women workers during his work on a project on small scale miners in Papua New Guinea.

In the last part, we take our readers through four chapters to show that if the mining industry is becoming globalized with its attendant problems, there are also innumerable and small ways women are using it to their advantage and resisting the oppression that connection to a globalized economy entails. Women's labour has been crucial in the expansion of capitalism and the reproduction of its modes of production in the mining frontier. Gender hierarchies in developing countries facilitate certain forms of capitalism, capitalism that is intersected by patriarchy. Jeannette Graulau, in her chapter on women's labour in 'peasant mining' of gold in the Brazilian Amazon, describes the structural vulnerabilities of organizing women's labour in mining communities. In particular she illuminates the gendered aspects of production and technology and their intersection with historical relations of class, colonialism, ethnicity and Westernization. She draws upon debates of feminist political economy and geography to show how women's work in the Brazilian Amazonia is functionally and spatially connected to the contemporary process of capital accumulation. The essential standpoint nature of Ingrid Macdonald's chapter is highlighted by her use of the global rights framework that has been put in place by the international agencies in discussing the steps that need to be considered in engendering the mines. This international perspective is again contrasted by local resistances such as those by women in Bhowra colliery in Jharia described by Lindsay Barnes in her chapter. Lindsay in her chapter provides examples of women building upon their collective strength to contest the powerful forces formerly beyond their control, and how small resistances can build up in yielding positive results for women. The last chapter is by Kuntala Lahiri-Dutt which throws light on yet another aspect of globalization of mining. It explores the attitudes to and perceptions of women truck operators in a large colliery operation in a remote part of East Kalimantan, Indonesia. It shows that the gender impacts of globalized mining cannot be put simply in black and white, seeing women either as victims of agents. It reveals the need to transcend the anti-mining bias of the development literature, an urgent theoretical task if feminists intend to understand the complexities of mineral production in the developing world. Gendering the mines might be the only way to generate the theoretical and policy tools to challenge the neo-liberalist doctrines that favour corporate mining capital and its global expansion.

Gender identities and inequalities are socially and historically constructed, and sustained in the mines in a multitude of ways. To what extent the experiences of workers and particularly women workers in advanced countries can be of relevance when examining pit women and others from the developing countries? Can we talk about an 'average' pit woman in all the mining practices in the world? Is there a representative 'mining woman'? Can we locate pit women's experiences as those mediated by class as a whole, internationally, or even in individual countries? By scrutinizing the 'pit women and others' of countries other than Europe and North America, both in the past and at present, this book not only brings out the roles and contributions of women from a range of countries but connects their histories with the realities of gender in contemporary mining communities. Through the chapters written by a range of experts with diverse experiences, *Women Miners in Developing Countries* as a volume illustrates women's productive roles as miners and engages with the attitudes and issues related to such roles. The chapters reveal those particular cultural, political and economic contexts in which women and men's work in the mines becomes gendered. They also dislodge the conventional notions of women in developing countries as victims of patriarchy and examine the various ways women express their agency. This book sheds light on the ways that women have participated in mining in the past as well as exploring the continuities into a present, where women, although still labouring against sexism in various forms, engage in a range of jobs in mining. In rendering women miners in developing countries visible, past and present, we hope also to generate questions about the nature of mining development itself. It brings forth the rights of women to be heard and to participate in the processes – from consultations, compensations and jobs, to closure (Macdonald and Rowland, 2002; Macintyre, 2002).

There is a considerable development literature arguing that the introduction of large-scale mining has so far adversely affected women more than men (among others, see Robinson, 1986, 1988; Tauli-Corpuz, 1997; Bhanumathi, 2003a, b; Bose, 2004). Similarly, mine closure pushes women into more vulnerable forms of labour. Again, the gaze tends to be focused upon large scale rather than small scale mining, without understanding the complexities of mineral production affecting women's livelihoods and work. Many of the social, economic, environmental and political ills associated with mining projects could be dramatically reduced if women's voices were heard and their participation sought (Ranchod, 2001). Historically, women's participation in the labour force has given them greater economic autonomy and has been the basis for politicization and demands for gender equity. Can participation in mining be empowering for women in developing countries? We hope that this book will stimulate further enquiry into the position of women in mining and create a path for consolidating a Gender and Mining approach in developing countries.

References

Allen, V. L., 1981. *The Militancy of the British Miners*, Moor Press, Shipley.

Amutabi, M. and Lutta-Mukhebi, M., 2001. 'Gender and Mining in Kenya: The Case of Mukibira Mines in Vihiga District', *A Journal of Culture and African Women Studies* (JENdA), vol. 1, no. 2, available from http://www.jendajournal.com/vol1.2/amutabi_lutta-mekhebi.html (retrieved on 22 August 2005).

Ballard, Chris and Banks, Glenn, 2003. 'Resource Wars: The Anthropology of Mining', *Annual Review of Anthropology*, vol. 32, pp. 287-313.

Bhanumathi, K., 2003a. *Women and Mining in India*, Report of National Seminar held from 25 April to 26 April, 2003, at India International Centre, New Delhi, Mines, Minerals and People, Hyderabad.

Bhanumathi, K., 2003b. 'Women and Mining in India', Paper presented at the Oxfam Community Aid Abroad *International Conference on Women and Mining*, Melbourne.

Bose, Sharmistha, 2004. 'Positioning Women within the Environmental Justice Framework: A Case from the Mining Sector', *Gender, Technology and Development*, vol. 8, no. 3, pp. 407-12.

Campbell, Beatrix, 1984. *Wigan Pier Revisited: Poverty and Politics in the Eighties*, Virago, London.

Carr, Griselda, 2003. 'Pit Women: Coal Communities in Northern England in the Early Twentieth Century', *Journal of Marriage and Family*, Feb 2003, vol. 65, Issue 1, pp. 268-269.

Coleman, McAllister, 1943. *Men and Coal*, Farrar and Rhinehart, Inc., New York and Toronto.

Connell, J and Howitt, R (eds), 1991. *Mining and Indigenous Peoples in Australasia*, Sydney University Press, Sydney.

Connell, R.W., 1987. *Gender and Power*, Allen and Unwin, Sydney.

Dange, S. A., 1945. *Death Pits in our Land: How 200,000 Indian Miners Live and Work*, Asia Publishing House, Bombay.

Douglass, David, 1972. *Pit Life in County Durham: Rank and File Movements and Workers' Control*, History workshop pamphlets, Ruskin College, Oxford.

Eddy, T. M., 1854. 'Women in the British Mines', *The Ladies' Repository*, vol. 14, no. 7, pp. 295-298 (available from The Making of America Journal).

Ericsson, Magnus and Noras, Penti 2005. 'A Note on Minerals-based Sustainable Development', *Minerals and Energy*, vol. 20, no. 1, pp. 29-39.

Espley, Samantha, Francis, Helen and Castonguay, Marie-Josee, 2004. *Hardrock Mining and the Evolution of the Roles for Women*, Inco Limited, Copper Cliff, Ontario.

Eveline, Joan, 1995. *Stories of Heavy, Dirty and Limp: Protecting the Institution of Men's Work*, Administration, Compliance and Governability Program, Research School of Social Sciences, The Australian National University.

Garside, W. R., 1971. *The Durham Miners 1919–1960*, George Allen and Unwin, London.

Gibson, Katherine, 1992. 'Hewers of Cake and Drawers of Tea: Women, Industrial Restructuring and Class Processes on the Coalfields of Central Queensland', *Rethinking Marxism*, vol. 5, no. 4, Winter, pp. 29-56.

Gibson-Graham, J. K., 1994. 'Stuffed if I Know! Reflections on Post-modern Feminist Social Research, *Gender, Place and Culture*, vol. 1, no. 2, pp. 205-224.

Goodrich, Carter, 1920. *The Frontier of Control: A Study of British Workshop Politics*, Harcourt, Brace and Howe, New York.

Goodrich, Carter, 1925. *The Miner's Freedom: A Study of the Working Life in a Changing Industry*, Marshall Jones, Boston.

Harrison, R. (ed), 1978. *Independent Collier: The Ccoal Miner as Archetypal Proletarian Reconsidered*, Harvester Press, Sussex.

Heemskerk, M., 2000. *Gender and Gold Mining: The Case of Maroons of Suriname*, Working Paper 269, Women and International Development, Michigan State University, East Lansing.

Hentschel, Thomas, Hruschka, F. and Priester, M., 2002. *Global Report on Artisanal and Small-scale Mining, Mines Minerals and Sustainable Development*, International Institute of Environment and Development, London.

Hinton, Jennifer, 2003. 'Summary Report of the Workshop on Women in Artisanal and Small Scale Mining', Report from *Women in Mining: Voices for Change Conference* in Madang, Papua New Guinea, August 6-10.

Hinton, J. J., Veiga, M. M. and Beinhoff, C., 2003. 'Women and Artisanal Mining: Gender Roles and the Road Ahead', in G. Hilson (ed) *The Socio-economic Impacts of Artisanal and Small-scale Mining in Developing Countries*, A. A. Balkema/Swets and Zeitlinger Publishers, Rotterdam, Chapter 11.

Hipwell, W., Mamen, K., Weitzner, V. and Whiteman, G., 2002. *Aboriginal People and Mining in Canada: Consultation, Participation and Prospects for Change*, Working Discussion Paper Prepared for the North-South Institute.

International Council on Mining and Metals/World Bank, 2005. *Community Development Toolkit*, URS, ICMM/Washington, DC.

Jennings, N., 1999. 'Social and Labour Issues in Small Scale Mines', Report for Discussion at the *Tripartite Meeting on Social and Labour Issues in Small Scale Mines*, International Labour Office, Geneva.

John, Angela V., 1980. *By the Sweat of their Brow: Women Workers at Victorian Coal Mines*, Croom Helm, London.

Kerr, C. and Siegel, A., 1954. 'The Interindustry Propensity to Strike', in A. Kornhauser, R. Dubin and A. M. Ross (eds) *Industrial Conflict*, McGraw Hill, New York.

Koczberski, Gina, 1998. 'Women in Development: A Critical Analysis', *Third World Quarterly*, vol. 19, no. 3, pp. 395-409.

Labonne, B and Gilman, J., 1999. 'Towards Building Sustainable Livelihoods in the Artisanal Mining Communities', Paper presented at the *Tripartite Meeting on Social and Labour Issues in Small Scale Mines*, Geneva.

Lahiri-Dutt, Kuntala, 2000a. 'Gendered Exclusion: Women Workers in Indian Coalmines', Paper presented at the *Workshop on Regional Histories of Women*, School of Women's Studies, Jadavpur University and Department of History, Calcutta University.

Lahiri-Dutt, Kuntala, 2000b. 'Gender Dimension of Mining: Women Workers in Indian Coalmines', *Mining Environmental Management*, London, September, pp. 19-21.

Lahiri-Dutt, Kuntala, 2001. 'From Gin Girls to Scavengers: Women in Raniganj Collieries', *Economic and Political Weekly*, vol. XXXVI, no. 44, pp. 4213-21.

Lahiri-Dutt, Kuntala, 2003. 'Not a Small Job: Stone Quarrying and Women Workers in the Rajmahal Traps in Eastern India', in Gavin Hilson (ed) *The Socio-Economic Impacts of Artisanal and Small-scale Mining in Developing Countries*, AA Balkema, a division of Swets and Zeitliger Publishers, The Netherlands.

Lahiri-Dutt, Kuntala and Robinson, Kathryn, forthcoming in 2006. 'Period Leave at the Coalface', in Lyn Parker (ed) *Women and Work in Indonesia*, University of Western Australia.

Library of Congress, 2001. *A Library of Congress Guide for the Study of Women's History and Culture in the United States*, Library of Congress, Washington DC.

Massey, D. and McDowell, L., 1984. 'A Woman's Place', in D. Massey and J. Allen (eds) *Geography Matters: A Reader*, Cambridge University Press/Open University, Cambridge, pp. 128-147.

Macdonald, Ingrid and Rowland, Clare, 2002. *Tunnel Vision: Women, Mining and Communities*, Oxfam Community Aid Abroad, Melbourne.

Macintyre, Martha, 2003. 'Women and Mining Projects in Papua New Guinea: Problems of Consultation, Representation and Women's Rights as Citizens', Paper presented at Oxfam Community Aid Abroad *International Conference on Women and Mining*, Melbourne.

Metcalfe, A., 1987. 'Manning the Mines: Organizing Women Out of Class Struggle', *Australian Feminist Studies*, vol. 4, Autumn, pp. 73-96.

Metcalfe, A., 1988. *For Freedom and Dignity: Historical Agency and Class Structures in the Coalfields of NSW*, Allen and Unwin, Sydney.

Mines, Minerals and Sustainable Development (MMSD), 2002. *Breaking New Ground, MMSD Final Report*, International Institute of Environment and Development, London.

Mitchell, W., 1975. 'Wives of the Labour Movement', in A. Curthoys, S. Eade and P. Spearritt (eds) *Women at Work*, Australian Society for the Study of Labour History, Canberra, pp. 1-14.

MBDA, 2003. *Ninja Gold Miners of Mongolia: Assistance to Policy Formulation for the Informal Gold Mining Sub-sector in Mongolia*, Final Report, Canada Fund Mongolia, Mongolian Business Development Agency.

Moore, Marat, 1996. *Women in the Mines: Stories of Life and Work*, Twayne Publishers, New York.

Moore, Robert, 1974. *Pit-men, Preachers and Politics: The Effects of Methodism in a Durham Community*, Cambridge University Press, London.

Nash, June, 1979. *We Eat the Mine, the Mines Eat Us: Dependency and Exploitation in Bolivian Tin Mines*, Columbia University Press, New York.

Noestaller, R., Heemskerk, M., Hruschka, F. and Drechsler, B. 2004. *Program for Improvements to the Profiling of Artisanal and Small-scale Mining Activities in Africa and the Implementation of Baseline Surveys*, Communities and Small Scale Mining, World Bank, Washington DC.

Orwell, George, 1937. 'Down the Mine', in *Collected Essays*, Secker and Warburg, London.

Parkinson, George, 1920. *True Stories of Durham Pit Life*, London.

Parpart, J.L., 1983. *Labour and Capital on the African Copperbelt*, Temple, Philadelphia.

Ranchod, Sarita, 2001. *Gender and Mining: Workplace*, Research Topic 3, Mines, Minerals and Sustainable Development South Africa, and African Institute of Corporate Citizenship.

Robinson, Kathryn M., 1986. *Stepchildren of Progress: The Political Economy of Development in an Indonesian Mining Town*, State University of New York Press, Albany.

Robinson, Kathryn M., 1988. 'What Kind of Freedom is Cutting your Hair?', in Glen Chandler, Norma Sullivan and Jan Branson (eds) *Development and Displacement: Women in Southeast Asia*, Centre of Southeast Asian Studies, Monash University, Clayton, pp. 63-78.

Robinson, Kathryn M., 1996. 'Women, Mining and Development', in Richard Howitt, John Connell and Philip Hirsch (eds) *Resources, Nations and Indigenous Peoples: Case Studies from Australasia, Melanasia and Southeast Asia*, Oxford University Press, Oxford, pp. 137-149.

Roonwal, G. S., Shahriar, K., and Ranjbar, H. (eds), 2005. *Mineral Resources and Development*, Daya Publishing House, Delhi.

Rothermund, Indira, 1994. 'Women in a Coal Mining Area', *Indian Journal of Social Science*, vol. 7, no. 3-4, pp. 251-64.

Tauli-Corpuz, Victoria, 1997. 'The Globalization of Mining and its Impact and Challenges for Women', Paper presented in the *International Conference on Women and Mining*, Baguio City, Third World Network.

PART I
Reconstructing Gendered Histories of Mines

Chapter 1

Women Miners: Here and There, Now and Then[1]

Gill Burke

Introduction

This chapter looks at women miners in Asia at a general and contextual level, mainly during the modern period. It suggests that in the majority of instances women have always participated in small scale, artisanal mining and, indeed, continue to do so in many countries today. In larger scale operations however, women's participation apparently occurred only at the early stages of industrialized mining. The reasons for this and the subsequent development of the masculinist model for mine workers will be examined.

Although the main focus of the chapter is historical it attempts to find linkages with contemporary practice in developing countries. The purpose is to present material at a general level, a *tour d'horizon*, that sets the scene for the detailed country-specific chapters that follow. I acknowledge that in the process I may raise more questions than I answer. I also acknowledge – and hope to avoid – the risk of falling into the 'universalism' rightly criticized by Chakrabarty (1989) with, indeed, the added risk of being accused of 'Orientalism' when testing whether Asian experience fits western constructs. In addition, as an Australian, I may be accused of 'cultural cringe' for giving so many European examples – what a risky business it all is. I could justify it but prefer simply to press on.

Women were and are involved in mining in three different ways. They can be miners – actually involved in the extractive process either below ground in large mines or in small scale artisanal mines. Or they can be workers at surface – involved in the sorting, crushing or preparation of coal or ore. Nowadays there are other women working at surface as well, as clerks, secretaries, nurses, but I am excluding them from this examination. Being a secretary in a mining company is not much different to being a secretary in any other company. The third

[1] This chapter is based on a paper given to the *Women in Asia* Conference, Melbourne, October 1993. The comments of participants are gratefully acknowledged and have generally been incorporated.

involvement for women is as members of the mining community – responsible for the reproduction of the workforce on a day-to-day as well as a generational basis. This will be touched on but not to any great extent.

Let's begin by setting the context for these women workers. It is crucial to emphasize that it is as *miners* that they are being examined. Thus mining provides the perspective, framework and parameters for any analysis. To have it otherwise is simply to reprise old debates by writers of women's history; to have yet another anthropological take on yet another area where women are (probably/usually) disadvantaged or to devolve into yet another study of women in development. In addition, mining as the contextual framework allows for comparisons over time and place in what otherwise is a very, very heterogeneous field. As Rounaq Jahan (1982) has warned:

> Any generalization about Asian women is hazardous. Asia is a continent of immense diversity and vast contrasts. Asia shows women in powerful positions . . . yet millions of the world's most oppressed women live in Asia. In major Asian cities one finds thousands of well educated and well trained professional women but the vast countryside is full of millions of illiterate, poor women working long hours to find and carry water, firewood and food for the family. Class cleavages and rural-urban differences result in wide disparity in life styles and opportunity structures for different groups of Asian women. Any report on Asia has to take cognizance of such differences.

Whilst it's true that women miners are indeed 'hidden from history', any attempt at 'recovery' needs contextualizing within the problematics of mining history. This is especially the case with Asian women miners. International comparisons and linkages can help especially because of the lack of visibility, both in public perception and most scholarly work, of women miners in Asia and, indeed, almost everywhere else. In particular, the low profile of women miners takes on a special quality due to the contrasting high public profile of male miners.

Mining is an ancient, global industry. Metal mining and its concomitant metallurgy was highly developed many thousands of years before the Christian era. Mining tin and copper, then smelting them to make bronze, afforded artifacts of a high order. Notable early examples are the bracelets and spearheads found at Ban Chiang in Northern Thailand and the large incense burners of cast bronze of Shang period China which all date from around 3000BCE (Raymond, 1984)[2]. Coal mining also has a long history but remained highly localized around areas where outcrops were easily accessible. Transport costs precluded wider use of bulky coal especially where alternative fuel sources were easily available. Precious and base metals had few substitutes, and far higher use-value. Thus metal mining, smelting

---

[2] In Raymond's triumphalist and occasionally mind-boggling account of the history of metallurgy the emphasis is indeed on 'man'kind. Neither a mention nor any picture of women can be found.

and manufacture was far more widely spread[3]. This remained true in both Asia and Europe from pre-modern times.

From early times it seems that miners were perceived as 'other' – as a group apart, independent and lawless. In pre-industrial societies as disparate as Ming China and feudal Europe special laws emphasized this separateness. The coming of industrialization did not fundamentally alter this perception; indeed, if anything, it heightened it. By modern times the miner as 'other' had become characterized by two, not necessarily exclusive, forms of behaviour – the one, reckless and given to drinking, gambling, fighting and so forth; the other as highly politicized proletarian in the forefront of industrial action. Metcalfe (1988), in his study of the New South Wales coalfields, terms the ideal types of these two forms the 'Larrikin' and the 'Respectable' mode of class struggle. Although it is possible to suggest – and I will give some examples, that women mineworkers often showed similar independence traits it is the male miner who provides the dominant image. Thus the Ying of invisible women is overshadowed by the Yang of only too visible men. Gender therefore is central to the business of unpacking miners' history.

How these images of the male miner developed will be discussed after first examining women miners since, I suggest, the one is indissolubly linked to the other. If indeed Metcalfe is correct in posing ideal types, these proletarian workers in an industrial context that he describes have come a very long way from the shared family enterprise that characterized earliest mining and still typifies much small-scale mining today.

Women as miners

The organization of the mining process has, inevitably, a strong geological determinism. The early period of mineral development, at different points of time in different countries, begins with outcrops or with 'streaming' – extraction of alluvial minerals from rivers. Then, unless the deposits continue as alluvial, mining goes underground, by various means, from there. With some exceptions it is not until the modern period that geological prospecting leads to shaft-sinking, open-cuts or dredge-pond building as the first step. Thus early mines were generally small.

---

[3] An additional factor was the diverse location of mineral deposits. For example the tin and copper needed to make bronze were not necessarily contiguous. Tin from Cornwall in Britain was reportedly exported to Cyprus (where there was a large copper mine) via Phoenician traders as early as 2500BCE. Tin from Nigeria was similarly widely exported – first north, by Arabic traders, and thence to south Asia, see Cline, Walter, (1937) *Mining and Metallurgy in Negro Africa*, Winsconsin, quoted in Freund, Bill, (1981) *Capital and Labour in the Nigerian Tin Mines*, Essex.

They were/are also, generally, family-worked operations[4]. It is in this context that most women become miners. In many instances the family work unit divides its time between mining and agriculture, with mining as a 'cash-crop' alongside food production. Lest this rural idyll seduces, let us not forget that even in this context mining (and quarrying[5]) is hard and dirty physical labour for all concerned. In addition, the family may not be free labour, gaining full benefit from their output, but instead be contractors, or corvée labourers or in debt bondage to others who own the mineral deposits.

Examining mining in Pre-conquest South America, Taussig (1980) describes a small-scale activity run by the Incas as a monopoly, with labour provided by a rotating corvée that was not particularly burdensome to the miners. It was a minor economic activity because the Incas valued silver and gold only as a form of ornamentation and not as a form of wealth:

> One of Pisarro's secretaries describes mines of 60 to 240 feet deep which were worked by a few score miners, both men and women… the miners worked only four months of the year … having to return to their villages to resume their agricultural responsibilities… Writing in 1588, Acosta stated that the Inca entertained the miners with all they needed for their expenses and that mining 'was no servitude to them but rather a pleasing life'.

It would be misleading, I think, to simply term this form of mining 'pre-industrial' since in many countries other more complex forms of mine and mining organization developed alongside these even in the pre-modern period. And, indeed, small mines exist alongside large, complex ones today. What is true however is that industrialization of mining operations changes the nature of women's work and ultimately forces them out of directly being miners.

Inca gold mining apart, I have so far found only one past example of women working as underground miners of precious or base metals. This was in the gold mines of Mysore, India, during the 19th century. The presence there of family work groups underground was commented on by the Inspector of Mines who was relieved to see that in most mines this practice was being phased out (BSP, 1894). Elsewhere, there are women gold miners aplenty in small scale mines, but as these mines deepen the women stay at surface. Quite why this is so is not clear, reasons of geology, of required skills, of belief and superstition come to mind[6]. Even today,

---

[4] I use the present tense here to allow for different periods of development in different countries/continents since the basic components/conditions of small scale mining are really little changed over the centuries save in location. Even the tools and equipment are basically the same.

[5] I treat quarrying – for stone and industrial minerals – as a sub-set of mining and do not discuss it in detail.

[6] The 'rape' of mother earth by metalliferous miners is frequently very clearly articulated and ritual actions taken to avoid the consequences. See Taussig above and also

in many countries with highly developed and mechanized metalliferous mining industries, there is a reluctance to allow women underground even as visitors let alone as workers[7]. It is coal mining that provides the many examples of women underground miners.

Unlike metalliferous mining, coal mining requires muscle rather than much skill[8]. In unmechanized mines almost anyone can be a collier. Indeed, Harrison (1978) has suggested this is a significant causal factor in the development of western coal miners' industrial militancy. The toughest muscle is needed at the actual coalface to hew coal from rock. This seems to have always been done by men and the 'Big Hewer' rapidly became the heroic face of mining in many countries. Even when women worked alongside it was in an inferior position, gathering the hewn coal, loading it into containers and tramming or carrying it to surface.

This distinction can be clearly seen in a drawing of a man and woman at the Chikuhō district coal mines, Japan[9] (Figure 1.1).

The gallery they are working in is too low to stand. He lies flat, hewing the coalface with a pick; she crouches kneeling, her bare breasts hanging down, and rakes the hewn coal into a basket. Despite the risk of explosive fire-damp, the only light comes from the smoking wick of a small oil lamp[10].

Nash, June, (1979) *We Eat the Mines and the Mines Eat Us: Dependency and Exploitation in Bolivian Tin Mines,* New York.

[7] The writer's personal experience backs this up. There is are some ex-miners in Cornwall, UK, who blamed the closure of Wheal Jane mine (1972) and South Crofty mine (1998) not on fluctuations in the international metal markets but on the fact that a woman (myself) went underground in each case a few months prior to closure. In addition, I was dropped from a consulting team for a project in Bolivia, at the request of the Bolivian mine management, as it would have involved my going underground.

[8] This generalization applied to mines prior to full mechanization. And, of course, even in completely unmechanized coal mines some basic skills and knacks were necessary. Worksongs, such as the 19th century British '*Jowl, jowl and listen lad and hear the coalface working*.' taught such skills. But the skill difference between coal and metalifferous mining is indisputable. One example is the vast numbers of Cornish miners who went overseas during the 19th century. I have argued elsewhere they did so not because they were in poverty at home (although they were) but because they had highly exportable skills. They went as miners to a mine not as migrants pushed by poverty and seeking whatever work they could (see Burke, Gill, (1984) 'The Cornish Diaspoa of the Nineteenth Century' in Marks, Shula and Richardson, Peter *International Labour Migration: Historical Perspectives*, London, pp. 57-76). A closer look at Chinese miners in Australia during the same period might, I suggest, show a similar pattern not least since some may have already worked in California during the 1849 gold rush.

[9] Original source unknown, but used as illustration in the German Mining Museum, Bochum, *Frauen und Bergbau: Zeugnisse aus fünf Jahrhunderten*, Germany, 1989, p. 55.

[10] The 'safety lamp' that puts the naked flame behind gauze was invented in early 19th century Britain by Humphrey Davy but clearly took some considerable time to reach Meiji Japan. Use of the lamp alone however is not enough to prevent gas explosions, it must be

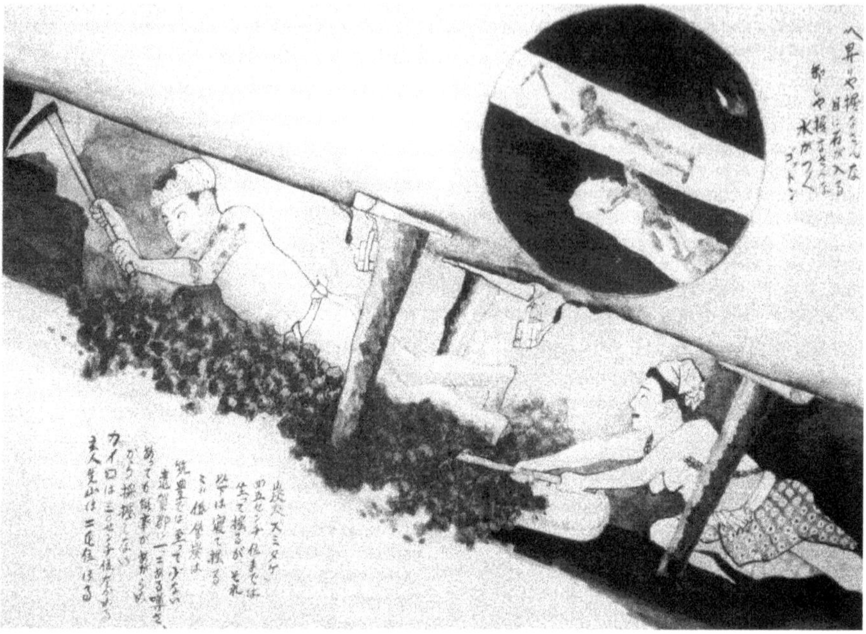

Figure 1.1 Male and female working team, Chikuhō, 1890 (Yamamoto Sekubei)

Source: Reproduced from Frauen und Bergbau, Deutsches Bergbau-Museums, Bochum

In 1910, a series of photographs taken underground at mines in Chikuhō (Murray, 1982, p. 334) showed near-naked men and women hewing, raking and tramming. As Hane (1982) quotes, in Japanese coal mines the heat underground was intense, the air stifling and difficult to breathe:

> The women's job was to transport the coal that had been mined. The coal was loaded into a 4-foot square wooden box. The bottom of the box had metal runners. We had to pull this box with a sash over our shoulders. It was hard work. Where it was uphill, there were wooden logs to serve as rails… Going downhill, when the angle of the slope was over thirty degrees, we would get on our hands and knees, grab the log rails firmly, hold back the box with our heads and slowly crawl down backward. With a lamp in our mouths and with our heads holding back the box full of coal we would feel our way down, inch by inch, with the tip of our straw sandals. If you slipped, it wouldn't be only you who got hurt, because there were others ahead of you. Some of the women had babies on their backs while hauling coal. Since water was dripping all the time it was

---

accompanied by proper ventilation methods. It is worth noting here that as late as 1963 four hundred and sixty-four miners were killed in a gas explosion at Miike Colliery, see Allen, Matthew, (1991) *Stories from the Coal Mouth: An Ethnographic Account of a Japanese Coalmining Community's Past and Present*, University of Sydney PhD.

> slippery ... It was really hard to get a loaded coal box moving. We'd push it with our heads and our rears.

This aligns closely with statements by British women and children miners to the Royal Commission of 1842. One woman commenting that the chains that harnessed a woman to a coal sledge hurt 'more if we are in the family way'. Illustrations to the Commissioners' Report showed children and women chained to heavy sledges or carrying half-hundredweight baskets up ladders. One shows a ragged old woman working a windlass that lowers a half-naked girl bestraddling her boy companion into a mine (BPP, 1842). What outraged Victorian society however, were not so much the working conditions as the proximity of naked bodies. As the Commissioners put it:

> That in the districts in which females are taken down into the coal mines, both sexes are employed together in precisely the same kind of labour, and work for the same number of hours; that the girls and boys, and the young men and young women, and even married women and women with child, commonly work almost naked, and the men, in many mines quite naked; and that all classes of witnesses bear testimony to *the demoralizing influence of the employment of females under ground* [my italics].

And that was the end for British women miners. The Mines Act 1842 prohibited women from working underground. Elsewhere in Europe women continued to work underground tramming the heavy boxes of coal, so vividly described by Zola, albeit as fiction, in his monumental work *Germinal.*

> ...to get a firm position by straddling the legs and planting the feet against the timbering on either side of the gallery. The body had to be bent forward and the arms kept stiff, so that one could push with all the muscles of the shoulders and haunches... she ran along with her behind so high and her hands so low that she seemed to be trotting on all fours like one of those dwarf animals in a circus. She sweated and panted and her joints cracked, but she never complained for familiarity had brought apathy and you would have thought that being doubled up like that was part of the normal course of human suffering.

But, whilst Second Empire France may have experienced a *frisson* reading about the explicit and violent sexual behaviour described in *Germinal*, there was no public outrage against the conditions depicted. Women continued to work underground in France and in Belgium they continued so working into the 20th century.

In Japan, women worked alongside men underground from well before the advent of large scale modern mining in the latter half of the 19th century, peaking at over 66,000 in about 1920. They stayed in the pits until 1933, when most female underground labour was banned. Exceptions were made for thin seams and mostly worked-out areas, and the ban was poorly enforced in seldom-inspected smaller mines. This ban was lifted for women 25 or older in 1939 amid a wartime labour shortage but reimposed by the postwar Occupation authorities. The last officially recognized women left the mines in 1947, although some women continued to work illegally in small mines until at least 1961 (Smith, 2005). Similarly in India, women worked underground in the smaller shallow mines, the *pukuriya khads*. In 1920 they accounted for 37 per cent of the total colliery workforce. Regulations to restrict female mine employment were drawn up in 1928, but in 1932 there were still over 11,000 women employed at the Bihar mines alone (John, 1980). Then an ILO ban and the Indian Mines Act forbade them. Like Japan, women were brought back underground during World War II but once again were totally prohibited after Independence in 1947[11].

In China, until the Cultural Revolution, women rarely went underground. In Chenxi county, Hunan, women worked underground as hauliers in the 1930s but this was a very rare phenomenon. Nor, in early times, did women work in mining at all. Wright (1984) attributes this to the prevalence of foot binding in northern China where the coalmines were mainly located. Nor were women miners found initially in Vietnam, where large-scale mining began under French colonial rule in the late 19th century – although, of course, they had been working in small scale mines over the centuries. However, the need to obtain sufficient and regular mine labour and to prevent rampant seasonal absenteeism – as men returned to their villages for Tet or for planting and harvesting times – forced the French mine companies to coerce workers into new villages at the mines. To achieve this they allowed women and children to become part of the mine workforce. By the 1930s mechanization had shifted the profile of the workforce more closely, but not entirely, towards male workers (Murray, 1982).

For geological reasons deep underground mining is not characteristic in Malaysia, Thailand or Indonesia. The geology of tin deposits there meant mining was by panning, hydraulic sluicing or dredge. In any case, in each of these countries tin mining was initially done by imported Chinese contract labour (Yip, 1969; Cushaman, 1991; Heidhues, 1992) so there was no established pattern of family enterprise. In later years women worked in the tin industry mainly as '*dulang* washers' sieving and panning the waste slimes. There were coal deposits in Malaysia and Indonesia, but the Malayan Batu Arang collieries which opened in 1913 did not use women as mine workers and the Indonesian coal mines at Ombilin and Pulau Laut used convict labour.

In the Philippines archeological evidence suggests that mining and smelting gold, copper and iron was undertaken from ancient times, fostered and assisted perhaps by trade and economic links with China. In the early days of Spanish colonialism in the 16th century when the Philippines became part of the Hispanic quest for 'Gold, God and Glory', output from the gold mines was shipped to the Americas and thence to Spain. Women and children were employed alongside men, but the mines, despite potentially rich deposits, remained underdeveloped and unproductive. In the 19th century numerous ventures were started, for mining a

[11] Thanks to Kuntala Lahiri-Dutt for this information.

wide spread of minerals, with a labour force that still included women and children. All were spectacularly unsuccessful. As Lopez (1992) states:

> ...mining was not a priority. Prospecting was initially undertaken by the *adelantados* [pioneers], followed up by some persevering Spaniards, but the mining industry never really got off the ground. Native miners using panning and smelting technology acquired centuries earlier were *the* [his italics] mining industry... It was in this state that the Americans found the industry in the late 1890s.

The Americans brought industrialized mining and over time the industry grew substantially and included major mines such as Benguet and Lepanto. Labour statistics were kept from the early 20th century. By that time 'manpower' meant 'men'.

Hane suggests that working alongside men and producing as much gave women in Japanese mines a strong sense of equal worth (Hane, 1982) even though they received far lower pay. A similar sense is characteristic of women underground miners in the United States – a modern development since the 1970s. Mechanization and later pit closures, rather than outrage, took women from underground work in major mines in Japan, France, Belgium, India and elsewhere, and paradoxically it is mechanization that now allows American women miners to participate in work at the coal face.

Following the passage of the Equal Rights Amendment and a sexual discrimination suit in 1977, coal companies in Kentucky, Pennsylvania and Virginia began, very reluctantly, to employ women as underground workers. Under the slogan '*Women miners can dig it too!*' these women have survived opposition and harassment from bosses and male miners to become an acknowledged part of the workforce (Disney, 1981):

> Whatever anyone does or says to us, the rule is we do our cryin' in our own bathhouse, in private. We ain't gonna let the men know they can get to us. It's important they think we can take it and give it right back, otherwise we couldn't survive here.

By 1981 there were some 2,500 women miners in the US. Almost all became miners for economic reasons – it was the only way to support a family in those parts of rural America. Lack of seniority, plus a degree of male malice, meant that initially they were allotted the hardest, dirtiest jobs. And when cutbacks came, they were fired (Disney, 1981):

> That's when we all got close – the men and women who were going to lose their jobs. Now I'm tryin' to get rehired. I really love mining, there's nothing so exciting to me. I like having grown these muscles, I like being able to handle the job and I like earning good money.

The above quotation is worth noting given the general assumption that a mine is not a woman's place. In rural areas around the world where there is little or no other well-paid work, for a woman to go mining is a sound economic decision – provided the mine itself is reasonably well run and mechanized. Where this is not the case the focus should be on making it so for *all* miners rather than stating, as many commentators do that the work is 'too hard' for women.

Over time women became more accepted by male miners. It also became clear that work crews with women have better safety records and fewer accidents than all-male crews (Disney, 1981):

> The men are always proving how brave and macho they are. We women don't have that problem, we want conditions to be safe and we refuse to work in a section if it is dangerous.

That is fine, but the above quotes were made over twenty years ago. All major mines today are highly mechanized, the 'Big Hewer' a figure of history, indeed increasingly it is the open pit rather than underground working that characterizes major operations. Open pit mines work by blasting, then shoveling the split rock (by machine) into dumper trucks that take it to the separators (machines). Like the pits, the diggers and dumpers are huge but it does not need muscle to operate them, simply a Heavy Goods Vehicle Licence and some training. Yet where are the women in large numbers?

Visit any significant coal or metal mining operation today and look for women workers. Almost always they will be back in the office as clerks and secretaries. Some will be 'professionals', especially geologists, anthropologists and environmental officers. They are most unlikely to be miners. Ask why there are no women mineworkers and the answer usually is 'cultural reasons'[12]. 'Culture' apparently allows women to drive trucks at Lihir mine[13], Papua New Guinea (PNG), but not at Porgera mine in the PNG Highlands (where the prevailing culture of male violence has been enhanced rather than lessened by the presence of the mine). Where are women miners at the Freeport mine in West Papua/Irian Jaya, the world's largest gold mine? Does 'culture' prevent women working underground in South Africa (where the Minister of Mines is a woman)? Women led coal-cutting teams in China during the Cultural Revolution, but the few working alongside their comrades underground now do so illegally since women were prohibited in 1992 from underground work. No, women miners are back where they always have been – in small scale mines.

---

[12] 'Culture' in this context is defined not as part of the long standing and often complex academic debate but as 'the local men wouldn't like it'.

[13] I was told at Porgera that this is the case, but I don't know for certain.

Women at surface

Women have been surface workers both at metalliferous and coal mines from the moment these began. In both their tasks were much the same, sorting the mined rock, breaking large lumps with a hammer to enable them to be easily crushed, then sieving the crushed mass to separate mineral from dirt. The first European mining textbook, Agricola's *De Re Metallica*, published in 1556, has numerous woodblock illustrations of women doing these tasks.

Industrialization brought no initial change to the surface worker's situation above ground other than to increase their workloads. Increased mechanization, such as coal sorting belts, also initially increased the numbers of women employed. In many places they faced constant opposition from miner's Unions who saw them as cheap labour, liable to divide the workforce and pull down wages. Over time, of course, mechanization displaced and displaces women at surface.

One notable characteristic of women surface workers, one that seems ahistorical and multilocational, is their bold independence. They are far from the docile, quiet ideal type that defines women in the domestic sphere and especially in Asia. These are working women getting on with their lives. This is splendidly illustrated by a wood block print in *Kudō Zuroku* (Illustrated Book on the Smelting of Copper) first published in Japan in 1801[14]. The copper was from the Sumitomo mines at Besshi, started in the 17th century.

Entitled 'Crushing and Sorting the Ore' the print shows a group of women kneeling on mats (Figure 1.2). Each has a large rock in front of her on which she puts the lumps she will crush. Each holds a curved hammer – similar to the 'spalling' hammer in Cornish mines and used by women surface workers there. Bits of broken ore lie all around. They are busy, but also clearly chatting. One woman turns her face towards her neighbour, another, with her baby on her back, is being addressed by a tough grey-haired old biddy smoking a pipe. A kettle and tea-bowls stand on a shelf nearby.

Their demeanour contrasts with that of the men shown, in other illustrations, smelting and refining. These look grim as, near naked because of the heat, they remove slag or pour molten copper into moulds without a tea kettle in sight. In another illustration of women sieving and washing ore, one has her pipe and tobacco pouch resting just behind her. There is no denying the work is hard, but the womens' demeanour indicates that it is bearable and they are not brought down or immiserated by it.

---

[14] Facsimile edition with English translation by Zenryu Shirakawa, Connecticut, 1983.

Figure 1.2 Women crushing and sorting ore (Kudō Zuroku, Plate III)

Source: Reproduced with permission from the Burndy Library

Figure 1.3 Women sorting ore (Agricola, De Re Metallica)
Source: Reproduced from the1950 facsimile edition of Hoover and Hoover, 1912

Women are chatting too in one of the illustrations in *De Re Metallica* – although maybe 'chatting' is not quite the right term (Figure 1.3). They are standing in a wooden building at a long table covered with lumps and pieces of rock. One woman, bosom to the fore, is pointing at and clearly haranguing another who looks crossly back. Beside her another woman leans forward apparently thrusting a sharp pointed piece of rock towards the face of a man who leans anxiously backwards. There's no kettle, but a large pitcher hangs from a hook on the wall. Quiet and mild-mannered these women clearly are not.

This lack of subservience observed across the centuries may be one of the reasons such groups of independent women (factory workers are another) have been largely ignored by male historians and others who seem to have inherited without question the comments of earlier writers about women's true place being in the home. In Europe during the 19th and 20th centuries this was frequently stated and many men argued that female employment before marriage should be of a kind to inculcate domestic skills:

> The hard work is not the greatest calamity of which we complain, that is a mere physical evil; what we most deplore is, that when called to take upon themselves the duties of wife and mother they are totally unfit for them. How can the moral standard of society amongst the lower orders be raised, with mother and sisters of such education and example?[15]

Writers may not put it quite so bluntly today. But even after the growth of feminist interpretations of history, studies of mining women surface workers are few.

One notable exception is the work of Angela John (1980) on the 'pit brow lasses' who worked at surface in British coal mines. The work was hard – wheeling heavy barrows, dragging coal trucks, shaking large heavy sieves, and it was dirty. The women were always black with coal dust. John draws attention to the way in which 'blackness' for many in the 19th century was equated with 'dirt, sexuality, sin, the Empire, slavery and chaos' – the underside of respectable white British society. The rugged independence and 'mannish' language of the pit brow lasses did little to calm such fears and there were frequent attempts – often supported by male mine workers – to get the women's work ended. But the women loved their work and spoke out strongly for it, comparing factory and domestic work unfavourably to their healthy life in the open air.

That women worked at surface also in Asia is undeniable. For example, in Vietnam in the 1930s some three to four hundred women and children still worked at nights for the *Société Françaís des Charbonnages do Dong Trieu*, sifting coal and preparing it for treatment (Murray, 1982). In the Asian context, studies such as John (1980) are harder to find[16], especially in Western languages. There are several collections of oral histories of women coal miners in Japanese, such as Idegawa Yasuko's *The Mothers who Gave Birth to Fire*, based on interviews in the 1970s[17]. In Malaysia the few remaining '*dulang* washers' – women who work over the tin wastes – have been the subject of study but not, I think, of wider publication. This lack of easily accessible studies poses the question, 'are women still there to find?'

---

[15] Henwood, George *Cornwall's Mines and Miners*, first published as series, *Mining Journal*, London, 1857 and 1859. Quoted in Burke, Gill, (1986) 'The Decline of the Independent Bal Maiden', in Angela V. John (ed) *Unequal Opportunities: Women's Employment in England 1800–1918*, Oxford, p. 189.

[16] And, of course, are not always accessible to one without the language. Whilst I would not for one minute suggest that 'nothing is good if not in translation', the variety of languages in Asia – plus of course those of earlier colonialism, makes gaining an overview quite hard. Thus case studies, such as those presented below, are invaluable.

[17] Idegawa, Yasuko, *Hi o Unda Hahatachi: Onna kōfu kara no kikigaki* (The Mothers who Gave Birth to Fire: Stories from Women Miners), Fukuoka. Quoted in Allen, 1976.

I suggest it would be unusual today to find women working at surface in other than small to medium mines. Larger operations, as mentioned above, are highly mechanized. Mineral is crushed, separated, sorted by machine. This can be exemplified by contrasting another drawing from the Chikuhō coal mine region with the *Kudō Zuroko* print of a century earlier. This (Figure 1.4) shows two lines of women sorting coal as it passes them on a mechanized belt. Unlike their sisters toiling and sweating underground to win the coal, they are trim and neat. Their heads are covered by identical scarves; their short patterned coveralls are fronted by tidy aprons. They have hammers ready to hand to break any recalcitrant lumps. But they are not chatting. There is no tea kettle. They all appear to be young women and clearly none would ever smoke a pipe. Their whole attention is downwards, every eye focused on the passing coal. They look neither independent nor happy.

Women in the mining community

Mining communities worldwide tend to be isolated, reflecting the location of mineral deposits. Thus miners are often industrial workers but rural dwellers. Isolated communities have their own dynamics and mores, conformity to these mores is in turn reinforced by the isolation. In any context, there are few places where life is so visibly lived than in small rural communities.

Within these isolated communities the miner's wife is further isolated by virtue of the sexual division of labour. The miner mines and his wife provides the means whereby he can continue to mine. Thus she undertakes the reproduction of the labour force on a day-to-day basis – washing the dirty work-clothes (and, where there are no pithead baths, also washing the miner), providing food and comfort. She also undertakes the reproduction of the labour force on a generational basis by bearing his children. True, in long established mining communities she will have the support of family and kin, but this support will also act as a break on any attempt to assert her individuality. In most cases, given the demands of this service and the added exigencies of the man's shift work she will have little time or energy for such attempts.

The requirements of the domestic sphere meant that in the past (and still in many cases today), once family based mining was replaced by industrially organized operations, a miner's wife did not usually go 'out' to work although they often had and still do have some very necessary income-generating activities:

> What the husband earns is too little and really we all have to help out, like my making *salteñas*. Some women help out by knitting, others sew clothes, others make rugs, others sell things in the street. Some women can't help out and then the situation is really difficult.
>
> Barrios de Chungara, 1978

Figure 1.4 Female coal sorters, Chikuhō, 1910 (Yamamoto Sekubei)
Source: Reproduced from Frauen und Bergbau, Deutsches Bergbau-Museums, Bochum

The women's mining surface work described by Barrios de Chungara was undertaken in the main by single women, widows and unmarried or deserted mothers. What signs of 'independence' had manifest in a working young woman seemed to rapidly shut down after marriage. Quiet, apparently conformist – and poor, for although miners in general earned more than most rural workers their wages were still low, subject to deductions and often paid in arrears. Thus the miner's wife carried the whole burden of family budget management. Accident, injury to her man or just if he drank too much could cause the whole fragile edifice to collapse. In very many cases there was the added stress of living in mine company housing. The death of a miner could lead to eviction of his widow. The threat of eviction was often a method used by companies to force striking miners back to work.

The dependent role of the miner's wife can also be said to help guarantee the continuation of poverty in the mining communities. As Emberson-Bain (1993) has pointed out in reference to the Vatukoula mine in Fiji:

> Through their unpaid labour, Fijian women have contributed to sustaining a cheap labour system at Vatukoula. Women supplement the earnings of their husbands (in turn subsidizing the cost of male mine labour) by growing food and fishing for domestic consumption, gathering firewood… (Also) the rigorous demands of mine work and the

24 hour shift system that operates in the mining town, have brought changes in the traditional division of labour. Today it is the women who perform most, if not all, the garden work and their work load is further extended because of the low wage system mining town.

When these pressures and constraints could be no longer contained they would boil over into violent outbursts. 'Housewives riots' over the price of rice or bread[18] could trigger male miners first into supporting the women then striking and rioting themselves for extra pay, or better conditions. Not all were such dramatic occurrences as in the protest described in Zola's *Germinal*. There, after the hated shopkeeper had accidentally been killed during the riot the women crowded round his body jeering until one old woman reached down and tore off his genitals – 'Then Ma Brûlé stuck the whole thing on the end of her stick, raised it on high and carried it like a standard down the street followed by a rout of shrieking women' – but they were nonetheless often violent and often with disastrous consequences. In Japan for example, in 1918 a riot begun by women longshore workers in Toyama was taken up by miners in Ube and elsewhere (Hayashi, 1990). When their demands were not met they went on the rampage, the army was called in and thirteen people were killed. A further three hundred and fifty were indicted and sent to prison or fined. As Hane (1982) comments, the fact that these activities 'remained merely outbursts of pent-up frustrations made them basically ineffective in improving the miners' condition'; true enough, but given the institutionalized violence at Chikuhō, that continued long after 1918 until the mines finally closed in the 1970s, no other forms of action improved the miner's condition either. This will be looked at more fully below.

At a more mundane level, the particularities of the miner's work and the isolation of his wife often resulted in their having little to do with each other beyond the basics of social and conjugal interaction. As the long-time classic sociological study of British coal-mining, *Coal is our Life* (Dennis et al., 1974) puts it:

> Just as the men in their clubs talk mainly about their work and secondly about sport and NEVER [their capitals] about their homes and families, so do their wives talk first of all about *their* [their italics] work, i.e. the home and family and secondly within the range of things with which they are all immediately familiar.

This could be said to match the model of the subservient Asian wife (which itself is open to question), but is this really the case? Are/were western miner's wives different to their Asian sisters in isolation and subjugation? Relativities aside, I suspect not all that much. It is possible to say that the reproduction of the labour

---

[18] I'm always amazed when historians of the Labour and Trades Union Movements refer to such events as 'merely' bread riots – presumably as opposed to 'proper' strikes. Being driven to riot by the high price of bread doesn't seem very 'merely' to me.

force in large mining enterprises necessitates that women stay home wherever such enterprises are located, but closer, more detailed examination needs to be done.

A more recent dimension to the isolation of miners' wives is the growing use by companies of 'Fly-in-fly-out' (FIFO) strategies rather than building towns in remote locations – Argyle Mine in the Kimberley region of Australia and Red Dog mine in the Canadian arctic are two 'developed country' examples. Thus the mine worker is away from home months at a time, back for a short period and off again – becoming in effect more like a sailor or long distance truck driver in terms of managing a relationship (Stoney and Shrimpton, 1989).

However, in very many contexts in Asia and elsewhere, it is often only the senior mine workers – management, engineers, geologists, etc., who are flown out. That these people are generally white puts a neo-colonial spin to FIFO. The 'natives' continue to live around the mine, usually in purpose-built towns – see, for example, Freeport mine in West Papua, the major mines of PNG – Porgera, Tolukuma, Ok Tedi, Lihir. The resentment this engenders amongst indigenous mine workers has been noted by Robinson in her study of the Soroako Nickel Project, Sulawesi, Indonesia (Robinson, 1991). Similarly, in PNG, following protests during the mine development stage, some limits were placed on FIFO. But, since the local people had hoped initially that the whole mine workforce would be flown in and out, thereby maintaining a village-based way of life, these gains were more apparent than real. The need of mine companies to retain the indigenous workforce at the mine site echoes that of the earlier French colonial mining companies in Vietnam mentioned above. In both cases what Gerritsen and Mcintyre have termed the 'capital logic' of a mine company – the source of its finance, need to make a profit, timing of loan repayments, etc., all force a particular outcome (Gerritson and Mcintyre, 1991). For miners' wives, FIFO, whether for their husbands or the white bosses, brings little that's good.

The male miner

> We are women, we are strong, we are fighting for our lives
> Side by side with our men who work the nations mines
> United by the struggle, united by the past
> And it's - here we go! Here we go!
> For the Women of the Working Class[19].

---

[19] Flamin' Nerve, *Here We Go! For the Women of the Working Class* Song, Women Against Pit Closures Benefit Concerts, Britain, 1985. The tape of Flamin' Nerve songs, recorded in Sheffield during the 1984–1985 Miner's Strike, contains some very funny as well as highly political numbers – songs about the strike but also about a woman's lot. For example, *Don't Get Married Girls* has a chorus: 'You get him off to work/The Mighty Hunter is restored/And you're left at home with nothing/But the dreams you can't afford'.

The British coal miner's strike of 1984–1985 saw the emergence of a previously hidden group. The 'Women Against Pit Closures' movement began amongst the women of the northern coal mining communities – the wives, mothers, sisters and daughters of miners. Highly politicized and highly effective they were conscientious fund raisers who nonetheless posed a challenge to the ruling male hegemony of the coalfields. Previously (and in the minds of many men, still) it was unheard of for women in mining communities to speak out, to participate in 'men's' activities such as being on the picket line.

Although Britain is a developed country with compulsory schooling to mid adolescence, a small country where isolation is relative, it nonetheless provides an illustrative example in the context of this discussion. Coal mining and concomitant industrial action by miners in Britain were no new things, localized mining unions had begun to be established early in the 19th century and the first Miner's National Union in 1875. But although in the past there were food riots, the women in coal mining communities did not overtly and publicly join in protest alongside the miners for a further hundred and fifteen years. Why did it take so long? To find some answers it is necessary to look at the men.

In Western Europe and North America the miner holds a particular place in public perception. Despite being a matter of history the image of the 'Big Hewer' remains powerful. This is particularly the case in terms of political activism. Consider some past images – in Poland, striking coal miners alongside shipyard workers in Gdansk in support of Solidarity; striking miners in the then Soviet Union backing Yeltsin and thus contributing to the fall of Gorbachev; striking coal miners in Britain bringing down one Conservative government in 1974 but being routed by another in 1985[20]. In these and many other contexts the miner's industrial and political militancy has raised them to cultural icons.

Certainly the image of the 'Big Hewer' is a masculine one, but I would argue that it is also masculinist and that masculinism underpins and informs most popular culture and academic work on the subject. Any examination of this masculinism shows how central gender is to the debate. Consider the following, by a Professor of History no less (Allen, 1981):

> Coal mining evokes popular images of hard unrefined men, distinct and separate from other workers, hewing in mysterious dungeons of coal: dirty, strange men, in some ways frightening and for this reason repellent, yet attractive because they are masculine and sensuous ... The power of miners, however, does not lurk behind images; it is real, there for everyone to see and experience, resting on the importance of coal as a source of energy.

---

[20] In the context of this present discussion there is a certain irony in the fact that it was a woman – Prime Minister Margaret Thatcher – who singlemindedly if not singlehandedly (she did have the support of her government colleagues and several thousand riot police) eventually brought down the miners and virtually terminated the British coal industry.

Such a view was firmly slapped down by writer Beatrix Campbell (1984). But the very passion of her critique indicates the power of the masculinism she is criticizing:

> The socialist movement in Britain has been swept off its feet by the magic of masculinity, muscles and machinery. And in its star system the accolades go to the miners – they've been through hell, earth, fire and water to become hardened into heroes. It is masculinity at its most macho that seems to fascinate men.
>
> Miners are men's love objects. They bring together all the necessary elements of romance … It is the nature of the work that produces a tendency among men to see it as essential and elemental, all those images of men down in the abdomen of earth, raiding its womb for the fuel that makes the world go round. The intestinal metaphors foster the cult of this work as dark and dangerous, an exotic oppression … it constructs the miner as earth-man and earth-man is true man. And it completes the equation between some idea of elemental work and essential masculinity. This romance is duly mirrored in working-class politics – miners are the Clark Gables, the Reds of class struggle.

Why should this masculinism be so strongly held? Do miners believe their own publicity? Campbell suggests that, crucially, its roots lie in the circumstances surrounding the exclusion in Britain of women from underground work in 1842. Certainly successive Mines Acts reconfirmed this ban. Even the 1979 Sex Discrimination Act – unlike the United States two years earlier – 'retained for the present' the exclusion of women.

Does the same or similar masculinism underpin writing on Asian miners? Are male mine workers the heroes of the labour movement, archetypal proletarians, 'larrikins'? A brief examination suggests the answer to be 'Yes and No – and mainly No, but not for want of trying'.

That Asian male miners were/are as given to larrikin behaviour as their western brothers is not in doubt. Wright (1984) in his study of the Chinese coal industry, pointing to the miserable and poverty-stricken life most Chinese miners endured, quotes Li Dazhao who wrote in 1919: '…how the miners at Tangshan tried to finish two weeks work in one week and spent the other week in wild sprees of eating, drinking, whoring and gambling.'

Allen (1981), commenting on statements by one of his interviewees, remarks:

> At this point it must be said that for a miner to give up drinking there has to be a formidable reason because there is tremendous social pressure incumbent on a miner to drink with his mates. In this there are close parallels to the Australian notion of 'mateship'. That is, if a man does not drink with his mates he is not a 'real' man.

Similar examples could be found throughout Asia despite the awful and immiserating conditions under which miners lived and worked. In many cases these conditions were exacerbated by the fact of colonialism but for most miners their most immediate source of repression was their overseer. So many miners were contract labourers. They were recruited by the overseer who often also supervised work, ran the company store and had, in effect, power of life and death over the men. The overseer's goal was to maximize output at minimum cost to himself. The consequences for the miners were often malnutrition and an appallingly high accident and death-rate. The miner's response – at a general level, was to drink and/or to riot. By early in the 20th century however, these riots took on an increasingly politicized character.

During the 1920s and 1930s there were waves of strikes all over Asia as miners formed unions and began protesting their pay and conditions. For example, what Murray (1982) has termed 'the coal miners' dramatic entry onto the political scene' occurred in 1936 with a spontaneous work stoppage that soon spread throughout the Vietnamese coalfields[21]. Such disputes were increasingly politically informed and underpinned by the principles and growing strength of the communist party. The prime target for activists was the contract system. Thus, when Liu Shaoqi went to Pingxiang in 1922, some of the earliest lessons he tried to teach attempted to break down the dependent relationship between workers and contractors.

During 1936 in British Malaya, the Malayan Communist Party, aware of discontent at Batu Arang collieries began to work with the miners – whose consciousness had already been raised by the growth of a strong labour movement in China. By 1937 the Chinese miners at Batu Arang had formed a Coal Workers Union that then struck for increased wages. Indian workers there also stopped and a general strike of almost 5,000 colliery workers began. The strike was violent – and was also swiftly and violently crushed. Two hundred police and two companies of the Malay regiment were sent in, two people were killed and one hundred and sixteen arrested (Kaur, 1990).

The above suggests that perhaps the process to 'quintessential' if not 'archetypal' proletarian amongst Asian male miners was not dissimilar to that of their European or American counterparts. However, the larrikin/respectable dichotomy is less clear there, primarily as Asian miners' leaders had Communism as their 'respectable' ideology, whose illegality undermined its respectability.

In China and Japan the process of proletarianization proceeded alongside mechanization of the mines, but in both countries the mine workers' strength and solidarity was initially hampered. In China, the activities of secret societies and in Japan the hold the yakuza maintained at the mines were significant in muting if not stifling protest. These factors would have lasting implications. After the Second World War the process diverged. In China, the reaction against secret societies had helped produce the Revolution and subsequently a society of workers and peasants

[21] It is worth noting here that in 1938 anthracite from Indochina began to be marketed in the United States. Although it was cheaper to mine the coal and transport it from Indochina than to mine in West Virginia, the United Mine Workers Union was able to pressure the US government into erecting a high tariff in order 'not to let the standard of living of the Virginia miners be attacked' (Naville, Pierre, (1949) *La Guerre Du Vietnam*, Paris, quoted in Murray, p. 586). So much for international solidarity.

where it might be said that in theory at least *all* workers were heroes of the labour movement. In Japan even though there continued to be strikes and violent protests, the miners remained powerless, part of what Hane (1982) terms the 'underside of modern Japan'.

The Japanese situation is worth looking at in more depth. I suggest its circumstances, whilst particular to Japan, encapsulate the principle reason why coal miners in Asia are not perceived as the labour heroes they are in the West. My main source is Matthew Allen's *From the Coal Mouth*, an examination of contemporary welfare dependency in the Chikuhō region, where the mines are now closed. Allen (1981) pinpoints two main issues:

> The first issue concerns the nature of history of coal mining in Japan and the violence and terror which has accompanied its rise and fall. *The coal industry is undoubtedly one of the most violent and repressive industries Japan has ever seen* [my italics]. The second issue is the power that underlies this violence.

The main locus of this violence lay with the coal company's use of *yakuza* as overseers (*rōmu gakari*). The practice, introduced during the war to keep workers controlled and productive, meant that company violence became a way of life for the miners. Retribution against miners who fled company housing (*tanjū*) and the mine but were subsequently caught by the *rōmu gakari* was so savage that the English word 'lynch' entered the local dialect. On a daily basis the *rōmu gakari* carried weapons that allowed them to assert their physical superiority over the workers without fear of recrimination.

Violence towards miners in Asia, under colonialism and elsewhere, was commonplace – any list of protests and strikes triggered by violent overseer behaviour would run to several pages. Violence was particularly prevalent during earlier periods where slave and convict labour worked the mines. Indeed, Hane (1980) suggests that it was the routine violence during the time the Meiji government ran the coal mines with convict labour that established a precedent for the future abusive treatment of miners. Very many similar western examples could also be found. What distinguishes the Japanese example is the *institutionalized* basis of the violence. The incorporation of the *yakusa* went far beyond their simply being employed – they were a fundamental part of the structure, of the power that lay behind the violence.

The other distinguishing factor was the weakness and fragmentation of the Union movement. Again, this was not unusual elsewhere but the situation of the Japanese coal mines was particular enough to almost be termed culturally specific. Miners had gained the right to form unions during the 1920s – a right which lapsed but was then revived by the US Occupation authorities after the war. By 1948 some fifty per cent of all miners belonged to a union.

But they did not necessarily belong to the independent National Coal Miner's Union (*Tanrō*) – there were unions at companies both large and small that were essentially company unions (Allen, 1981). They operated on the premise that they represented individual member's interests and that disputes could be settled internally with management.

As a consequence there was no cohesion in the union structure. The only area of consensus amongst union leadership was that antagonizing management would serve no positive purpose. It was not until the year-long strike at Miike mine in 1960 that *Tanrō* was demonstratively effective. By then, however, the Japanese government had made the decision to switch its energy policy from coal to oil, the coal companies were virtually abandoning mine maintenance – with horrific consequences, and it was only a matter of time before the mines closed (Allen, 1981):

> The legitimating bodies – the government, police, courts, gangsters and the mine management – were able, through their manipulation of the political agenda, to prevent the development of the miners' consciousness as a class, as an oppressed group and as a potentially powerful labour force.

Thus, although the proletarianization of the male miners in Japan could be said to have developed after the exclusion of women from underground work, this exclusion neither helped nor hindered a masculinist model developing. It was other factors that did this – effectively disempowering the miners in public perception and in reality.

Women in small scale mining

Nowadays, when women in mining are considered it is rare for them to be seen as workers in medium or large mines. Rather they become part of family units working small scale ones. Small scale/artisanal mining itself is far from homogeneous. So womens' experience and participation may vary quite dramatically depending on the nature and type of operation.

Artisanal mining covers a wide range of activity. From the very long standing traditional practices described below by Evelyn Caballero, where mining forms part of a holistic pattern of village management, to the exploitative and deadly work in asbestos mines in India. In some countries small mining operations are legal and so training and safety programmes can be introduced – as Geoff Crispin describes. In others they are illegal and the miners are at risk of police or military intervention, plus the 'ordinary' risks associated with the mining activity. They cover coal, industrial minerals and some metals – mainly precious ones. They also cover gems, as the recent focus on 'conflict diamonds' attests.

Even in this context where women are numerous they often become invisible[22], often subsumed under the general heading of 'family'. As such academics and

[22] A recent example of this is the comment 'women and children are usually not directly involved with the mining or mineral processing in Chinese artisanal mining', in Gunson and

commentators generally examine them on a country-specific case study basis. Thus they may form part of discussions about peasants, they may be used as examples in discussions of women's work, but rarely as miners per se. Indeed this is one of the reasons the approach of this book is so welcome – it provides an opportunity to relocate women back into debates about mining.

But having relocated them, is there anything more to be said? Can we pull together some common threads, some notes towards the development of theory? I suggest this will not be easy. And one of the reasons may be because at present small-scale mining itself rarely figures in mining debates. The history, the constructs, the debates, all focus on large mines and their work forces. So it is up to us, the historians, to bring out of the pit and into the light our doubly disadvantaged subjects.

References

Agricola Georgius, 1556. *De Re Metallica*, Translated from the Latin by H. and L. Hoover, facsimile edition, Dover Publications Inc., New York.

Allen, V. L., 1981. *The Militancy of British Miners*, The Moor Press, Shipley.

Barrios de Chungara, Domitila, 1978. *Let me Speak! Testimony of Domitila, Woman of the Bolivian Tin Mines*, Originally published in Mexican, English Edition, London.

BPP, 1842. *Report of the Commissioners on the Labour of Women and Children in Mines*, British Parliamentary Papers.

BSP, 1894. *Report of the Inspector of Mines for India, vol. 1*, British State Papers. India Office, (26) 2028.

Campbell, Beatrix, 1984. *Wigan Pier Revisited: Poverty and Politics in the Eighties*, London, p. 97.

Chakrabarty, Dipesh, 1989. *Rethinking Working-class History: Bengal 1890–1940*, New Jersey, pp. 219-230.

Cushman, Jennifer W., 1991. *Family and State: The Formation of a Sino-Thai Tin Mining Dynasty 1797–1932*, Oxford.

Dennis, Norman, Henriques, Fernando and Slaughter, Clifford, 1974. *Coal is Our Life: An Analysis of a Yorkshire Mining Community*, London, pp. 203-204.

Disney, Anthea, 1981. 'Where a Woman's Place is Down the Mine', *Observer Magazine*, 27th July, London, pp. 14-17.

Emberson-Bain, Atu, 1993. *Catching a Common Vision? Mining, Development and the Perspectives of Pacific Women*, Research Paper for the International Women's Development Agency, Victoria, p. 10.

Yue Jian, 2002, see also www.iied.org/mmsd/mmsd_pdfs/asm_china.pdf, p. 11. The author's acceptance of the 1992 prohibition of women working in mines was shown subsequently to be mistaken. Mine accidents continued to reveal women working illegally both in small and in state-owned mines – as they had in the preceding years.

Gerritsen, Rolf and Macintyre, Martha, 1996. 'Dilemmas of Distribution: the Misima Gold Mine, Papua New Guinea', in John Connell and Richard Hewitt (eds) *Mining and Indigenous Peoples in Australasia*, Sydney, pp. 35-55.

Gunson, A. J. and Yue Jian, 2002. *Artisanal Mining in the People's Republic of China*, Mines, Minerals and Sustainable Development, International Institute of Environment and Development, London.

Hane, Mikiso, 1982. *Peasants, Rebels and Outcasts: The Underside of Modern Japan*, New York.

Harrison, Royden (ed), 1978. *Independent Collier: The Coal Miner as Archetypal Proletarian Reconsidered*, Sussex.

Hayashi Eidai, 1990. *Yami o horu onnatachi* (Women who Dig Darkness), Akashi Shoten, Tokyo, p. 196.

Heidhues, Mary F Somers, 1992. *Bangka Tin and Mentok Pepper: Chinese Settlement on an Indonesian Island*, Singapore.

Idegawa, Yasuko, 1980. *Hi o Unda Hahatachi: Onna kōfu kara no Kikigaki* (The Mothers who gave Birth to Fire: Stories from Women Miners), Fukuoka. Quoted in Allen, 1981.

Jahan, Rounaq (ed.), 1982. *Women in Asia*, Minority Rights Group Report, London.

John, Angela V., 1980. *By the Sweat of their Brow: Women Workers at Victorian Coal Mines*, London, p. 237.

Kaur, Amajit, 1990. 'Hewers and Haulers: A History of Coal Miners and Coal Mining in Malaya', *Modern Asian Studies*, vol. 24, no. 1, pp. 75-113.

Lopez, Salvador P., 1992. *Isles of Gold: A History of Mining in the Philippines*, Oxford, p. 42.

Metcalfe, Andrew, 1988. *For Freedom and Dignity: Historical Agency and Class Structure in the Coalfields of New South Wales*, Sydney.

Murray, Martin J., 1982. *The Development of Capitalism in Colonial Indochina (1870–1940)*, California, p. 334.

Raymond, Robert, 1984. *Out of the Fiery Furnace: The Impact of Metals on the History of Mankind*, Australia.

Robinson, Kathryn, 1991. 'Land Rights, Labour Relations and Fertility in the Soroako Nickel Project, Sulawesi', in John Connell and Richard Hewitt (eds) *Mining and Indigenous Peoples in Australasia*, Sydney, pp. 107-119.

Storey, Keith and Shrimpton, Mark, 1989. *Impacts on Labour of Long-distance Commuting Employment in the Canadian Mining Industry*, Newfoundland.

Taussig, Michael T., 1980. *The Devil and Commodity Fetishism in South America*, North Carolina.

Tsuna, Masuda, 1983. *Kodō Zuroku: Illustrated Book on the Smelting of Copper*, Colour woodcuts by Tokei, Niwa Motokuni. First published Japan, 1801. Facsimile edition with English translation by Zenryu Shirakawa, Connecticut.

Wright, Tim, 1984. *Coal Mining in China's Economy and Society 1895–1937*, Cambridge, p. 162.

Yip, Yat Hoong, 1969. *The Development of the Tin Mining Industry of Malaya*, Kuala Lumpur.

Chapter 2

Japanese Coal Mining: Women Discovered

Sachiko Sone

Introduction

The subjects of my chapter, the coal mining women of Japan, are unknown even to most Japanese. Very few Japanese text books of history refer to this group of women – coal mining women – despite the fact that, at the turn of the 20th century when Japanese industrialization was in the full swing, coal mining was the second largest single occupational category for women.

Until 1960, when miners went on strike at the Miike mine in the biggest labour dispute in post-war Japan, miners were traditionally viewed as the most conservative and inactive segment of the labour force, and they were seen to be of low status by historians. Within the context of a patriarchal society, female miners were placed even lower than their male counterparts. This division of labour in the industry, although it has been underestimated by labour and economic historians, was crucial before the mid-20th century when rationalization and mechanization occurred. Until female miners were prohibited from working underground by law in 1928, a male hewer (miners) and a female haulier (carriers) worked together underground as a minimum unit and extracted as much as a ton of coal per shift in most coalfields. In this chapter, I examine the history and role of women in the Japanese coal mining industry, with particular reference to Chikuhō, Japan's largest coalfield, illustrating their special characteristics by comparison with the female factory workers of the textile industry, the largest occupational category for women in the modern era. I follow their fate in the course of the many technological, political and economic changes that affected the industry, through government and company records and statistics as well as local accounts. These sources can reveal important social and economic patterns and trends in the women's lives. The ordinary Japanese women's lives such as female coalminers who left not a single letter, let alone a diary need alternative sources to understand their way of life and mentality. The absence of written records of such women's lives in Japan is partly related to the traditional definition of history in a patriarchal, male-defined world, as is the case throughout Northeast Asia. The alternative sources such as oral records of interview and miner's songs reveal levels of complexity in these women's lives not hinted at in standard historical texts, but which is outside the scope of my chapter.

Though women were legally prohibited from working underground in the large scale mines in the late 1920s, I show that the demise of the female underground miners had more to do with family labour practices, technology and geology, than it did with legal sanctions.

The Japanese coal mining industry

Coalmining has a long historical lineage in Japan, though until the 19th century the main use of coal was restricted to the production of salt in the coastal area of the Inland Sea. However, even from the outbreak of the Opium War (1840–1842) Japan had supplied some coal to foreign vessels, the product being purchased by Chinese and Dutch ships in Nagasaki, the only port open to the outside world before 1858. The country's increasing contact with ocean-going vessels caused Japan to re-evaluate its mineral wealth, and production expanded rapidly following the Meiji Restorations, the government recognizing the value of the mineral as a source of foreign exchange. Thereafter, Japan's coal industry was developed in response to the demands of industrialization associated with militarism. The industry also played a crucial role in the period of reconstruction and development in the post-1945 period. Coal was Japan's chief energy source, one of the few mineral resources in which the country was self-sufficient, and Japanese coal reserves (see Figure 2.1) were ample for its early needs except for anthracite and coking coal (Lockwood, 1968: 11). This would have proved a problem for the burgeoning metals industries had Japan not been able to obtain coking and blending coals from Manchuria that could be mixed with local low- and medium-grade bituminous resources. The coal mining industry is thus closely related to Japan's advance into Manchuria.

The important contribution of the coalmining industry is illustrated by the rapid development of Japan's mines, with five million tons being produced in 1897, 20 million in 1913, and 42 million by the eve of the Manchurian foray in 1936.

During the Meiji period (1868–1911), significant reserves were discovered in Kyushu (the southern Island) and Hokkaido (the northern Island), and smaller reserves in Jôban (in Fukushima and Ibaragi prefectures) and Ube (in Yamaguchi prefecture) in Honshu. All were intensively developed. The largest coalfields were located in Kyushu, but although the area contained one of the few large Japanese mines, the Miike mine, the district was generally characterized by smaller mines than those of the less extensively mined Hokkaido fields.

Looking at the outside world around the turn of the century when the *Zaibatsu* (giant financial cooperate bodies) established their foundation, it cannot be denied that producing coal at minimum cost was crucial for an industry where the world markets were so competitive, especially when the low quality of Japanese coal is taken into account. Of crucial importance was that Japan was able to establish its markets in East Asia while Western powers were preoccupied with the political and economic affairs in their own countries around the turn of the 20th century (Yamashita, 1980). However, the industry could not continue to function unchanged in the face of competition when other nations' problems had been resolved and this inevitably led to a process of capitalization and mechanization.

Figure 2.1 The Japanese coalfields

With the shift towards rationalization in the larger mines in the 1920s, women were dismissed, together with many of their fellow male workers, to be replaced by even 'cheaper' Korean labour. There is no doubt that enormous profits were made from the coalmining business by keeping miners' wages low, but the post-World War I recession and increasing competition forced a change in the mode of production. By the mid-1930s, there were 110 mining firms that controlled the bulk of production with Mitsubishi and Mitsui controlling 50 per cent of the capital

(Lockwood, 1968). The story of the *Zaibatsu* is therefore an integral part of the story of coalmining, as are their policies associated with labour recruitment in which family labour and women played an important role. A major reason for this was that women were cheaper than men and thus represented a vital part of the workforce until mechanization took place from the mid-1920s. Before this, they amounted at times to nearly one third of the mining workforce, with two thirds of them working underground alongside men up to 1928. At the coal mining industry's peak, in the post World War II boom, there were about 900 coalmines in Japan, more than half of them in Kyushu (Yamada, 2003). In the 19th and early 20th century in Chikuhō[1], a large proportion of these mines were small endeavours employing less than 100 workers, many of which worked in family teams, including wives and children (Sone, 2003: 14).

Historical records of women miners

There is no record of when women first took up work in the pits of northern Kyushu, but there is early physical evidence of the presence of female miners. When the Mitsui-Yamano mine was first developed in 1899, a wooden memorial inscription dating back to 1723 was discovered. It marked the site of a mining accident in which four male and two female miners perished. The inscription also revealed interesting details about the history of coalmining in northern Kyushu. It recorded that the memorial had been placed there by the village headman. This confirmed other evidence that headmen of villages had been involved as entrepreneurs in operating local coalmines since the early 18th century, when profit made from these mine sites provided additional revenue to the village economy (Higaki, 1991).

By the mid 18th century, coal was increasingly being used as domestic fuel in the cities like Hakata and Fukuoka in the northwest of Kyushu as well as for salt production in the coastal area of the Inland Sea. Consequently, there were more reports about females working in the coalmines. For example, a record from Kajiyama village in the Karatsu domain in 1862 (now Saga prefecture, where

[1] The Chikuhō coalfield occupies the northwest part of Fukuoka prefecture. The field is 48 km in length and 12 to 28 km in width, covering an area of 816 square kilometres. It is set in a low basin surrounded by mountain lands 300 to 600 m high on the south and northwestern sides, with a number of smaller parallel ranges within the basin. The Onga-gawa (river) runs through the central valley in the basin and empties into the Japan Sea. Coal from the area was transported via the Onga-gawa and its tributaries until railways were developed in the late 19th century. The population began to increase in the late 1880s when the central capitalists advanced in to Chikuhō. By 1896, coal production on the Chikuhō coalfield exceeded 2,580,000 ton, which accounted for 51.2 per cent of total national production. With the boom in the industry, the population in Chikuhō doubled from 260,000 to 520,000 between 1897 and 1915. It is claimed that two thirds of employed miners had their families with them.

coalmines were developed by the local lord in the early 18th century) tells the background of the miners in three pits (Higaki, 1991):

Pit 1: 36 miners, including a family of five, one family of four, five couples and 15 singles including one female.
Pit 2: 33 miners, including one family of five, five families of four, one family of three and one single.
Pit 3: 14 miners, including two families of four, two families of three and one single.

It is of note that 13 out of 83 miners, accompanied by their 28 family members, came as seasonal labourers from the Chikuhō coalfield where coalmining had started in the late 17th century.

As coal mining expanded in the second half of the 19th century, this part-time work turned into a full-time occupation for both men and women, who had been mostly tenant farmers. In 1897, an investigative journalist, Kônoe Kitarô made a pioneering visit to 113 large and medium scale mines in four districts of the Chikuhō region, at a time when the coalmining industry in the region was about to expand. Apart from these larger mines there also existed numerous small ones; in fact, in 1897 as many as 330 small mines were recorded in Chikuhō. In the course of his observations, Kônoe (1898) counted more than 23,000 workers, of whom women accounted for well over 30 per cent of the labour force. He described miners' work as follows:

> A *Sakiyama* (male hewer) who works at a pit with roof less than 1.3 meters high has to crawl while extracting coal... As for an *Atoyama* (female haulier), she too crawls, holding a 30 to 60 centimeter long wooden stick in each hand while carrying coal in a basket from the working face to the loading point for coal wagons... An *Atoyama* carries a stick with coal baskets at both ends on her shoulder. Once these men and women step underground they wear very little and soon are covered with black coal dust while groping around. Indeed their labour is beyond imagination as human work... Surprisingly, some women bring their babies into the pits and nurse them there.

At the turn of the 20th century, this traditional family labour was fully utilized by the mine owners through the *Naya seido* (sub-contracting system) to meet the ever-growing demand for labour by tapping the large labour reserve of women[2].

The first national survey on miners employed in the large coalmines with an annual output of over 50,000 tons in 1906 reveals that four out of six coalfields,

---

[2] Naya means barracks, which were built to accommodate migrant workers when the coalmining industry began to grow rapidly in the latter part of the 19th century. The Naya boss who was employed by the owner of the mine, would usually recruit workers in numbers from as few as five to several hundred through his personal network of connections. The Naya boss would also supervise workers in pits and control their life in the Naya community. This system changed early in the 20th century when the Naya boss was slowly replaced in the newly centralized management system.

Chikuhō (25 large mines), Jôban (7), Karatsu (6) and Miike (1) employed women in mining operations. Three coalfields – Chikuhō, Jôban and Karatsu – used a working pair system at the pits, and the ratio of female miners to male miners was approximately 1:1. In Chikuhō the ratio of working pairs of married couples to single miners was approximately 4:6. By contrast, the Hokkaido and Nishisonoki coalfields indicated very low employment of female labour, partly because of their geographical isolation from the mainland (Ogino, 1993).

Family work in the Chikuhō coalmines, the so-called *hitosaki* (a pair working team for mining), consisted of a *Sakiyama* (male hewer; literally meaning 'at the working face') and an *Atoyama* (a female haulier; literally meaning 'backstage in the pit'). The extensive use of family labour was directly related to the topography and condition of the coal seams. The mines of the Chikuhō coal field, where about half of the coal seams occurred between 300 and 1,200 meters below ground, had a particularly high incidence of ingress of gas and water due to crustal and volcanic activities in the region (Takahashi, 1961). Although better mining techniques should have been developed to cope with this situation in Chikuhō, cheap labour was, instead, increasingly sought in order to produce more coal to meet the growing demand. Consequently, extracting coal as well as its transport relied solely on human labour. Until the mid-1920s, when the more efficient method of the long-wall mining was introduced, the traditional stall-and-pillar system of mining was employed in most mines.

Owing to the long history of coalmining in northern Kyushu, a particular mode of working had developed which proved to be more adapted to the use of female and child labour than in other coalfields such as Hokkaido, where single male migrant miners were extensively employed. In 1924 there were more than 68,000 female labourers in the coalmines nationwide, of whom over 48,000 (71 per cent) were working underground[3]. More than 20,000 female labourers were employed in the Chikuhō pits. In other words, every third miner was a woman in Chikuhō. About 10,000 of the Chikuhō women worked above-ground, mostly on the coal-sorting belts.

After the mid-1920s, mechanization of the production process and rationalization of the work force took place in large mines resulting in mass dismissals in the industry. This was critical in Chikuhō, where the labour force fell by about a half between 1925 and 1930, whereas the comparable national decline was only 35 per cent[4].

These rationalizations affected female workers more than males, with the proportion of females slumping markedly by 1930. It marked the end of family work in the mines and many women lost their jobs. Despite the large reduction in female labour nationally between 1920 and 1930 (from 94,870 to 38,440), 63 per

---

3 'Saikin 15 Nenkan Kakutoshi 6 gatsumatsu Genzai ni Okeru Sekitan Yama Kôfusû Shirabe', in *Sekitan Jiho* (Time Signal on Coal), vol. 11, no. 4, pp. 46-7.

4 *Chikuhō Kôgyô Kumai Geppô* (Monthly Report of the Chikuhō Coalmining Industry Association, 1904–1941), 1928–1931.

cent of female workers were still sent underground, which means that re-organization of the workforce proceeded slowly.

After 1937, when the Japanese economy increasingly shifted on to a war footing, the demand for coal as the only reliable fuel grew rapidly and a severe shortage of domestic labour within the industry was soon felt. One way of overcoming this was to employ more Koreans; another was to recall women into the mines. The labour protection law, which was revised in 1928 and had already been undermined by a growing number of special exemptions (Iwaya, 1999: 113-8), was finally put aside in 1939. By 1944, 70,000 women (or approximately one-seventh of the total labour force), were working in the coalmines, of whom 10,000 or so were sent underground to take haulage work alongside Korean hewers. Chikuhō records from 1944 show women miners/hauliers were mainly employed in the medium or small scale mines, which accounted for only 6.65 per cent of the total underground workforce (Tanaka, 1984). Women miners, as a unique characteristic of the Chikuhō coalfield, had all but faded away during the period of rationalization.

After World War II, the government and allied occupation authorities again banned women from underground work. Women employed in the large mines had to make space for men who were returning from the front. No official records exist on women miners after 1947, but oral accounts tell that some of those in the small mines in Chikuhō and Jôban, where such mines had traditionally been operated, continued to work underground discretely up until the 1960s, when most coalmines closed down due to the shift in national energy policy from coal to oil.

Characteristics of female labourers

From 1922 onwards, national statistical surveys were instituted by the government. Among the few governmental records and statistics about female coal miners, the 1924 survey on industrial labourers provides some information about such women whose numbers at the time were at their peak.

The characteristics of female labourers in the coalmines can be illustrated by comparison with female workers in the textile industry, the largest occupational category, which absorbed 80 per cent of female labourers up until the 1930s.

Age and marital status

The age structure of female labourers in the coalmines was in striking contrast to that of female factory workers in the early 1920s as Figure 2.2 shows[5].

---

[5] The sources of data for this and the following charts are compiled from *Rôdô Tôkei Jicchi Chôsa Hôkoku* (Labour Statistics Survey Report), Office of the Cabinet, Bureau of Statistics, 1924 and *Kôfu Rôdô Chôsa* (Statistics of Labour in Mines),Tokyo: Bank of Japan, Bureau of Investigation, 1924, Nishinarita Yutaka, 'Sekitan Kôgyô no Gigyutsu Kakushin to

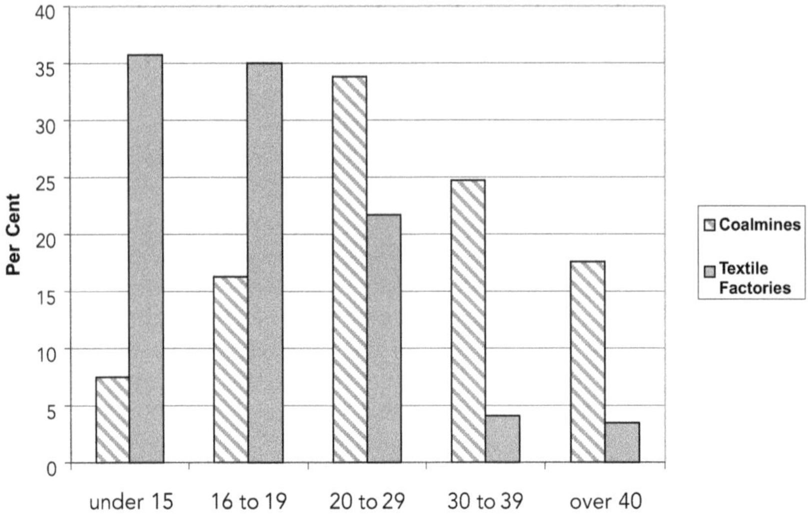

Figure 2.2 Age ranges of women in coal mines compared with those in textile factories in 1924

Women under 20 years of age constituted 24 per cent of female mining labour, compared with 71 per cent in the factories[6]. On the other hand, 58 per cent of mining women fell into the 20 to 39 year bracket, compared with only 28 per cent in the factories. It is therefore not surprising to find that in 1924 75 per cent of women in the mines were married, while just 12 per cent of women in the factories were married.

The age structure of women in the mines, however, differed greatly depending on the type of work in which they were engaged. Women miners were in general older than coal-sorting women: 64 per cent of female miners were between 20 and 39 years of age, the remainder being mainly older, whereas 36 per cent of coal-sorting women were under 19 years of age. The marital status of the women differed also: 80 per cent of women miners were married, compared with only 38 per cent of coal sorting women. These figures demonstrate that up until the mid-1920s mining was still carried out as family labour in most mines.

Joshi Rôdô' (Technological Innovation on the Coalmining Industry and Female Labour), in Nakamura Masanori (ed.), *Gijutsu Kakushin to Joshi Rôdô* (Technological Innovation and Female Labour), Tokyo: Tokyo University Press, 1985, pp. 81-5.

[6] *Rôdô Tôkei Jicchi Chôsa Hôkoku*, 1924. Nishinarita Yutaka, 'Sekitan Kôgyô no Gijutsu Kakushin to Joshi Rôdô', pp. 80-1.

Length of employment

The length of employment of female labourers in the coalmines also forms a remarkable contrast to that of female factory workers (Figure 2.3).

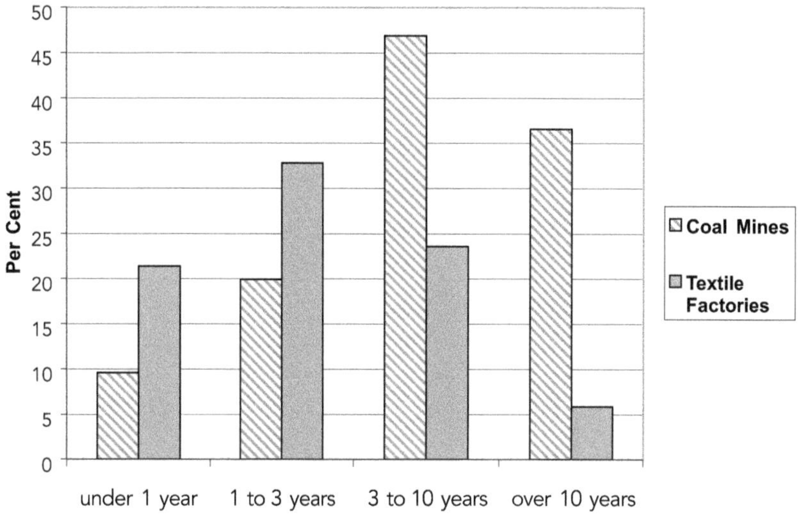

Figure 2.3 Length of employment of women in coal mines compared with those in textile factories in 1924

53 per cent of women in the mines worked for more than five years, whereas 54 per cent of women in the factories worked for less than three years. The trend of higher stability among female workers in the mines did not apply to all types of work: 57 per cent of women miners worked more than five years, whereas only 37 per cent of coal sorting women worked more than five years, and 45 per cent of them worked less than three years. In other words, work such as coal sorting, which required less skill and was paid less, was not attractive to workers and resulted in lower stability.

School education

Female labourers in the mines received very little education (Figure 2.4). 35 per cent of the mining women had no schooling at all and another 35 per cent did not finish the four-year compulsory education required by law for all children from 1871. In this context, it must be noted that the rate of attendance of school-age girls nationally in 1882 was less than 20 per cent, whereas that of boys was 60 per cent. The girls' attendance rate rose gradually and reached 87 per cent nationally by 1901 (Fukaya, 1980). Thus in 1924, when the survey was conducted, the female workers in the textile factories, of whom 70 per cent were under 20 years old, would most likely have had a superior educational standard than the mining women, of whom only 24 per cent were under 20 years old. Indeed, 93 per cent of factory girls completed their primary school education, whereas among female coalmine workers the proportion of primary school graduates was 26 per cent. This indicates that the majority of the mining women had been denied the opportunity to have primary education. Poor educational background can be seen not only among female mine labourers but also male labourers in the coalmines before 1945[7].

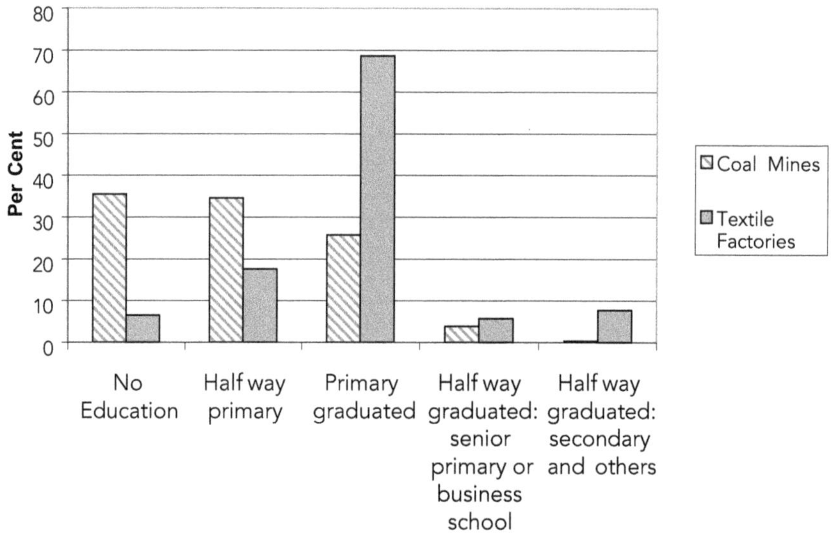

Figure 2.4 Educational levels of women in coal mines compared with those in textile factories in 1924

Origin and recruitment

The origin of mine workers was not a subject of the 1924 survey. However, a survey on the origins of labourers in the Chikuhō coalfield in 1928, conducted by the employment office in Fukuoka prefecture, casts some light on those of females working in the mines. The data indicate where female workers came from, that more than two-thirds of them were married, and that many of them worked with their husbands. The statistics show that the ratio of local to migrant workers had

[7] According to the survey on the large coalminers in Chikuhō conducted by Osaka Chiho Shokugyô Shôkai Jimukyoku in 1925 (the Mitsui Shinnyû mine and the Mitsui Tagawa mine) was no better than in the 1924 national survey: 29.3 per cent of females had no education, whereas 10.5 per cent of males had no education.

not changed greatly since 1906, when the first survey on workers in the mines was conducted. In 1928, 28,280 (45 per cent) of the 63,646 workers were from Fukuoka prefecture. The rest came from various prefectures such as Kumamoto (9 per cent), ôita (9 per cent), Hiroshima (7 per cent), and Korea (9 per cent). This shows that the labour market for the Chikuhō coalmines had gradually been extended from Fukuoka to other parts of Kyushu Island, and from beyond the island to the Inland Sea regions such as Hiroshima prefecture. It should be noted that these prefectures were known for their high rates of tenant farming and also the low rice yields in overpopulated lands. As a consequence, they sent many migrant labourers not only to other parts of Japan but overseas.

By the mid-1920s most major mines had replaced the *Naya Seid*, or sub-contract system, by direct company hiring, but some boss stewards were still responsible for the recruitment of labour in addition to the company-employed recruiters. The recruiters would visit small villages in remote regions where farming conditions were harsh to smooth-talk the desperate tenant farmers. The stories would spread from mouth to mouth among farmers and some would decide to take up a new working life as contract labourers in the mines. For example, five villages in a region in Shimane prefecture sent 458 workers to the mines in Chikuhō, and another 405 came from a single village in a region in Hiroshima prefecture[8]. The high numbers from certain villages suggest that recruitment from them was not a matter of chance, but was because those villages had previously established a good reputation for providing diligent mine workers. This practice, called *Enko Boshû* (recruitment through personal connections), was most common in the recruitment of labour in the coalmining industry before 1945[9]. It persisted even though the seven national government employment offices had been established after 1923, when the public employment office system was proclaimed. The *Enko Boshû* system would have been most effective in controlling as well as recruiting employees.

Working conditions: Accidents, illness and death

Although female labour played a crucial role in the development of the coalmining industry in Japan, especially in the Chikuhō coalfield, once the women lost their value as cheap labour they were dumped under the name of rationalization. During their 50 years work in the industry between the 1880s and the 1930s, women continuously suffered as much as men from illness, injury and the fear of death. Women, however, were more affected by diseases than men. A national survey of diseases among miners in 1927 found the highest sickness rates to be in the coalmines. The percentages of men and women registered as having been sick and

---

[8] Chikuhō Tankô Rôdôsha Shusshin-chi Chôsa, 1928: 5-10.

[9] Ibid: 25-30. In the 67 coalmines in Fukuoka prefecture, about 70 per cent of the total labourers or over 68,000 came by personal connections through the employers. At Nagasaki (13 mines) this was 93 per cent, while at Saga (12 mines) it was 64 per cent.

absent from work for least 3 days were 145 per cent and 200 per cent respectively. In other words, on average every woman fell ill twice in a year. Harmful effects of heavy mining labour on maternity are strongly indicated by the fact that about 20 per cent of women suffered from diseases of the urogenital system, the majority being diseases of the womb, whereas the urogenital rate for men was 3 per cent. The most common diseases for both men and women were those of the digestive organs (men 21 per cent, women 37 per cent), followed by those of the respiratory organs (men 21 per cent, women 31 per cent). Skin diseases and nervous disorders were also very common[10]. The world of pitch darkness, with its dust, fierce heat, humidity, and lack of basic sanitation was the cause of these diseases.

When the 1927 figures are compared with the first national survey of diseases among coalminers conducted in 1913 (sampling 22,807 men and 8,960 women in 26 large mines), it is clear that working conditions had significantly worsened over the period. Although relief systems were well established by the late-1910s, and most large mines had built their own hospitals and clinics as well by the mid-1920s and 1930s as welfare facilities, these improvements above-ground did not prevent miners from falling ill. Furthermore, accidents occurred most frequently in coalmines and as a consequence the death rate of coalminers was the highest among workers in the metal, non-metal and oil categories throughout the history of the industry. In fact, up until their closure in the 1960s, the coalmines were the most dangerous of all Japanese industries in which to work.

In the late 1920s the large mine owners such as Mitsui, Mitsubishi and Kaijima finally began to take safety measures underground and as a result the accident figures decreased, only to rise again during the Pacific War. Fluctuations in the number of coalmining accidents were mostly related to rises and falls in the total labour force. For example, there was a huge increase of accidents between 1914 and 1919 when the number of workers increased abnormally from 182,000 to 348,000. The highest number of fatal accidents in coalmines (992, which was 90 per cent of the total number of deaths in the entire mining industry) before World War II occurred in the following year, 1920, when the total number of miners had increased by only 5,000, to 353,000. This resulted from a decline in working conditions underground due to a cut in safety measures during the post-war recession[11]. The most victims were male workers.

---

[10] ILO, 1933: 248-9.

[11] The largest number of accidents in the mines including the coalmines arose from the collapse of gallery roofs. A large number of deaths also occurred from accidents in connection with the haulage systems, and also through explosions. Electric shock, defective machinery or the use of defective tools produced many accidents in mines. In the large mines of the Chikuhō coalfield in 1926, 318 miners died from roof-falls, and there were 32 explosions, killing 33 miners outright and fatally injuring a further 154. See Nishimuta, 1928: 29.

Wage levels and economic contribution

Despite hard labour and dangerous working conditions, coal mining provided women of peasant origin higher wages than any other occupation. Female miners in Chikuhō, who accounted for 60 per cent of the female labour force nationally, received the highest daily wage not only among female workers in the mines but also among female workers in other industries at the beginning of the 20th century[12] (Figure 2.5).

The first national survey in 1906 reveals common features with regard to females in the large coalmines in Chikuhō, Jôban and Karatu coalfields. First, they were employed in the same jobs as male workers, except for carpenters, blacksmiths and mechanics. Second, their wages were lower than men's, ranging from 88 per cent to 61 per cent of male wages depending on the type of work.

The largest occupation category for female labourers in Chikuhō was as underground miners, where more than 5,000 women worked as *Saitanfu* (miners) and in excess of 3,000 women worked as *Teko* (miner's helpers). This means that 70 per cent of women labourers were engaged in actual mining in 25 large mines in the Chikuhō coalfield at the beginning of the 20th century. The average daily wage for female miners was 55 Sen, which was 81 per cent of the male wage, and for female miner's helpers the wage was 49 Sen, which was 88 per cent of the male wage[13]. The third biggest occupation for women in the mines was as *Sentanfu* (coal sorters), with over 2,000 women working above-ground. Female coal sorters received 28 Sen, which was 73 per cent of the male wage, and equal to about half of a female miner's wage. The lowest paid job in the mine was miscellaneous work above-ground. Six hundred women were engaged in such work, where their wage was as low as 16 Sen, which was 39 per cent of the male wage.

The major work in the mines was of course mining, and mining was carried out by teams. Under the stall and pillar system, a team included a miner who extracted coal, and a miner's helper who put coal into a basket or tub and, harnessed by a rope which was put over both shoulders, and pulled it from the working surface to a coal wagon at the loading point. Because the work was so hard, the wages of female miners and miner's helpers underground were nearly equal to those of the men. As wages were on piece rates, with payment according to the amount of coal as well as its market price, this wage system restricted and strained the wellbeing of members in a mining team. The quantity and quality of coal produced depended largely on the skills as well as the cooperation between hewer and haulier. It is therefore not surprising that nearly 40 per cent of mining pairs in Chikuhō were married couples.

---

[12] Source data from *Kôfu Taigû Jirei*, Nôshômushô Kôzankyoku, 1908 (for 1906), *Fukuoka, Saga, Nagasaki and Kumamoto Kakukenka ni okeru Rôdô Jijyô*, Chûô Shokugyô Shôkai Jimukyoku, 1925 (for 1924), and *Kôdzan Rôdô Tôkei*, Nihon Ginkô Chôsakyoku, 1924.

[13] *Kôfu Taigu Jirei*, ibid: 57.

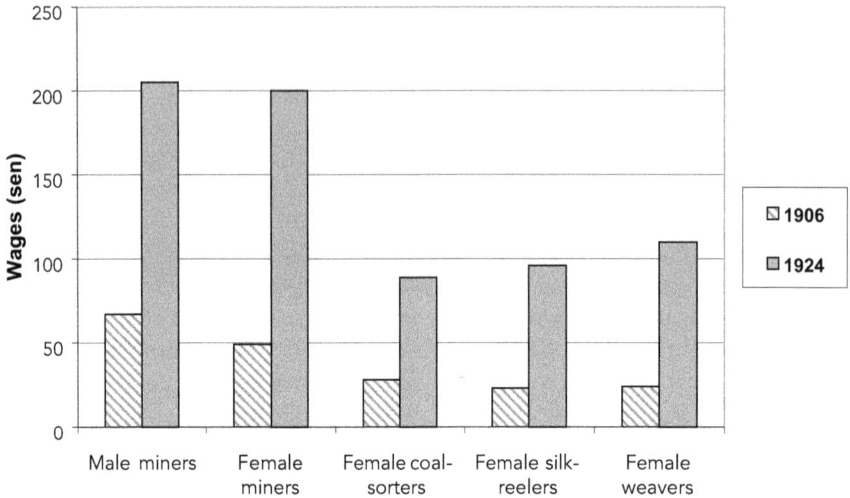

Figure 2.5 Average daily wages for male and female Chikuhō mine workers 1906 and 1924 in Sen (100 Sen = 1Yen), compared with those of female textile workers

Although female miners received higher daily wages than other female industrial labourers such as weaving factory workers, their monthly reward was no higher, because their working rate was quite low, around 65 to 70 per cent, or 20 days a month at maximum. The work underground was as strenuous for women as it was for men, yet there was no less amount of domestic work to be done once they returned home. Women had to work 20 hours a day, including labouring underground for 12 to 16 hours, then above ground in the domestic sphere at home.

According to the first national survey on 386 households of labourers (with families including one to five dependent members) in the large coalmines in 1925, conducted by the Bureau of Social Affairs, there was no household which was able to manage solely on the income of the head of the household. The advantage of having double incomes can be seen in the slightly higher expenditure in clothes, leisure and education. The high level of savings was forced on the miners by the companies as a means of holding them to the mine. The households' economic contribution rate from women was 27 per cent. Although their income remained as a supplementary income for the households, their contribution cannot be neglected. This is especially true in Chikuhō where the average monthly income of 218 households in 14 mines made by the single breadwinner was 51 Yen, which was 7 Yen lower than the national average, and so an additional income was needed. Female hauliers in Chikuhō earned as much as 30 Yen per month, a significant supplement to the 58 Yen wage of a hewer as the head of the household. On the other hand, the household income contributions made by the wives of mechanics and carriers were only 8 Yen and 7 Yen respectively.

The economic contribution by women employed in the smaller mines would have been even more important, since the wages for male miners were lower than in large mines.

Rationalization and the banning of female miners

Notwithstanding the economic contribution made not only to their households but to the mining industry, women were eventually expelled from underground. The basic legislation governing the general question of women and child labour consisted of the *Kôjô Hô* (the Factory Act) of 1911 and the Amendment Act of 1923. The Factory Act that came into force in September 1916 contained special rules for so called *Hogo Kôfu* (protected workers, including women and children under 15 years of age) concerning working hours, shifts, night work, rest periods and minimum wages. A month before the enforcement of the Factory Act, the first law to be applied for the protection of the labourers in the mines, *Kôfu Rôeki Fujo Kisoku* (Regulations for the Aid of Mining Workers) was promulgated and came into force in September 1916. The working hours for women and children (under 15 years) were restricted to 12 hours a day. Night work between 10 p.m. and 4 a.m. was permitted only under a two-shift system. Women were prohibited from working for the first 35 days after childbirth, though this could be reduced to 22 days with a doctor's certificate. However, this ordinance, like the Factory Act, included no penalties and contained so many exemption clauses that it was of little effect.

Between 1916 and 1928 (with a break in September 1925 because of the big earthquake in the Kantô region) heated discussions were continuously held between the industrial capitalists and the Ministry of Agriculture and Commerce (until 1922), and the Bureau of Social Affairs in the Ministry of the Interior (from 1923 to 1928) about the need for protective labour legislation. Two factors finally made the industrial capitalists give up the traditional labour-intensive coal production utilizing family labourers in the pits. The first was political, being pressure from the government, which was sensitive about its international reputation. Pressure was then put on the industry to take measures in response to the labour regulations contained in the Versailles Peace Treaty of June 1919 and to the ILO resolutions of October 1919, which prescribed eight working hours and a prohibition on the night shift by women and by children less than 16 years of age.

The second factor was economic. The mining industry underwent a serious decline during and after the Great Recession starting in March 1920. Demand for coal decreased, and by 1921 unsold coal stocks had risen to 4 million tons. The first industry-wide organization was formed to restrict coal production per mine in 1921. By 1924 this cartel system was taken advantage of by large firms such as Mitsui and Mitsubishi to control the production among mines they owned, leading to a situation where a few large mines monopolized the production of quality coal at less cost. This form of production control by individual mine owners was soon revised and changed to regional control. There was also a growing threat from cheap coal imports, which increased by a factor of almost three, from 0.7 million tons in 1919 to 2 million tons in 1924. Negotiations were soon undertaken between the *Rengôkai* (the Large Coalmine Owners Association most of whom were central capitalists) and the South Manchurian Railway Company, a company half owned by the government and also the owner of the Fujun coalmines, to restrict the imported coal. In other words, capitalists and the military government negotiated in order to co-exist.

From 1926 to 1928 the coal market became active again as domestic demand increased. By this time regulations restricting the rate of coal production had been removed, so the industry began mechanizing underground mining operations, resulting in huge cuts to the labour force. The mechanization further increased the gap in the scale of operations run by the large mines owned by a handful of large enterprises, and the smaller ones owned by local entrepreneurs who had a little or no capital to invest on new machinery. In the mid-1920s and 1930s about 200 small mines were forced to cease their operations whilst large mines managed to enhance their productivity. For example, in the Mitsui–Tagawa mine in Nôgata, Chikuhō coalfield, where mechanical cutters and conveyors were introduced in the pits in mid 1928, the daily productivity of a miner nearly doubled from 2.21 tons to 4.19 tons (Ogino et al., 1980: 176); the cost of production, moreover, decreased by 14 per cent. Having proved that mechanization was profitable, the industrial capitalists finally took the step of banning women from working underground, though not so much for their welfare as to smooth the path of rationalization and reduce labour costs. In September 1928 the revised regulation was promulgated under which night work for women and children (under 16 years) between the hours of 10 pm and 5 am was forbidden, and underground work by women and children was to be banned after a five-year period of grace. In fact, rationalization of the underground mining workforce commenced in the large mines in the late 1920s, and the removal of female labour was well underway by 1933 when the Amendment Regulation came into force (Figure 2.6).

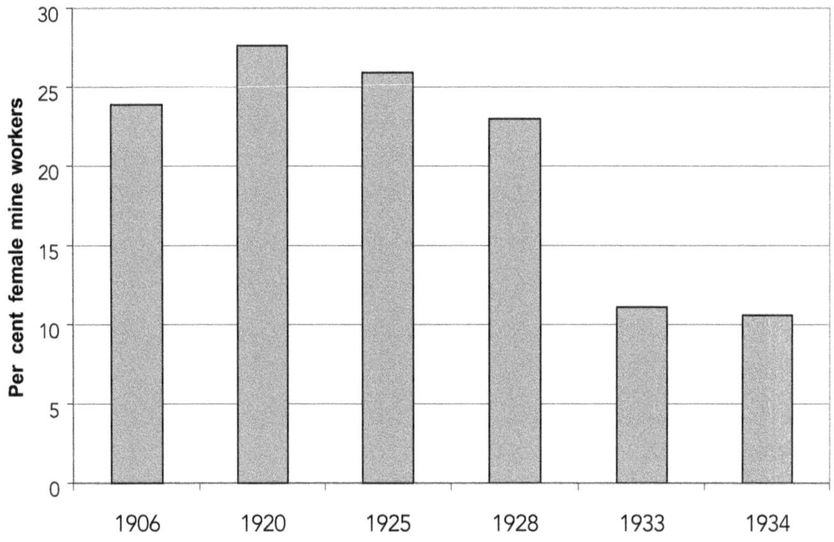

Figure 2.6 The trend of female labour in the coalmines 1906–1934

Source: Compiled from Nôshômu Shô Kôzan Kyoku (Ministry of Agriculture and Commerce, Office for Mines), *Kôfu Taigû Jirei* (Case-Study of Labour Conditions of Miners), 1908 (for 1906); Tsûsan Sangyô Daijin Kanbô Chôsa Tôkeibu (Research and Statistics Division, Minister's Secretariat, Ministry of International Trade and Industry), *Hompô Kôgyô no Sûsei 50 Nen Shi* (History of the Mining Trends of Japan), republished, Tokyo: Ryukei Shosha, 1963 (for 1920–34)

The smaller mine operators, on the other hand, were unable to modernize their operations. Moreover, their business relied heavily on cheap (i.e. family) labour. In 1932, the *Chikuhō Sekitan Kôgyô Gojokai* (Mutual Aid Association of medium and small mine owners), which was founded in 1930 and representing mines employing 20,000 workers including 5,000 females, made a plea to the Ministry of Interior for an exemption, that would allow women to work in the mines where coal seams were too thin (below 60 centimeters) to be mined by modern methods, and also for mines where remnant coal was extracted. The exemption was granted in June 1933, just two months short of the expiry of the original five-year period of grace. The large mine owners also insisted that night work by coal-sorting women be done under a three-shift system. Thus the protective value for women in this Amendment Regulations for the Aid of Mining Workers was doubtful. The proportion of female labourers changed little in the following year. The doubt was confirmed as the war economy grew and, when the industry was in need of miners, women were again 'allowed' to work underground.

Discharge of the female miners

The Revised Miners' Regulation in 1928 exerted influence not just on female miners but on all workers as Table 2.1 below shows. More than 55,000 men and 38,000 women were dismissed nationally between 1928 and 1933. In Chikuhō, within just four years between 1928 and 1931, the number of workers was nearly halved about 100,000 to 50,000 in which about 25,000 males and a similar number of females were dismissed[14].

Table 2.1 Number of workers in coal mines in Japan 1920–1944

	1920	1924	1928	1933	1940	1944
Male	248,003	182,617	183,213	127,592	289,140	353,574
Female	94,870	68,452	54,677	16,010	39,973	70,577
Female u/ground	66,396	47,948	37,730	6,573	11,083	10,897

Source: Tsûsan Sangyô Daijin Kanbô Chôsa Tôkeibu (Research and Statistics Division, Minister's Secretariat, Ministry of International Trade and Industry), *Hompô Kôgyô no Sûsei 50 Nen Shi* (History of the Mining Trends of Japan), republished, Ryukei Shosha, Tokyo, 1963

Large companies such as Mitsui, Mitsubishi and Kaijima were ostensibly well prepared for a smooth transition, at least in the public eye. Firstly, three to four month's severance payment of around 200 Yen was paid soon after the dismissed workers names were announced[15]. Secondly, knowing that the male workers' wages would not cover family expenses, companies offered side jobs to unemployed women. The Kaijima mine, for example, began such a scheme in 1929, and by 1933 more than ten types of jobs, such as knitting gloves, making bamboo baskets and making clay packing (for explosives), were offered to these women, the products being sold at company stores[16]. The daily wage from this work was, however, around 40 Sen, which was one-third of their former wage as a haulier or timberer. Moreover, the number of side jobs was about 350 per day on

[14] Chikuhō Kôgyô Kumiai Geppô, 1928–31.

[15] Large mine operators tried to cope with the so-called Showa Recession (1930–32) by reducing the labour force. For example, in 1930 two mines (the Asô-Sannai and the Aso-Yoshikuma) dismissed all 699 members of their female underground workforce, as well as 37 males. Four mines dismissed 1,680 workers, and four others suspended operations and discharged 951 workers. The situation worsened in the following year. In 1931, five mines (Kaijima-Ônoura, Nippon Seitetsu-Futatse-Uruno, Chûô, Takao No. 1 and No. 2) discharged all female underground workers, totalling more than 2,000, as well as 500 men. See *Chikuhō Sekitan Kôgyô Shi Nenpyô*, 1973: 354-58, 371.

[16] *Fukuoka Nichinichi Shinbun*, November 26, 1993.

average, equivalent to only ten per cent of the total number of women seeking work[17].

The Kaijima mine conceded that side jobs were offered to the women in order to 'prevent them from being idle or corrupting public morals', adding that the measure would also discourage male workers from leaving the mine. The company's concern clearly was not solely with the welfare of the females, but was also aimed at avoiding adverse reactions from the male workers.

As many as 17 labour disputes and strikes took place in Chikuhō in 1931, marking the peak of labour union activities before World War II. The main issue for the disputes was the mass dismissal of male workers, including union members, who were targeted by the management in the large mines (Ogino, 1993: 181-4). Of five strikes that took place in 1931, details of four are recorded in the diary of a union – the Nippon Sekitan Kôfu Kumiai. Claims on behalf of female workers were listed as items in only two out of the four recorded instances, and then at best ranked only fourth out of 15 items[18]. From the way the women's claims are recorded in the diary, they seem to have been added as an afterthought, along with provisions relating to non-mining company functionaries.

Did female workers stand up and protest against the prohibition of women working underground and at the mass dismissals? Considering the atmosphere in small company towns, where mine officers constantly checked the movement of workers, and were patrolled by police who often had connections with the *Tokkô* (the Secret Police Service), and the low priority placed by male union leaders on the claims of female workers, it is highly doubtful that women took on a prominent role in strike actions. The Nippon Sekitan Kôfu Kumiai diary tends to confirm this when it records that some women assisted by cooking rice for the male strikers, and by joining in the singing of miners' songs at the strike headquarters[19].

Women miners themselves, of course, did not want to give up their work underground, but from the late 1920s they were persuaded by both government and mine officials to come above ground and return home to raise their children, the nation's future soldiers. While revision for the Miners' Relief Regulation was underway, the Home Affairs directed mine owners to establish coal miner wives associations in 1925. Numerous coal miner wives associations were soon formed

---

[17] Ibid.

[18] The author inspected the original of the diary, written on rice paper with a brush, at the Research Centre for Coalmining Materials, Kyushu University, in November 2000. The fourth ranked claim was among demands made on the Sumitomo Tadakuma mine, and called for some work to be given to discharged female miners. The answer, 'will consider', was tantamount to refusal. Another claim, on the Inatsuki mine, ranked 12 out of 23, called for women to be offered work in stripping remnant coal, and for women discharged in the last six months because of pregnancy to be reinstated. The first claim was answered 'consider', and the second 'can't do'.

[19] Record during the Takao No. 1 Mine dispute in May 1931. See Tanaka Naoki and Nakadate Kô, *Chikuhō Tanden Sôgi Jôhô*.

with goals such as to 'improve home life', 'support security movement' and 'cultivate quality as good wives, wise mothers and upstanding members of society'. So women in large and medium mines were finally legally bound by the reformist-nationalist law of the state policy in 1928.

Conclusion

The role of mining women at the family level was, from the earliest days of the industry, as breadwinners undertaking arduous work in partnerships with men. They were paid less than men for doing the same work, and they faced the same dangers as men and probably suffered more than they did from the effects of the harsh sub-surface environment; but their rates of pay were substantially higher than those of women in other industries, such as textiles. For the employers, women were relatively cheap labour, but as part of a family unit they were also valued as they tended to help stabilize what was a fairly mobile workforce.

As in most industrialized countries, and sometimes in response to international pressure, Japan legislated to protect women and children in the workplace. Regulations intended to prevent women from working underground were, however, modified or suspended in response to economic cycles or the demands of wartime production. The main factors involved in taking women away from the coalface had more to do with mechanization and less labour-intensive mining than with legislation. Women in fact provided a convenient reservoir of cheap labour, to be drawn on in times of need, but as quickly discarded when a cheaper supply, such as imported Koreans or forced labour, came along.

Coalminers demonstrated a high level of militancy, and their unions initiated 138 disputes in the Chikuhō coalfield, including 17 in one year, before the coal mining union movement was crushed by the *Zaibatsu* (the giant financial cooperate body) and the government in the late 1930s. It is doubtful, however, that women took a leading part in these actions, or even that particular claims were made on their behalf. Since two thirds of women coalminers in Chikuhō at that time were miner's wives, their interests were probably considered to be inseparable from those of their husbands. Though crushed organizationally, the miners, men and women did not passively accept their fate but protested with their feet, taking advantage of the availability of alternative employment in the many small mines of Chikuhō coalfield.

References

Government Publications

ILO, 1933. *Industrial Labour in Japan*, Studies and Reports Series A (Industrial Relations) no. 37, Geneva: International Labour Office.

Nôshômu Shô Kôzan Kyoku (Ministry of Agriculture and Commerce, Office for Mines), 1908. *Kôfu Taigû Jirei* (Case-Study of Labour Conditions of Miners).

Nôshômu Shô Kôzan Kyoku (Ministry of Agriculture and Commerce, Office for Mines), 1913. *Kôfu Chôsa Gaiyô* (Summary of Survey of Miners).

Shakai Kyoku (Bureau of Social Affairs), 1924. *Kôgyô Rôdô Jijyô Chôsho* (The Report on Labour Condition in the Mines).

Shakai Kyoku Rôdô Bu (Bureau of Social Affairs, Section of Labour), 1934. *Kôzan ni okeru Fujin no Chika Rôdô* (Female Labour Underground in Mines).

Shakai Kyoku Rôdô Bu (Bureau of Social Affairs, Section of Labour),1926. Sekitan Kôgyô Rôdô Jijyô Chôsakai, *Tankô Kôfu Kakei Chôsa* (Survey on Households among Labourers in the Coalmines).

Shôkô Shô Kôzan Kyoku (Ministry of Commerce and Industry, Office for Mines), 1906–1939 and Tsûshô Sangyô Daijinkanbô Chôsatôkeibu (Research and Statistics Division, Minister's Secretariat, Ministry of International Trade and Industry), 1946–1988. *Hompô Kôgyô no Sûsei* (The Trend of Mining Industry in Japan).

Newspapers

Fukuoka Nichinichi Shimbun, 26 November 1933.

Local government and company reports and publications

Chikuhō Sekitan Kôgyô Kumiai, *Chikuhō Sekitan Kôgyô Kumiai Geppô* (Monthly Report of the of the Chikuhō Coalmining Industry Association), 1904–1941.

Chikuhō Sekitan Kôgyô Shi Nenpyo (The Chronological Record of the Coalmining Industry in the Chikuhō Region), 1973. Fukuoka: Nishishimbunsha kyokai.

Chikuhō Tankô Rôdôsha Shusshin-chi Chôsa, 1928. 'The Survey on the Origin of the Labourers in the Chikuhō Coalmines' in *Chikuhō Shokugyô Shôkai Jimukyoku Chôsa Hôkoku* (The Report of the Surveys by the Local Employment Office), pp. 5-10.

Sekitan Kôgyô Rengôkai, *Sekitan Jihô* (Time Signal on Coal), 1926–1940.

English Books and Articles

Iwaya, S., 1999. 'Work and Life at a Coal Mine: The Life History of a Woman Miner', in Wakita Haruko, Anne Boundy and Ueno Chizuko (eds), *Gender and Japanese History* (English version) Osaka: Osaka Uni. Press, vol. 2, pp. 413-445.

Lockwood, W., 1968. *The Economic Development of Japan. Growth and Structural Change 1868–1938*, Princeton: Princeton UP, expanded edition.

Japanese Books and Articles

Fukaya Masashi, 1980. *Ryôsai Kenbo Shugi no Kyôiku* (The Education for Good Wife and Wise Mother), Nagoya: Reimei Shobô, pp. 204-5.

Higaki, G., 1991. *Kinsei Hokubu Kyushu Shohan Shi no Kenkyû*, (A Research on the History of Domains in the Northern Kyushu prior to the Meiji Period), Fukuoka: Kyushu UP.

Kawahigashi, E., 1994. 'Nihon Shihon Shugi Kakuritsu Ki no Joshi Rôdô' (Female Labour in the Period of Establishment of Japanese Capitalism), in Takenaka Emiko (eds), *Shin Joshi Rôdô Ron* (New Essays of Female Labour), Tokyo, Yûhikaku, pp. 46-90.

Kônoe, K., 1898. *Chikuhō Tankô Shi* (The Record of Coal Mines in the Chikuhō Region), Fukuoka, Nakamura Kinkokô, reprinted 1975, Tokyo: Bunnka shuppan.

Nishinarita, Y., 1985. 'Sekitan Kôgyô no Gigyutsu Kakushin to Joshi Rôdô' (Technological Innovation on the Coalmining Industry and Female Labour), in Nakamura Masanori (eds), *Gijutsu Kakushin to Joshi Rôdô* (Technological Innovation and Female Labour), Tokyo: Tokyo UP, pp. 70-105.

Ogino, Y., 1993. *Chikuhō Tankô Rôshi Kankei Shi* (The History of Capital-Labour Relations in Chikuhō Coalmines), Fukuoka, Kyushu UP.

Sone, S., 2003. 'Exploitation or Expectation? Child Labour in Japan's Coalmines before World War II', *CAS* (Critical Asian Studies), vol. 35, no. 1, pp. 33-58.

Takahashi, M., 1961. *Chikuhō*, Fukuoka: Kyushu UP.

Tanaka, N., 1984. *Kindai Nihon Tankô Rôdô Shi Kenkyû* (Research on the History of Coalmining Labour in Modern Japan), Tokyo: Sôfûkan, p. 549.

Tanaka, N. and Ogino, Y., 1980. 'Hogo Kôfu Mondai to Saitan Kikô no Gôrika: Kôfu rôaki fujo kisoku o chûshin ni shite' (The Issue of Protection of Labour and the Rationalization of Mining System: Mainly about the Division of the Miners' Relief Regulation), in *Enerugî to Keizai Hatten* (Energy and Economic Development), Fukuoka: Nishinihon Bunka Kyôkai, pp. 345-412.

Yamashita, N., 1980. 'Nihon teikoku shugi seiritsuki no higashi ajia sekitan shijô to mitsui bussan' (The Coal Market in East Asia during the Establishment of Japanese Imperialism and the Mitsui Company), in *Enerugî to Keizai Hatten*, Fukuoka: Nishinihon Bunka Kyôkai, pp. 235-306.

Yamada, H., 2003. 'Demolishing of Coalmines under Energy Policy after World War II in Japan', Proceedings, the 6th International Mining History Conference, Hokkaido-Akabira city, 26-29 September, pp. 321-4.

Chapter 3

Race, Gender and the Tin Mining Industry in Malaya, 1900–1950

Amarjit Kaur

Introduction: Political and economic frameworks

Britain's imperial drive in Malaya after 1870, which was consistent with an agenda of competitive state-building overseas by rival European powers, resulted in the redrawing of the political map of Malaya. Building on the earlier acquisition of the East India Company's Straits Settlements, Britain placed the Malay states under formal protectorate status between 1874 and 1914. The new colonial state had precisely delineated boundaries, a new style of administration, and institutional and legislative structures to oversee the various aspects of government. In the mining sector the colonial administration introduced a series of mining regulations, including titles and leases to mining land, and codified the 'mutual rights and obligation of the State and miners' (Wong, 1965: 15). The details of these regulations need not detain us here. Suffice it to say that these regulations impacted on tin mining in three main ways. First, the state assumed control over the development of the tin industry. Second, the commodification of mining land meant that miners obtained security of tenure which enabled them to undertake mining operations on a more 'permanent' basis. Third, a 'new' class of independent operators, the *dulang* workers, was created alongside the other miners. The role of these *dulang* workers must be understood in the context of the structure of the tin industry after 1850 and the mining methods and labour systems.

The tin industry, 1850–1950

Three phases may be distinguished in the history and development of the Malayan tin industry, and these coincided with increased world demand for tin and the growth of Malayan tin supply; the role of Chinese enterprise and labour; technological advances in mining methods and Western enterprise; and improvements in transport and communications (Lim, 1967; Yip, 1969; Wong, 1964; Kaur, 1985).

During the first phase (until c. 1850), the industry was in the hands of the Malays who controlled tin production. Output and demand were limited since mining was largely subsidiary to Malay subsistence agriculture and was carried out as a part-time occupation by the Malays. Apart from sluicing or *lampanning*, *dulang* washing or panning for tin was a common practice among the Malays. Its popularity as a mining method stemmed from the fact that it required little capital outlay and there was a lack of digging tools to overcome flooding of mines. The *dulang* (pan) is a large shallow dish, usually made of wood. It is very light and the *dulang* washer swirls the *dulang* such that the particles of sand and gravel are carried away, leaving behind the black grains of tin sand.

Panning was essentially a method of recovery and concentration of ore rather than an actual extractive method. It was used by miners to recover alluvial tin ore from the rivers and streams of mining areas (see Metcalfe, 1999) in the foothills of the Western Malay states. Additionally, panning was also used to recover tin from the abandoned tailings of mines. During this period, there was no strict identification of this method with gender. It is estimated that tin output during this phase averaged 500 tons annually (Yip, 1969: 19). Most of this tin was used for soldering and for making funerary accessories in China.

During the second phase (c. 1850–1920), soaring demand for tin in the West, consistent with the increased use of tin plate in the food canning industry, and the discovery of extensive rich tin deposits in the Western Malay states of Perak, Selangor and Negri Sembilan resulted in a great expansion of the Malayan tin industry. This expansion coincided with the passing of control of the industry to Chinese enterprise and labour and the adaptation of mining technologies such as the chain pump, hydraulic sluicing and the gravel pump to the Malayan situation. Additionally, colonial government measures, including the provision of rail and road networks; the regulation of the industry through mining codes and land legislation (Ooi, 1961: 379-382); and a liberal policy on migrant labour recruitment also aided the expansion of the industry. In 1874 Malaya had accounted for around 10 per cent of world tin production. By 1895 the figure was around 55 per cent and Malaya emerged as the world's largest supplier of tin (Wong, 1965; Yip, 1969).

Turning to mining methods, the Chinese introduced the chain-pump (*chin-chia*), adapted the Malay method of ground-sluicing, and subsequently embraced European mechanical technologies, including hydraulic mining (or hydraulicking) and gravel pump mining. Chinese miners were a superstitious lot and had prohibited women from working in the mines prior to the early 20th century. Women were considered 'unlucky' and since mining was regarded as a gamble, they had to be kept out of mining areas. Use of the new methods and technologies radically altered the mining landscapes and there was a shift to inland areas, since miners were no longer restricted to sites where there was good natural drainage. Moreover, these methods produced large amounts of tailraces and tailings and subsequently introduced segmentation of work in the industry. Panning was 'revitalized' since it was imperative that there should be no wastage. Additionally, since the tailings were some distance away from the mines, it probably became acceptable to allow women to pan for tin. At any rate, panning had few prospects, and the work was tedious and tiring. Panning was thus identified as an occupational category by the colonial state owing to the fact that the tin-bearing ground (*karang*) had to undergo 'two stages of treatment, the major process performed in the mines, and a minor recovery by *dulang* washers, working over abandoned tailings' (Ooi, 1961: 354). These developments thus coincided with a system of regulation of *dulang* washers. The first *dulang* washing passes were issued by the Perak State Government in 1907 (Ooi, 1961: 354). The panners or dulang washers had to apply for a *dulang* pass from the Mines Department which permitted them to recover a fixed quota of tin-concentrates per month. The *dulang* washers, who were predominantly Chinese women, were not regarded as mining wage workers but rather as independent and self-employed persons. Most of them were based in Perak, which had the richest tin deposits in the country.

During the third phase, European enterprise, capital and mining technologies (especially the dredge), enabled European companies to displace Chinese as the leading tin producers. While in 1910 European companies accounted for only 10 per cent of the total Malayan tin output, by 1928 these companies produced almost as much as the Chinese miners. In 1929 the output from European mining companies comprised 61 per cent of total Malayan tin output, compared to 39 per cent for Chinese output (Yip, 1969: 20). European dominance has been attributed to both political and economic factors. The European companies had at their disposal large amounts of capital, whereas most Chinese companies were family-owned or were partnerships. Second, the practical effects of mining legislation tended to favour western mining methods. Third, colonial government land regulations; the abolition of revenue farming (an important source of capital for Chinese miners); the control of opium and secret societies and legislation against the 'truck' system of payment (all of which had contributed to Chinese dominance over mining labour) enabled European companies to take control of the tin industry. Western enterprise also subsequently gained control over the smelting of tin in Malaya (Wong, 1965: 273-8; Yip, 1969: 20-22). Politically too, colonial government legislation introduced in the 1930s to restrict Chinese male immigration into Malaya also impacted on Chinese mining labour supply. The Europeans had no qualms about hiring workers from other racial groups or women and there was both a diversification of the labour force and increased employment of women. Crucially, these processes resulted in increased Chinese female immigration into Malaya and foreshadowed the growth of Chinese women's participation in the tin industry.

Race and gender in the mining labour systems

No explanation of the differentiation of men's and women's economic roles and the peripheral role of women in mining would be complete without an examination of production structures and labour systems. Tin mining in Malaya rested on a complex production structure that was shaped by the pioneering nature of the industry. This in turn was a consequence of three conditions: the location of mines; the risks involved; and the different tasks involved in mining operations.

A number of factors accounted for Chinese dominance of tin mining from about 1850 to 1920. One was that the traditional methods of working, where the emphasis was on labour, and communal organization, coupled with an aptitude for hard work, helped the success of the Chinese. The miners also gathered in a form of organization (*kongsi*) that was a self-contained unit. Equally significant was the influence of Chinese clans or societies, which often organized and controlled migration; facilitated such matters as the provision of capital; as well as regulating relations with local rulers. Migrant numbers were small, and Chinese labourers worked the mines under the direction of Malay chiefs, but the capital was advanced by Chinese merchants/financiers in the Straits Settlements.

The discovery of rich tin fields in the Larut district of Perak and Ampang in Selangor in the 1840s led to changes in this arrangement and the Chinese merchants started making advances directly to Chinese 'mines advancers' and mine managers, while the Malay chiefs drew tribute and certain taxes from the tin produced. The mines advancer functioned as middleman and had a multiplicity of roles – tin miner, shopkeeper, tin-ore dealer, and operator of revenue farms. He could even be a source of capital to smaller mines advancers (Yip, 1969: 90). The new arrangements led to an expansion of tin production in the 1850s and 1860s and there was a tin rush which coincided with large-scale Chinese migration into the western Malay states.

The Chinese government did not support Chinese emigration, and this opposition made it impossible to utilize open, regulated recruitment arrangements. It was not until 1893 that the imperial Chinese government lifted its ban on Chinese emigration. Consequently, until the early twentieth century Chinese migration comprised two main networks: the kinship-based migration network and the credit-ticket system network. The kinship-based migration network involved recruiter-couriers who recruited migrants from their own villages/regions whilst the passage money and travel expenses were commonly guaranteed by relatives or friends from the migrants' hometown. The credit-ticket (steerage) system, which the bulk of the migrants relied on, involved the passage money and travel expenses being met by labour brokers, captains of junks or labour agencies. Upon arrival at their destinations, the migrants' employers paid the passage money owed by the migrants, and the migrants entered into verbal or written contracts for the repayment of their debt in the form of labour service. When they had repaid their debt, the workers were released from their obligation and were free to choose their employer and place of employment. This system of indirect recruitment later gave way to a direct system whereby labour recruitment agencies, foremen and other middlemen in Malaya (such as Chinese officials closely associated with the Malayan mining industry), were entrusted with the job of recruitment.

The recruitment of these workers was consistent with a rather elastic use of migrant labour. They had many characteristics in common. They were young, predominantly unskilled adult males who emigrated as individuals and thus had no dependents. The Chinese government also took 'precautions' to prevent women from being taken overseas. This prohibition was later abolished in stages for the different dialect groups (Purcell, 1965: 254).

Labour control

Labour control was achieved in three main ways. The first was through the mechanism of the indenture contract. Mine owners or *coolie* brokers paid the travel costs and expenses of the new migrant (*xinke*) in exchange for a contractual obligation on the *xinke*'s services for a specified period. The prices offered for *xinke* varied with labour requirements and employers used the enforceable labour agreements to manage their labour costs and supply.

The second form of control was through the distinctive Chinese form of social organization, the triad or secret society (*hui*). Trocki has argued for the classification of both the secret societies and clan associations (which combined dialect, regional and occupational groupings) under the general category of *kongsi* (Trocki, 1990: 11). Crucially, the *kongsi*, apart from being a business co-operative, also maintained social control (law and order) and fostered social solidarity (Yen, 1986: 117-18).

Mining was organized through the *kongsi* which functioned as a resilient organization in a frontier society, and was based on bonds of brotherhood and partnership in economic activity. The isolation of the tin mines from settlement areas or towns meant that the *kongsi* had to provide a multiplicity of ancillary services needed by the workers. The *kongsi* also provided an institutional framework into which new arrivals were inducted. More significantly, it allowed mining operations to continue unhampered by the fluidity of workers. The *kongsi* thus offered a sense of security and identity – it relied on a variety of mechanisms including a personal recruitment system, kinship links, clan ties, and provincial connections. Membership of secret societies was often obligatory – the society offered protection and established its own law and code of conduct while receiving complete loyalty from the workers. The *kongsi* also provided a basis for the emergence of leadership in the organization. In mining areas the *Kapitan Cina*, with whom the colonial authorities dealt with, was usually the head of the most powerful local secret society as well as the leading Chinese entrepreneur in the area.

The third method of labour control was the 'truck' system whereby workers usually received their wages at long and irregular intervals. Workers were not paid until the ore had been smelted and sold and this normally occurred every six months. In the interim period the mine owner provided the workers with provisions and other items for personal consumption (usually including opium), deducting the cost from their wages. The mine owner in turn entered an agreement with the mine advancer who provided the provisions (and even mining tools) on credit in exchange for guaranteed sale of all the ore at fixed prices. The price of ore was fixed below its market value while the items obtained through the mine advancer and supplied to the workers were charged well above market rates. This practice, which was distinct from indenture, often bound the workers to employers in perpetual debt and strengthened employers' control over labour. Workers also became indebted to mine owners through their consumption of opium and involvement in gambling. In some areas, opium was often provided in lieu of wages, further tying workers to employers (Kaur, 2004: Ch. 3; Yip, 1969: 82-6; Wong, 1965: 43-9).

Wage structures and payment

Workers in the tin mining industry in Malaya were normally employed under two categories of remuneration. These were a system of wages and a tribute system. Wage-earning labourers were paid either on a time-rate (*kongsi-kung*) or on a piece-rate (*naik chiang*) basis. The piece-rate workers were considered unskilled while the time-rate workers were regarded as skilled. *Kongsi-kung* labourers were normally employed when a mine was first opened and usually worked a seven-hour day. *Naik chiang* workers were employed to remove the overburden of earth that had first to be removed before the actual mining was carried out. They were paid by the *chiang* of earth removed. Both the *kongsi-kung* and *naik chiang* workers received free food, but had to purchase their own opium, tobacco and other provisions, usually from their employer. In contrast to the *kongsi-kung* workers, the *naik chiang* workers received their wages on a monthly basis.

By the early 20th century, these two categories of wage workers represented about 50 per cent of the total mining labour force in Malaya. The rest were employed under the tribute (or *hun*) system. *Hun* is the Chinese word for share and the workers who came under this category were all shareholders in the mining enterprise, with the headman usually holding two or three shares in contrast to the single share held by a worker. Workers employed under this system received no fixed wages but were entitled to a share of the mining enterprise's profits (Kaur, 2004: Ch. 3).

Compound accommodation was provided by mine owners in the vicinity of the mines. It consisted of a *kongsi* or communal house, and was a thatched, raised structure divided into three divisions: the centre (hall/store); and two wings which served as dormitories. The dormitories were divided into rectangular sleeping blocks for workers. A (male) cook was employed to prepare the food for workers. In smaller mines, the kitchen was part of the *kongsi* house whereas in larger mines, the kitchen was separate. No women were allowed in the *kongsi* house (Jackson, 1961: 86-87; Blythe, 1947: 112-3).

Gender and the mining sector

As noted previously, women's participation in the tin mining industry increased in the 1930s. This increased participation was consistent with four interrelated developments: Western dominance of, and increased mechanization of, the industry; expanding economic opportunities in the country; colonial immigration policy; and changing economic conditions in China.

The highly capitalized European mining sector was not only less reliant on Chinese labour but was also associated with new and specialized tasks. First, the large capacities of the dredges meant that European companies had to acquire extensive areas of mining land to operate their equipment on a regular basis. Although the emphasis was on mechanical operations, there was increased demand for workers in ancillary establishments to provide, operate, and repair machinery and carry out related tasks. Second, since the major mining areas in the FMS had evolved into urban centres, some with vastly increased administrative and commercial functions, the isolation of mining areas was also broken down, facilitating migration. This coincided with new employment opportunities and changing labour relations in the mining sector, since control of the labour force was no longer of overriding concern. Third, the establishment of labour departments, introduction of labour codes and an increase in the number of labour officials facilitated employer-employee relations. Declining employment in Chinese mines was offset to some extent by increased opportunities in the western mining companies. From a low figure of 228 workers employed on dredges in 1913, numbers rose to 16,800 in 1929 (Jackson, 1961: 144-5; Yip, 1969: 233). The racial composition of the mining population also changed, as shown in Table 3.1 below. Unfortunately, there are no detailed statistics on the breakdown by gender.

Turning to border controls and colonial immigration policy, new legislation introduced in the 1930s placed restrictions on the entry of Chinese. The new legislation represented the first attempts by the colonial state to use 'borders' as a means of keeping out a specific racial group. The rationale to exclude entry was based both on economic and security/political motives, as discussed below.

Briefly, three phases may be distinguished in colonial immigration policy and migration goals. These goals were consistent with patterns in world trade and the demand for Malaya's export commodities; economic conditions in the country; and labour activism among migrant workers. During the first phase, 1900–1927, the country witnessed the expansion of the tin and rubber industries and the entry of thousands of migrant workers to labour in these industries. For all three groups: Chinese, Indian and Indonesian, entry was completely free and unrestricted. There was, nevertheless, repatriation of some groups of unemployed workers during the depressed economic conditions of the 1920s.

Table 3.1 Employment in tin mining in Malaya by ethnic group, 1911, 1931–1965 (annual average)

Quinquennial Average	Chinese		Indian		Malay		Others		
	no.	%	no.	%	no.	%	no.	%	Total
1911	189,100	96.2	4,630	2.4	2,580	1.3	210	0.1	196,520
1931–5	44,530	85.1	5,130	9.8	2,270	4.3	430	0.8	52,360
1936–40[a]	61,310	81.3	9,420	12.5	4,010	5.3	650	0.9	75,390
1941–5	n.a	–	n.a	–	n.a	–	n.a	–	n.a
1946–50	30,980	68.5	6,660	14.7	6,690	14.8	880	2.0	45,210
1951–5	n.a	–	n.a	–	n.a	–	n.a	–	41,350
1956–60	n.a	–	n.a	–	n.a	–	n.a	–	30,443
1961–5[b]	26,700	60.9	5,060	11.5	11,470	26.2	610	1.4	43,840

Notes: [a] Average for three years, 1936–8; the figures for other years are not available
[b] Average for two years, 1962 and 1964

Source: Yip Yat Hoong, 1969, p. 384

In 1928 the British enacted the first piece of restrictive legislation in the Straits Settlements (SS), namely the Immigration Restriction Ordinance (IRO). This legislation empowered the Governor of the SS to regulate or prohibit immigration 'for the purposes of performing domestic or manual labour whenever the influx of immigrants threatened unemployment, economic distress or was not in the public interest' (Parmer, 1960: 12). The legislation, though applicable to all immigrants, was directed at the Chinese, particularly those of a 'criminal type'. It was viewed as an important instrument to deal with the Chinese since the state lacked legislation to manage Chinese migrants following the abolition of Chinese indentured labour in 1914. This legislation had two major ramifications. It enabled the colonial government to establish a basic framework for border controls and empowered it with the means to control the entry of labour deemed surplus to the requirements of the country.

Although no immediate measures were taken initially to restrict immigration, the state adopted a restrictive immigration policy from 1930 onwards. This policy was consistent with worsening world trade as a consequence of the Great Depression, rising unemployment and depressed economic conditions in the country. The closure of some tin mines and rubber estates prompted the government to impose a monthly quota on adult Chinese male immigration from August 1930 (Saw, 1989: 15). For the Indians, the British implemented a policy of repatriation of the unemployed and the destitute (unemployed Chinese were also repatriated).

Notwithstanding this, it soon became apparent that there were several loopholes in the IRO legislation. The IRO could only be resorted to during emergency situations, and it did not include provision for internal enforcement measures over immigrants once they had landed in the country. By this time too there were growing calls among Malay nationalists for increased restrictions on the immigration of 'alien' Asians. These calls not surprisingly coincided with competition for jobs, especially in the public sector.

In January 1933 the IRO was replaced by the Aliens Ordinance (AO) in the SS. This legislation was essentially designed to 'regulate the admission of aliens in accordance with the political, social, and economic needs for the moment of the various administrations in Malaya' and 'to provide a means of registering and controlling aliens resident in Malaya' (Parmer, 1960: 3). This legislation too was directed at the Chinese since Indians were classified as British subjects.

The AO subsequently led to the creation of an Immigration Department in the SS. Complementary legislation enacted in the Federated Malay States (FMS) and the Unfederated Malay States (UMS) saw the establishment of similar departments in these administrative units. All matters pertaining to Chinese immigration were transferred from the Chinese Protectorate to the Immigration Departments. Crucially, the AO provided the necessary mechanism for the registration of aliens resident in Malaya and was the first step in the development of internal enforcement measures over immigrants in Malaya.

The AO was significant for three reasons. First, it hastened the demise of the system of recruitment of Chinese labour either through intermediaries or through lodging houses. Second, the quota system which restricted the entry of Chinese males (based on the demand for their labour services in the country) resulted in increased passage costs for the *xinke*. This was principally due to the higher prices charged by the shipping companies to compensate them for their reduced trade. Third, since Chinese women were not affected by the quota system, their passage costs were cheaper than those of men. They were also encouraged to buy 'non-quota' tickets by ticket agents and lodging house keepers since ticket brokers would only sell the latter 'quota' tickets on the basis of one quota ticket to three to four non-quota tickets. According to Blythe, this was consistent with the fact that many women had now become the breadwinners since their spouses were subject to quota restrictions under the AO (Blythe, 1947: 102-3). Moreover, Chinese male migrants were also encouraged to bring their wives and children.

Conditions in China also favoured women's emigration. The prolonged effects of the slump in the silk industry of Canton, which has been attributed to the growing popularity of rayon, was a contributory factor, as was the impending war with Japan. Between 1934 and 1938, there was a net migrational gain of over 190,000 women migrants (Purcell, 1948: 100). These women were largely drawn to the tin industry, the traditional source of employment for the Chinese. Increased female immigration led to an improvement in the gender ratio, as shown below in Table 3.2.

Table 3.2 Malaya: Chinese females per 1,000 men, 1911–1947

Year	Females/1000 men
1911	215
1921	371
1931	486
1947	815

Source: Census Reports 1911–1947

The improved gender ratio contributed to the stabilization of the Chinese community, the reproduction of the Chinese labour force and increase in the proportion of women in the formal sector.

As noted previously, increasing regulation of the mining industry in the first decade of the twentieth century had resulted in the creation of a new category of independent mining workers who were not included in mining labour statistics. The term independent mining labour needs some clarification. Independent workers were those who were not paid on a daily/monthly basis or at the rate of *kung* removed per fixed number of hours. The *dulang* washers thus formed an independent category of miners and this term also included those paid at piece-rates or contract rates. The specific nature of the *dulang* washers' job description stemmed from the fact they were not allowed to *mine* tin, but only to recover it from streams or worked-out mines and to *concentrate* the tin ore recovered. Panning was favoured by the Chinese women miners because it could be carried out with very little capital, did not require specialized skills, and could be combined with other jobs.

The *dulang* pass system

Under the *dulang* licensing system, the women had to obtain a *dulang* pass or license from the Mines Department of the respective state in the FMS, upon payment of an annual fee of M$1.00. No age restrictions were specified, but child labour was prohibited. The pass allowed the panner to recover a fixed monthly quota of tin-concentrates. This quota varied according to the price of tin. The pass was not intended to allow women to raise their productivity or amass wealth. Rather they were seen as a means of increasing state revenue when the price of tin was high. The pass was also intended to provide some form of a livelihood for women in order to alleviate poverty among them.

The women were required to carry their passes on their persons at all times, and could be stopped for inspection at any time. Failure to carry the pass was liable to a fine, and repeated infractions resulted in a cancellation of the pass. Although the pass was non-transferable, it was possible to loan or 'rent' it out to another woman in return for either a percentage of the tin recovered or a daily 'rent'. Usually, the women were able to meet their quota in a fortnight and had to seek other part-time employment in order to survive.

The women worked on a co-operative basis, normally in small groups of five or more. They often lived in shacks or leaf hovels near the mines. The work of recovering the ore was tedious and time-consuming since the women had to stand up to their waists in the stream to wash for the ore (often with babies or infants strapped to their backs) (Siew, 1953; Siew, 1961: 07). The following comment in the FMS Mines Department Report for 1909 is fairly typical of the attitude of the state towards these women: '...Of the alien races who live in the States ... there are none to be compared with these [Chinese] women, who for sobriety, morality and honesty are not to be beaten...' (cited in Jackson, 1961: 46).

The *dulang* washer was permitted to sell the recovered ore to a licensed ore dealer only. Moreover, since *dulang* passes were registered, the ore dealers had to record the women's registration number against the purchases of ore. This was principally to ensure that the women did not exceed their prescribed quota. The women therefore were constrained from increasing their production since their fortunes were tied to the global market for tin and the particular tin restriction schemes of the colonial government. Thus, although the women were very small players in tin production, they were subject to similar restrictions as the big tin producers, despite the fact that they had none of the privileges of the latter. The colonial government effectively utilized the *dulang* pass system to regulate tin output.

The number of *dulang* passes issued in the FMS during the first 12 years (1908–1920), when the price of tin was high, and from 1925 to 1932, when tin prices fell, is shown in Table 3.3 below.

Apart from the ability to control output through the number of passes issued, the monthly quota system was also used to regulate output in keeping with tin's fortunes. During the depressed economic conditions of the 1930s, for example, the quota was reduced, as shown in Table 3.4 below.

As shown in Table 3.4, the *dulang* washer's monthly quota was much lower in 1932–4 compared to the period 1927–30, and the 1940s. It would appear that the *dulang* washers 'tend[ed] to thrive in crisis' (Siew, 1961: 106), that is, when the price of tin was high. In effect, manipulation of the permissible quota further served to marginalize *dulang* washers. Although their output varied and ranged from 0.5 to 1.50 per cent of total tin production (Siew, 1961: 105), they were subject to similar tin restriction controls as the big producers, as noted earlier.

Table 3.3 Malaya: Number of *dulang* passes issued by the FMS Mines Department, 1908–1920 and 1925–1932

Year	Number of Passes
1908	8,278
1909	9,596
1910	10,257
1911	10,807
1912	12,031
1913	14,155
1914	n.a
1915	15,859
1916	14,007
1917	13,870
1918	15,774
1919	15,553
1920	12,867
1925	7,792
1926	5,923
1927	7,536
1928	10,409
1929	8,947
1930	7,784
1931	8,738
1932	9,000

Note: n.a = not available
Source: (1) 1908–1920, Jackson 1961: 146; 2. 1925-32. (2) 1925–32, Khalid, 1992, p. 96

Prior to the commencement of the first International Tin Agreement, it appears that there were no fixed limits on the number of *dulang* passes issued. During the First Tin Restriction Scheme (March 1931 to December 1933), which coincided with a fall in rubber prices, the Malay peasantry was also affected by the depressed economic conditions and there were proposals that Malay women be given *dulang* passes without the usual fee. The response of the Senior Warden of Mines FMS to this proposal is particularly illuminating since it indicates that Chinese *dulang* washers were 'not easily controlled'. He argued that supervision of the Malay *dulang* washers would be difficult and that there was nothing to prevent them from 'evading actual *dulang* washing' and selling their passes to the Chinese. Despite these objections 30 passes were given to Malay women in the Kinta District with a maximum limit of 100. In Pahang too, passes were given to Malay women. In fact, in one mining area in Perak,

Batang Padang, the Inspector of Mines introduced *dulang* passes with the pass bearer's photograph attached to prevent the sale of passes (Khalid, 1992: 15-97)!

Table 3.4 Monthly quota allocated to *dulang* workers in the FMS, 1927–1949 (in *katis)**

Year	Monthly Quota
1927–30	24
1932–34	15
1935–38	20
1941	25-30 (estimate)
1948–49	25-45 (variations in the different states)

Note: * 1 *kati* = $1^1/_3$ pound
Source: (1) 1927–30, A. Azmi Abdul Khalid (1992) 'The social organization of the mining industry during the depression, 1929–1933 in Malaya', *JMBRAS*, 65 (Pt 2) (December 1992), p. 95; and (2) 1932–49 Siew Nim Chee (1961) 'Labour and Tin-mining in Malaya', in T. H. Silcock (ed.), *Readings in Malayan Economics*, Eastern University Press Ltd, Singapore, p. 407

The numbers of *dulang* passes issued to Chinese, Malay and Indian women in 1938 and 1950 are shown below in Table 3.5.

Table 3.5 FMS: *Dulang* passes issued to ethnic groups, 1938 and 1950

Year	Chinese	Malay	Indian
1938	10,108	806	115
1950	18,167	445	90

Source: Siew Nim Chee, p. 407

As noted earlier, the increase was a result of postwar need to raise revenue.

Adaptation and change

In the immediate post-World War II period, the mining industry required massive amounts of capital for rehabilitation. It is not surprising therefore that production by *dulang* washers increased as shown in Table 3.6 below.

Table 3.6 Tin production in Malaya by mining method, 1929-1962 (per cent)

Mining Method	1929	1939	1949	1962
Dredging	38.9	48.0	50.4	48.7
Gravel-pumping	41.5	36.5	35.0	37.7
Hydraulicing	8.4	5.7	4.0	2.2
Open-casting	3.9	3.8	1.2	3.5
Underground	4.6	4.0	4.0	4.0
Others	1.2	0.1	0.5	1.7
Dulang washing	1.5	1.9	4.8	2.2

Source: Lim Chong Yah, 1967: 63

The number of *dulang* washers also rose during this period as shown in Table 3.7. Indeed, by 1948, *dulang* workers comprised approximately 29 per cent of the total workforce engaged in tin mining.

The Chinese contribution to the development of tin mining in Malaya is remarkable for four main reasons. First, the Chinese provided the capital, labour and technology for the early growth and expansion of the industry. Second, Chinese dominance was associated with Chinese organization of mining and the type of labour systems in place in frontier regions. Through *kongsis* and secret societies the Chinese were able to import and deploy a very large labour force, and were able to operate large units. Third, Chinese enterprise contributed to the growth and spread of wage labour in Southeast Asia during this period. Fourth, though Chinese enterprise was labour-intensive, the Chinese were able to secure their capital requirements from the Straits Settlements merchants. Gradually, however, new technologies were adopted which both led to a reduction of the labour force and the passing of the industry to Western interests. But the lowly *dulang* washers continued to function as independent operators and thus ensured their survival under changing circumstances.

Table 3.7 Number of *dulang*-washers in the tin mining industry, FMS 1936–1950

Year	Total Number of *Dulang*-Washers	Total Workers in Tin Industry (exclusive of *Dulang*-Washers)	Total Product of *Dulang*-Washers (tons)
1936	11,809	80,218	565
1938	11,327	71,487	1,345
1946	22,973	23,026	2,859
1948	20,281	48,861	3,217
1950	18,702	47,224	3,292

Source: Siew Nim Chee, 1953: 406

Overall, since *dulang* washers were constrained by the terms of their employment category through the use of the quota allocation, they were forced to resort to other occupations to maintain themselves. Since most of them were located in the vicinity of mines, they also worked on daily or piece-rate wages when their *dulang*-washing quota was reached. Others turned to rubber tapping or weeding on estates. Others became market gardeners or pig farmers. A large number of them had left their spouses behind, and World War II and the Emergency period from 1948 to 1960 made it difficult for them to return. In order to ensure their survival, they often 'purchased' unwanted female babies whom they brought up as their own children to look after them in their old age (Siew, 1961: 110). These girls became the next generation of *dulang* washers. Single women and widows also found partners among the miners, had children and thus contributed to an improvement of the gender ratio and the reproduction of the mining labour force. Compared to the 19th century when women were barred from approaching mines because they were considered 'unlucky', Chinese women miners had made their presence felt in the tin mining industry by the 1950s.

References

Blythe, W. L., 1947. 'Historical Sketch of Chinese Labour in Malaya', *JMBRAS*, vol. XX, Pt. 1, pp. 64-114.

Jackson, R. N., 1961. *Immigrant Labour and the Development of Malaya*, Government Printer, Kuala Lumpur.

Kaur, A., 1985. *Bridge and Barrier: Transport and Communications in Colonial Malaya*, Oxford University Press, Singapore, Ch. 2.

Kaur, A., 2004. *Wage Labour in Southeast Asia since 1840: Globalization, the International Division of Labour and Labour Transformations*, Palgrave/Macmillan, Basingstoke.

Khalid, A. Azmi Abdul, 1992. 'The Social Organization of the Mining Industry during the Depression 1929–1933 in Malaya', *JMBRAS*, vol. 65, Pt. 2, December, pp. 85-98.

Lim Chong-Yah, 1967. *Economic Development of Modern Malaya*, Oxford University Press, Kuala Lumpur, Ch. 2.

Metcalfe, Ian, 1999. 'Geological Origins and Natural Resources', in A. Kaur and I. Metcalfe (eds), *The Shaping of Malaysia*, Palgrave/Macmillan, Basingstoke, pp. 11-41.

Ooi Jin Bee, 1961. 'Mining Landscapes of Kinta', in T. H. Silcock (ed.), *Readings in Malayan Economics*, Singapore Universities Press, Singapore, pp. 345-403.

Parmer, Norman J., 1960. *Colonial Labour Policy and Administration: A History of Labour in the Rubber Plantation Industry in Malaya*, J. J. Augustin for the Association for Asian Studies, Locust Valley, New York.

Purcell, Victor, 1948. *The Chinese in Malaysia*, Oxford University Press, London.

Purcell, Victor, 1965. *The Chinese in Southeast Asia*, 2nd edn, Oxford University Press, London.

Saw Swee Hock, 1989. *The Population of Peninsular Malaysia*, Singapore University Press, Singapore.

Siew Nim Chee, 1953. *Labour and Tin Mining in Malaya*, Cornell University Data Paper no. 7, Cornell.

Siew Nim Chee, 1961. 'Labour and Tin Mining in Malaya', in T. H. Silcock (ed.), *Readings in Malayan Economics*, Eastern Universities Press, Singapore, pp. 404-439.

Trocki, Carl A., 1990. *Opium and Empire, Chinese Society in Colonial Singapore, 1800–1910*, Cornell University Press, Ithaca.

Wong Lin Ken, 1964. 'Western Enterprise and the Development of the Malayan Tin Industry to 1914', in C. D. Cowan (ed.), *The Economic Development of South-East Asia: Studies in Economic History and Political Economy*, George Allen and Unwin, London, pp. 127-53.

Wong Lin Ken, 1965. *The Development of the Tin Mining Industry of Malaya*, University of Arizona for the Association of Asian Studies, Tucson, pp. 273-8.

Yen Ching-hwang, 1986. *A Social History of the Chinese in Singapore and Malaya, 1800–1911*, Oxford University Press, Singapore.

Yip Yat Hoong, 1969. *The Development of the Tin Mining Industry of Malaya*, University of Malaya Press, Kuala Lumpur.

Chapter 4

Patriarchy, Colonialism and Capitalism: Unearthing the History of *Adivasi*[1] Women Miners of Chotanagpur

Shashank S. Sinha

Introduction

While much has been written about mining and human displacement, studies on women miners still remain a relatively less-negotiated domain. *Adivasi* women form a marginalized category even in the emerging studies on women and as such remain 'hidden from history'. Based on an extremely scattered source material, this chapter attempts to bring a visibility to a section of population that has largely gained attention from the field of anthropological or developmental studies. The exercise is difficult on account of the absence of a coherent employment database and the fact that *adivasi*s were never treated as a distinct separate category even in

[1] The word *adivasi* originated in Chotanagpur in the 1930s and literally means 'original inhabitants'. *Adivasi* here is being used interchangeably with tribe in a very neutral sense to distinguish a group of people so socially organized from caste (though some of the tribes have also become castes or are in the process) sharing some kind of a group-sentiment, and generally (though not as a rule now) a common and specific territory. The socio-economic levels associated with them have never been fixed and permanent and should not convey a sense of complete isolation from the mainstream of Indian life: for apart from some isolated and primitive food gatherers, the tribes or *adivasi*s are, and have been, very much a part of Indian society as the lowest stratum of peasantry subsisting through plough cultivation, slash and burn agriculture, or as agricultural labourers, plantation coolies or (later) workers in factories and mines. As far as the cultural characteristics go, though some tribes continue to practice their specific rituals, traditions, religion and taboos, most of them have become sufficiently Hinduized or Christianized over a period of time. Exploited both by the immigrating *dikus* (non-tribal outsiders) and the colonial rule (and later by the capitalists), they also share some kind of a common experience of dispossession and socio-cultural displacement and therefore a resistance to the 'external' intervention.

the different census operations conducted during the colonial rule[2]. Centering on the general experiences of *adivasi*s, this chapter tries to give a historical insight into the lives and service conditions of women miners of colonial and post-colonial Chotanagpur [3]. An important concern of the study is to also show how the impacts generated by growth of mining enterprises were intersecting with those of colonialism to create new patterns of gender and social differentiation in *adivasi* patriarchies.

Canvas Chotanagpur

The crux of the problem lies in the fact that the region has simultaneously been the heartland of a large *adivasi* population and India's mineral base. Known as the 'Ruhr of India', Chotanagpur is one of the most mineral-rich regions of India with an abundance of coal, limestone, iron ore, mica, manganese, etc.; an attribute that accounts for the massive growth in mining activities in the region over a period of time. The majority of the population in this region is rural, including a considerable tribal population. The *adivasi*s now account for around 28 per cent of the total population of the state of Jharkhand. There are around thirty different scheduled tribes living in the region having their respective dialects and distinctive ways of life; notable among them being the Mundas, Santhals, Oraon, Ho, Pahariya, Birhor, and Asur (Singh, 1990: 82). The tribes of Chotanagpur are generally patrilineal, patrinymic, patriarchal and patrilocal and the lineage is defined in terms of descendents in the male line.

Crisis in Chotanagpur

The colonial land and forest legislations laid down 'legally' defined institutional foundations leading to marginalization of the *adivasi*s. The impacts included among other things, breakdown of the communal mode of production, emergence of private rights in land, commercial exploitation of forests, ban on shifting cultivation, penetration of market relations, intrusion of middlemen and moneylenders (*dikus*), social stratification and emergence of feudal relations in

---

[2] They were enumerated under different heads and names under different census reports: from a tribal form of religion (1891), to animists (1901), to tribal animists or tribal religion (1911), to hill and forest tribe (1921), to primitive tribe (1931), to tribe (1941) and scheduled tribe (1951 onwards).

[3] The districts that form the canvas of this study include Dhanbad, Singhbhum, Palamau, Ranchi, Hazaribag and Santhal Parganas. On account of concentration of tribal population, this region is also known as the *adivasi* heartland and currently forms a part of the state of Jharkhand that came into existence in 2000 as the 28th state of India.

land (Singh, 1988: 1227-28). Colonial administration also severely curtailed customary forest rights of the *adivasi*s. In the post-permanent settlement period, the battle between the *zamindar*s (landlords) and the colonial administration to secure large areas of jungle land had already played havoc with the forest acreage and the traditional common rights of the people (Damodaran, 2002: 143-45). The legislations of 1894 (Forest Policy) and 1927 (Forest Act) restricted traditional access to forests and forest produce. The efforts towards conservation of forests came only after the forest cover in Chotanagpur had been massively denuded. The National Forest Policy of 1952 was hardly any better, for what had become 'rights and privileges' in 1894 became 'rights and concessions' and then only 'concessions' in 1952. The development of industrial-capitalist enterprises in the colonial and post-colonial period contributed to further marginalization and loss of land. *Adivasi* lands were also acquired for the construction of railway stations, schools, colleges, roads, dams, hydroelectric projects and thermal power stations (Vidyarthi, 1978: 209). According to the Committee of Jharkhand Matters (COJM) Report, up to April 1989, there were 90,847 cases of land alienation filed from the region, of which 44,775 were accepted, 34,447 rejected and 13,595 cases were pending. Land had been restored in only 2,994 cases. (COJM, 1990). The phenomenon of dispossession took place both directly through deprivation of their land and indirectly through denial of benefits of development, of their rights (Areeparampil, 1989: 19).

Faced with problems of survival, the *adivasi*s started migrating to mines (and plantations) coming up in Bengal (and Assam) and after 1900, with the opening of mines at Jharia and Hazaribag, the migrations were mainly directed to the mining settlements in the region (Mohapatra, 1985: 252). The development of mines in the tribal heartland was paralleled by a massive influx of outsiders to cater to the requirements of skilled and later unskilled manpower. In terms of employment, the local men and women were disproportionately directed towards temporary, underprivileged and unskilled jobs. In due course of time, the local tribes even began to be evicted from the unskilled jobs. According to the 1921 Census, the tribes accounted for only 10 per cent in the total number of unskilled labourers in the coalmines of Hazaribag and Manbhum (former Dhanbad) districts. During the World War II period, the owners of the collieries began to recruit more labourers from outside Chotanagpur in spite of suggestions by successive government committees in 1896, 1906 and 1919 calling for recruitment of local labourers (Areeparampil, 1989: 24).

Situating women in the crisis

While the entire society felt the impact of such a crisis, women were affected more deeply. They suffered both as integral members of the society and in their exclusive capacities as women. Though they had never enjoyed proprietary rights in land they had played an active economic role under the traditional division of labour. In agriculture, with the exception of ploughing, all activities relating to production had traditionally been carried out by women including clearing fields, removing pebbles, sowing, weeding, harvesting, transplanting, etc. Similarly in hunting-gathering societies, hunting was the job of men while collecting fruits, tubers, timber, firewood and foliage was traditionally done by the women who also looked after the children and affairs of the household (Sinha, 2005: 35-60; 80-95)[4]. Land alienation and restrictions on the use of forest resources thus led to a diminution in women's traditional socio-economic space. Opening of mines therefore opened up alternative space for women[5]: for although the *adivasi*s had traditionally had an institutionalized sexual division of labour, it was the women who were often sent to the ore fields when the family was in need of some cash (Majumdar, 1937: 136).

In the early days of mining, female and child labour was in huge demand for carrying coal underground since no substitute male labour was available as cheaply. Female labour was not considered incompatible with the routine of the mines; in fact quite the opposite. It was cheap, relatively docile and well equipped to undertake many unskilled tasks in essentially huge open cast quarries (Corbridge, 1982: 48). The highpoint of women's employment in the mines was around 1920 when they formed about two fifths of the total workforce. This dropped to one tenth after the complete ban was enforced in 1939 and around 1967, women formed only about 6.6 per cent of the labour force. Besides prohibition, technological improvisations also displaced women workers employed in unskilled occupations (National Commission on Labour, 1969: 380). While there are some studies on the economics of women's employment in mines, what has gone relatively undocumented is their socio-cultural experience as miners. Their lives, as we shall see in the ensuing discussion, were encapsulated within 'layers of subalternity'.

In the first place, as subjects of *adivasi* patriarchies, they were required to conform to a strict taboo regime and other customs and traditions even when the families migrated to mining settlements in the urban regions[6]. The colonial rule played havoc with *adivasi* patterns of livelihood. The concurrent and subsequent development of mines and industries opened up new areas of exploitation. As

---

[4] For women's socio-economic roles in *adivasi* patriarchies see Shashank S. Sinha, *Restless Mothers and Turbulent Daughters: Situating Tribes in Gender Studies,* Stree, Kolkata, 2005.

[5] For the most part *adivasi* women miners were employed as unskilled casual labourers. Only a very small percentage of this section got regular jobs.

[6] Some prominent taboos included ploughing, roofing, levelling, weaving, using weapons, partaking of sacrificial meat offered to the family spirits. Thus, in spite of significant labour contributions, traditions kept women excluded from some core production activities, while keeping them perennially dependent in relation to the 'prohibited' activities. Sexual liaisons with non-tribal persons was also prohibited. For details, see Sinha, *Restless Mothers and Turbulent Daughters.*

workers they had to live at the whim of the *sirdars* (labour contractors/recruiters), petty capitalists and mine owners who would subject them to inhuman and exhaustive work regimes.

The division of labour in the mines somewhat represented an extension of the traditional division of labour requiring women to do tasks involving expenditure of tremendous manual labour. Gender discrimination was pervasive not only in the division of work but also in the disbursement of wages. During the colonial period, there was no established concept of regular wages and women were largely employed at very low, often dismal wages. They were invariably paid less than males doing the same jobs: in the 1850s, they earned two thirds of the daily wages of the male workers and by mid-1930s at the height of depression and the overproduction crisis in coal mining, they were drawing, in some cases, less than half the male wage (Simeon, 1998: 24). Further, the fact that such women belonged to socially stigmatized and lowly-rated 'tribal' groups made them easy targets for sexual abuse by the *sardars* and male co-workers, especially the up-country male immigrants. The outsiders often mistook sexual freedom enjoyed by *adivasi* women in their own societies as license for sexual harassment.

Colonialism and women miners

The Burdwan District Gazetteer points to the presence of a large number of Mundas, Santhals, Oraon, Kols and Bauris among the labourers migrating from the neighbouring districts of Bankura, Manbhum and Santhal Parganas to cater to the demands of labour in the mines and factories of Asansol and Raniganj (Peterson, 1910: 134-35; Coupland, 1911: 179; Imperial Gazetteer vol. I: 78-79). The number of Santhal and Bauri women employed in mines was high; high caste Hindu women worked in the mines only if they had been socially ostracized and abandoned by their families. From 1890 to 1919 the numbers of workers in the coal mines of Bengal increased from 25,000 to 175,000 – child employment was constant while number of women employed increased more rapidly then men (Rothermund and Wadhwa, 1978: 187). According to estimate, more than 15,000 women worked in Bengal mines in the early 1920s. Over 9,000 of them carried coal in the pits making about 35 per cent of the underground labour force. This ratio was valid for the surface workers as well (Engels, 1999: 230-31). The situation was largely similar at the iron mines in Singhbhum. J. L. Keenan, the Irish-American general manager at TISCO (Tata Iron and Steel Company, Jamshedpur) also saw 'endless streams of Kol, Santhal and Ho women, the inhabitants of dirt walking in a steady line from the mining faces to the narrow gauge track with baskets of ores on their heads' (Corbridge, 1982: 48). *Rezas* (female labourers) in fact accounted for around 36 per cent of the unskilled workforce in Jamshedpur at the time of the 1921 Census and accounted for the only women in industrial employment (Simeon, 1995: 22).

Nearly half the women who migrated to Bengal coalfields were single and employed and settled in so-called 'family units'. They might have migrated from the same village, but had done so as single women, abandoned wives or as widows and their relationship with the males in the work gang was based on carrying coal for them. Economic determinism was thus not the only factor behind women's migrations. They were migrating also because their position in the patriarchal family was crumbling (Engels, 1999: 226). According to official data, around 30 per cent of the women, working in 'family units' in the 1920s, were not formally attached to the men in their unit. A further 10 per cent had established informal liaisons with the men in their 'family units' (Engels, 1999: 231). In contrast, nearly 80 per cent of Santhal women worked alongside their husbands or male relatives in the Dhanbad coalfields. It was also learnt that: 'Santhalis in particular are so jealous concerning their women that they will not allow them to carry coal for other coal cutters'. The 1896 inquiry reported that men generally refused to carry and load coal in the case of up-country miners. This created a special difficulty, as many of them had come to coalfields as singles and needed the assistance of women and children of other castes to do the loading (Simeon, 1998: 24).

Gender was an important factor in determination of jobs and workforce management. The common parlance of the work place was imbued with gender – women coolies were called *reza*s and in their capacity as loaders attached to male hewers were known as *kamins* – a term signifying the performance of service (Simeon, 1998: 23). They were required to dump the coal over the belt (for the want of steam engines) but were also asked to carry it to the surface. In what was an extension of the family system of production so characteristic of the coal mines, the male workers at iron mines also arrogated to themselves the more interesting and better paid mining and drilling operations, while the women were more or less limited to loading the cut ores into baskets and carrying them to waiting tubs or aerial ropeways. Buchanan was struck by the 'somewhat strange relic of the old days' whereby the loading of iron was done from the baskets on women's heads rather than by steam shovels'. But the American mine manager explained the situation saying 'rice to feed women costs less than coal to feed the steam shovel' (Corbridge, 1982: 48)

The tasks performed by women also included gin winding, screening coal, pushing tubs and cleaning boilers. With only rare exceptions, they were generally employed in the labour intensive sections of the production process. At the Tata iron mines (Noamundi), women loaders doing most of the despatching of the ore were frequently on duty at night and had to leave infants in the care of small children near the workings (Simeon, 1995: 37). Physical exertion was not the only problem however. With the ban on children working in mines after 1924, there was the added problem of lack of supervision of children. In the absence of crèche or nursery facilities, too many children would stay unattended in the mining compounds always vulnerable to accidents and mishaps (Corbridge, 1982: 48-49).

The effects of hard labour were compounded by long working hours. There were no weekly holidays and at times the workers were made to do compulsory overtime. Also the daily-rated workers had no paid leave till 1945 while the *reza*s had to work in night shifts till 1920 (Simeon, 1995: 37). Despite the system of shifts, the working hours were mainly irregular and legal limits were often exceeded. Provision for weekly holiday was only made by the Mines Act of 1923 (*Royal Commission on Labour*, 1931: 123-24). The miners actually used to work for 18-20 hours, and only in 1934 were the weekly hours for both the surface and underground workers fixed to 54 (Roychaudhury, 1964: 307). The working conditions in the Singhbhum iron mines were equally tedious and exhausting for the contractor's labourers. They would start at the crack of the dawn and work continuously for twelve hours under the zealous supervision of the *sirdar*s.

Yet such hard labour was never compensated by the prevalent wage rates. Women miners received their wages as a part of their men's wages and their wage was seen as a contribution to the family budget and not as the income of a single person. There were marked variations in wage-rates between different collieries. In general, wages were the lowest in Indian-owned mines, higher in joint-stock concerns, and highest in the railway collieries (Simeon, 1995: 152-54). The coal cutters in Bengal earned from 8 to 10 annas (16 annas = 1 Rupee) a day for the underground work, while unskilled labourers on the surface earned only 4 to 5 annas a day. Women workers earned from 1 to 2 annas a day (Peterson, 1910: 134-35; Coupland, 1911). Similarly, the daily wage rates paid by the East Indian Railway Company for coal cutters were 6 to 8 annas and for women working underground 3 to 4 annas. Of the coolies working above ground, the men would get 2.5 to 4 annas, women 1.5 to 2 annas and children 1.25 to 1.5 annas (*Imperial Gazetteer*, vol. II: 325). Gender discrimination in wage rates was also reinforced by the growing inequity in wages paid to the skilled and unskilled workers and the permanent and casual workers: while the unskilled workers were weekly paid (daily rated), the skilled were monthly rated; while the permanent employees were entitled to other allowances and benefits, casual labourers were struggling to get even the basic wages regularly. Women, accounting for a large part of the casual unskilled workforce, happened to be the worst sufferers. In 1929, the piece rated *kamin* at the Dhanbad mines earned between 18 annas to 24 annas per week while the labourers performing mechanical equipment, pumps, boilers, carpentry would be paid on a time rate ranging from Rs 16.00 to Rs 54.00 per week. Some unskilled women workers such as the boiler *kamins* would also earn *hazira* (shiftwise earning), which was more demeaning than the piece rate (Simeon, 1995: 151).

A comparison of the wages paid to the clerical and female components of the workforce at Singhbhum iron mines between 1931–61 reveals that the absolute gap in earnings between the two had increased by 517 per cent. Further, the wage doled out to the typical tribal female quarry worker in 1963 was no greater than that paid to a non-tribal foreman and mates in 1931 (Corbridge, 1982: 55). Things did not improve much in the post-colonial period. According to figures provided by the Annual Report of the Chief Inspector of Mines in 1961, the average weekly cash earnings of overmen, sirdars, miners, loaders, and women working in open cast mines (at Dhanbad) respectively were Rs 34.15, Rs 21.68 and Rs 18.81 while in the case of labourers working above ground, the corresponding figures for clerical/supervisory staff, men workers and women workers respectively were Rs 45.47, Rs 21.84 and Rs 19.07 (Roychaudhury, 1964: 392-93). The limitations imposed by inadequate wages looked particularly glaring in view of the living conditions in the mines.

Most mines lacked ventilation facilities and the air was sticky and filled with toxic fumes. This was particularly true of the Raniganj and the mines worked by East India Railway at Giridih (*Royal Commission on Labour*, 1931: 115). At TISCO, continuous exposures to extreme heat, gas and temperatures were largely responsible for diseases like pneumonia, tuberculosis, cataract, hernia, rheumatism, migraine and silicosis of the lungs. The unhygienic conditions prevailing underground further complicated matters. Almost all mines were infested with hookworm. According to the Jharia Mines Board of Health, 90 per cent of the adult labourers were infected with hookworm. At Raniganj, nearly three quarters of the underground and over half the surface workers suffered from the disease (*Royal Commission on Labour*, 1931: 124). The safety measures at mines further complicated matters. In 1925, an average year, 38 women died in Bengal from collapsing walls or from accidents caused by explosions (Engels, 1999: 235). The decade 1930–40 in fact witnessed around 890 serious and fatal accidents in which 9,710 miners were injured (Simeon, 1998: 36).

Overcrowded housing and lack of sanitary facilities also caused much hardship for the *adivasis*. The miners used to live in lines (single room tenements called *dhowras* in Dhanbad) built like back-to-back barracks with no ventilation facilities. They were allotted to the labourers irrespective of the size of their families. In spite of Jharia Mines Board's byelaws restricting occupation of a quarter by a maximum of two adults and two children, the dhowras were generally providing accommodation to five to ten persons. In the absence of sufficient space, most miners had to carry their belongings to their work place in baskets. According to a survey conducted (under the auspices of Coal Mines Labour Welfare Organization) in 1959, only 20 per cent of the total working population was housed, 84 per cent of which lived in one-room tenements (Roychaudhury, 1964: 409). The housing conditions were extremely bad even at TISCO as only a minority of the skilled permanent workers was strictly entitled to the company housing while the unskilled and semi-skilled 'privileged' workers were herded into one-room tenements. The miners also suffered from a lack of drinking and washing water. There were no provisions for bathrooms and defecation and women used filthy pit water for bathing and washing. There were no latrines and the pigs acted as scavengers. On account of their meager income, the miners continually suffered from hunger and disease. There was no labour legislation worth the name to protect the interests of the labourers (Roychaudhury, 1964: 262). The Jharia Mines Board of Health and Jharia Water Board were formed only in 1914.

Lack of health and sanitary conditions caused regular outbreaks of cholera, smallpox and malaria and it was in fact frequent outbreaks of these diseases and epidemics which led to the constitution of Boards of Health at Jharia and Asansol in 1915–16 (*Royal Commission on Labour*, 1931: 133). Many miners also died on account of respiratory problems caused by humid underground conditions and bad housing. Women's health was in an even more precarious state. Reproductive and child-care facilities were lacking and women started receiving maternity benefits only in 1929 (Simeon, 1995: 37-38). TISCO got their separate Public Health Department only in 1932 (Roychaudhury, 1958: 152). Expectant mothers would be attended by *dais* (midwives). In 1937, there were only 11 certified midwives for the entire belt (Simeon, 1995: 37-38). It was only with the formation of the Coal Mines Labour Welfare Board in 1944 that regular provisions were made for maternity and child welfare centres (Roychaudhury, 1964: 273-74).

Added to poor housing and wages, squalor, poverty, poor diet, malnutrition, and waterborne diseases there were problems of drunkenness. Even the meager wages would thus trickle into the hands of the trader cum contractor who operated illegal liquor shops in the mining settlements. At Noamundi, contractor liquor shops competed with the government outlets by selling liquor at 2 annas/bottle in 1935–36 and 4 annas/bottle in the late 1940s. At Gua, liquor prices were double of those of Noamundi but alcohol remained perhaps the chief means of relaxation (Corbridge, 1982: 60-61). The liquor menace was a marked feature in other mines as well. On many occasions, husbands would surrender their vouchers to the moneylenders to raise cash for liquor. To cope with conditions of unending miseries, Santhal women increasingly started taking jobs as *kamin*s (loading coal on the trucks) while others allegedly started living as concubines to non-tribal workers. The latter could be particularly associated with the Bauris (Banerjee, 1978: 187-91). Dagmar Curzel's Report to the Government of Bengal in 1923 is dreadfully revealing. She writes: 'respectable Bengal women do not undertake industrial work and practically all such Bengalis found in the mills are degraded women or prostitutes' (Curzel, 1923:12).

In 1929 a notification was issued which required progressive elimination of employment of women working underground in the course of ten years[7]. In 1928, 29 per cent of the underground coal labour was female; each year 3 per cent was to be retired so that after July 1939, women were completely excluded (*Royal Commission on Labour*, 1931: 127). But actually, within the next year a quarter of the underground women were eased out of their jobs. The immediate effect of the ban was loss in wages and employment for women and where there were wives or connections of male workers, there was a corresponding reduction in family income (*Royal Commission on Labour*, 1931: 127-28). Estimates point to a 40 per cent decline in family income. Observers felt that the ban led to an increase in cases of prostitution. The Jharia Mines Board of Health pointed out that venereal diseases had increased in the mines (Roychaudhury, 1964: 651). Wherever women

---

[7] The Bill conferred the power to exclude women on the Government of India rather than the local governments. In fact there was such vigorous opposition to the proposed legislation measure in Bengal, Bihar and Orissa that the government had to declare that the same would be carried out in a phased manner.

remained for sometime, there was a change in their job profile; from loading coal tubs underground to working as unskilled labourers on the surface, loading wagons (*dust-kamins*), sorting various grades of coal and also working as labourers for manufacturing domestic coal. However even after being prohibited from underground work, women continued to subsidize the costs of colonial energy extraction by exerting a downward pressure on wages on the surface (Simeon, 1998: 27).

Added to the ban on underground work were problems related to the mechanization of mines. Women started getting marginalized from the mining sector with the technical up-grading of coalmines in the 1920s: when shafts and galleries began to replace the surface and shallow mines, tramlines were introduced for internal haulage and dynamite blasting was introduced. The prognosis linking the advent of machinery with the replacement of females was accurate, but took a longer time than the author might have surmised. Till the time of the 1921 Census, around one fifth of the coal hewers were women, as were nearly half of the coolies, loading and carrying coal above and below the ground.

Following the introduction of new devices, many *adivasi* miners left the coalfields either because they were operating as family units and mechanized haulage made many women and children redundant or because they were unaccustomed to dynamite blasting or because they found that their women were increasingly being subjected to molestation underground (Simeon, 1982: 224). In fact there was a general decline in women recruitment rates after the late 1920s. Women formed 37.5 per cent of the workforce in 1920, which declined to 25.4 per cent in 1929, the year central government ordered the gradual exclusion of females from underground work. This further went down to 13.8 per cent in 1935 and 11.5 per cent in 1938. Contributory factors also included a slump in the coal prices in the 1930s and the resultant closure of several small and under-mechanized enterprises employing a large number of women. Women's marginalization from the workforce was also caused by a general recession in sectors employing women. The tea industry underwent a recession in the depression years of 1931. The lac industry in Chotanagpur also slowed down because of a decline in European demand during the two World Wars. During World War II, labour shortages caused the government to lift the ban in 1943, leading to the employment of some 19,000 women in pits in 1944. In 1946, the ban was re-imposed (Simeon, 1998: 25).

Though the uproar that greeted the exclusion of women from underground work in coalmines had dissipated once the likely deficiency was made good by an influx of male labour, no such sanguine outcome could be expected in the open cast mines. Female labour remained an essential component of the mines' workforce. 'Easy to recruit when the demand for iron ore boomed, women were equally easy to lay off in the slumps' (Corbridge, 1982: 49). They remained a convenient source of cheap and periodically expendable labour. Escalating career prospects carried no meaning for them.

Post-colonial trends in women's marginalization

Adivasi women labourers continued to be employed, though in drastically reduced numbers, in the mines (surface work, particularly mineral extraction) but more comprehensively as unskilled casual labourers in agriculture, manufacturing and construction activities. Female involvement in the mining sector went down considerably in the post-colonial period. Between 1951 and 1969, the percentage of women working in coalmines went down from 15.7 to 5.8; in iron mines, it went down from 38.2 to 23.2; in the case of manganese mines, it declined from 43.9 to 39.7 (Corbridge, 1982: 47). Similar figures are available for other regions in Chotanagpur. At the time of the 1951 Census in Palamau, only 10 women were engaged in mining and quarrying activities in the coal mines and none in the case of iron ore (Roychaudhury, 1961: 229). The Annual Report of the Chief Inspector of Mines also indicated that women (working either on surface or in open castings) constituted only 14 per cent of the total employment in coal mining industry in the Dhanbad district in 1962 (Roychaudhury, 1964: 413). Scattered references also reiterate that there was not only a decline in the overall percentage of women working in mines but also in the male-female workers' ratio. According to the 1951 Census, the total number of women working in coal mines in the whole of Bihar was 24,097 while the corresponding figure for men was 108,135 (*Census of India, Bihar,* 1953: 133). Similarly according to the 1961 Census, the total number of tribal female workers per 1,000 males working in mining, quarrying, livestock, etc., was 389 (*Census of India, Bihar*, 1965: xvii).

Economic hardships were accompanied with attendant social and psychological strains. In the early years of mining and industrialization, there were numerous cases of migrant *adivasi* women workers being sexually harassed or abducted or even forced into prostitution and concubinage. There was no way out: they could not complain for the fear of losing their jobs; they could not return to their homelands either in the absence of alternative survival options. The ban on women's work in underground mines created fresh grounds for sexual exploitation. It resulted in an increased numerical disparity in the sexual structure of the workforce and gave encouragement to commercial sex. Such instances became a commonplace with growth of the mines and industries in Chotanagpur. In 1957, a survey undertaken by the Research Programme Committee, Government of India underlined a heavy disparity in the sex ratio in Jamshedpur as a result of a huge influx of single male labourers; 84.4 per cent females for every 100 males. A large section of the unskilled labourers in the city comprised the tribes living in surrounding areas. Cheaply available *adivasi* women were preferentially recruited on account of the fact that mineral exploitation in the district did not envisage the employment of underground labour. The simultaneous presence of a large number of unattached poor women labourers and single male workers having greater disposable incomes to 'seek pleasures' was, however, not found to be 'helpful for the maintenance of a certain moral standard.' The unmarried percentage among the males and females was particularly intriguing: 53.81 per cent of the males and

47.12 per cent of the females living in the city were unmarried. The reason for a huge percentage of unmarried *reza*s could also be explained by the high incidence of bride price among the Hos of Singhbhum. Rapid development of mines and industries had not only upset the sex ratio of the Jamshedpur city but Singhbhum district as a whole. In 1911, there were 1,035 females to 1,000 males as against 968 women to every 1,000 males in 1951. According to the Census of 1951 there were 6,213 unmarried males and 3,110 unmarried women in the age group of 15-24, a popular age group for marriage (Roychaudhury, 1958: 319-20). Studies on other mining regions of Chotanagpur unravel other dimensions of women's exploitation.

At the suggestion of the Dhebar Commission on Scheduled Tribes, a project was commissioned by the Bihar Tribal Research Institute, Ranchi to assess the impact of mining and industrialization on tribes living in and around Ranchi, Noamundi and Jamshedpur (Prasad and Sahay, 1961). This showed that at the Noamundi iron mines (Singbhum), the Santhal, Ho and Oraon boys and girls were found to be getting initiated into wage work at the age of 12. Working under the strict supervision of the contractors they were getting daily wages ranging from Rs 0.62 to Rs 1.37 (Prasad and Sahay, 1961: 13). Sometimes the women labourers would get even less. Many cases of *reza*s getting married to immigrating *Punjabis* and *Biharis* were also reported from the site. Interviews revealed that women preferred marrying the outsiders settled in Noamundi for business. In several such cases of marriage however, the 'husbands' had subsequently fled leaving behind scores of abandoned women and 'illegitimate' children. Women so deserted by their 'husbands' (non-*adivasi*s) were particularly helpless, as they had also lost the support of their clan by marrying outside. Their children were either found working as servants in shops and homes or had turned delinquents and were earning their living through pick pocketing (Prasad and Sahay, 1961: 19).

Incidents of marriage between *adivasi* women and non-*adivasi* men were widely reported from Jamshedpur region as well. The migrant *adivasi*s however would insist on rigid formalities of the marriage ceremonies. The acceptance of mixed marriages in the city was contrary to the attitudes prevalent in the interior villages where it continued to be regarded as sin (Ibid: 23-32). But other revelations from Noamundi and adjoining regions were quite disturbing. There were reports of the operation of a secret traffic in flesh under which many *adivasi* girls were being sold in other states. Flesh trade and prostitution was also prevalent in Jhinkpani and Gua (Roychaudhury, 1958: 249). Many women and men labourers working in the mines were found to be suffering from venereal diseases. They would not go to hospital for the fear of becoming publicly known in the event of which they might lose their jobs (Prasad and Sahay, 1961: 19-20).

The links between mining, prostitution, and patriarchy could be very visibly seen in the Bokaro region. Prostitution among the Bauri society in Bokaro was occasioned by the presence of the British army during World War II but more specifically by the development of mines and industries. The development of capitalist enterprises brought thousand of workers from far away places to this region. However, while the government and the capitalists did build factories to exploit this labour, other basic requirements of workers were not taken care of. In the absence of families or recreational facilities, many would go to the prostitutes (or in some cases take concubines) to relieve their boredom and loneliness (Sengupta, 1982: 84). It was noticed that a colony of prostitutes was developing in the region and many women were migrating and settling there (Sengupta, 1979: 100-102).

There was a fundamental economic dimension in such sexual liaisons. The workers were decently paid and hence could afford this odious luxury, while the uprooted, displaced and poverty-stricken women found an alternative source of livelihood in this profession. In some cases, the deserted concubines were taking to prostitution rather than going back to their villages (Sengupta, 1979). Sexual liaisons outside the tribe are taboo among the *adivasi*s. In the past, the Santhals had an excommunication ritual called *bitlaha* (Archer, 1974: 96-7). The persons cast out (the offenders and their families) were debarred from eating with others and especially from getting their children married. The victims were publicly ridiculed and hunted down. Though theoretically *bitlaha* applied to all members of the Santhali society, it was often the women who were the victims. Though the ritual ceased to exist in the post-independence period, a diluted version of the practice was still existent in the 1980s; the taboo, however, operated primarily against concubinage. Unlike rape, seen as an external phenomenon occurring either in situations of reprisal or isolation, concubinage (requiring the consent of the woman) was seen as an internal betrayal of the community by the Santhals. If the couple in question remained within the easy reach of the villagers, their house would be ravaged by an all male delegation and the couple would be driven away from the village. If the woman returned to the village, she was thrown out unless she confessed to repenting her act in front of the Panchayat and swore never to be a concubine again. In addition, the concerned woman also lost rights to remarriage (Kumar, 1982: 207-208).

The development of mines and industries however was not only affecting women exclusively but also the very structure of family. The institution of family was disintegrating and family life was becoming more and more individualistic (Roychaudhury, 1964: 157-58). A study on Jhinkpani cement plant (Das Gupta, 1978: 30)[8] reveals that in maintaining the agro-industrial economy, even the nuclear Ho family was breaking into two households; the husband, an industrial worker living in the town with school-going sons, while the wife, with minor and unmarried daughters, staying back in the villages to take care of the agricultural land. On many occasions, the husband had taken a second wife to manage his household in town while his other wife was struggling with the pressures of agriculture in the countryside. In protecting the dual interests in agriculture as well as the factory job, polygamy was becoming an adoptive measure within the limits of social sanction.

---

[8] The cement plant at Jhikpani was built during World War II and started production from 1947.

Though these studies are confined only to some mining regions of Chotanagpur, it can be safely stated that similar trends were observable elsewhere. The interaction between tribal and the immigrant caste cultures so visible in these regions was subtly generating new patterns and symbols of subversion. The appropriation of caste-Hindu customs and symbols by the *adivasi*s was to have profound consequences for women. Thus in the process of getting urbanized, learning new languages (like Hindi and Bengali), eating meals and wearing non-traditional clothes, the tribal patriarchies were also conceding women's conventional social space. Consequent to the screening of women in the presence of *dikus* or foreigners, traditional uninhibited dressing was giving way to covered clothing: the old one-piece handloom *lahanga* (long skirt) was fast being replaced by mill-made sari, one end of which covered the breasts. Like their caste counterparts, women were also found to be wearing petticoats and underwear and were adorning themselves with hairpins, ribbons, plastic bangles, and gold and silver ornaments (Prasad and Sahay, 1961: 23-32; Roychaudhury, 1964: 153; Roychaudhury, 1958: 244-45). The changes were quite noticeable among the Ho and Santhal women. Bride–price was disappearing but dowry was surfacing alongside. In fact, it was observed that dowry was an important factor in raising the marriageable age of *adivasi* girls. Similarly, *dhumkuria* (youth dormitory), which had already started disintegrating, had become non-existent and community dances were fast giving way to cinema as a mode of entertainment (Ivern, 1969: 17; Prasad and Sahay, 1961: 23-32; Roychaudhury, 1964: 153, 158, 166). In *adivasi* villages, these institutions had once formed not just physical manifestations of women's social space but were also important components of women's sexual and cultural autonomy.

The conflicts between the traditions and modernity were particularly marked at the site of mines and industries; in some cases the former paving the way for the latter, but more generally, women were living in conditions of conflictual coexistence. On one hand, they were struggling with 'age-old' patriarchal traditions, on the other; they were also trying to adapt to the new situations. Socio-economic dislocation was also accompanied by keen and unfair competition with the dikus, high aspirations, and changes in living styles, all of which were taking place in conditions of shrinkage in resource base. The *adivasi*s were also becoming vulnerable to new kinds of diseases and emotional disorders. Feelings like devaluation of identity and deprivation were aggravated by the not so pleasant attitude of the interacting caste societies towards their customs and traditions. With increasing frustration, marginalization and alienation, excessive alcohol consumption was becoming widespread and the state government was quite obliging.

It is argued that British government quite thoughtfully spread liquor shops in this part of the country to keep the rebellious *adivasi*s 'in check'. The colonial administration started imposing excise tax on *handia* (rice–beer), and giving licences to the *thikadars* (contractors) for manufacturing liquor (Jha, 1987: 61)[9]. In Jharia, liquor outlets were legalized in 1932 to encourage localized production. While cheapening the costs of liquor by up to 75 per cent, the move also augmented the state's resources by generating more revenues through the auctioning and licensing process. From 1929 to 1933, the number of liquor outlets in Dhanbad rose from 20 to 121 effecting a twelve fold increase in consumption. The proportion of family income spent on liquor consumption correspondingly increased from 13 per cent to 20 between 1929 and 1934 (Simeon, 1998: 31-2). Things did not change much under the 'nationalist' government, which earned about 8 crores a year leave alone the gains arising out of maintaining a low potential for struggle by keeping the *adivasi*s addicted (Sengupta, 1982: 27). Expenditures on liquor played havoc with the economies of families already living under huge debt burdens. In 1959, a survey (conducted by the Coal Mines Labour Welfare Organization) on the coal mine workers residing at Bhuli township (Dhanbad) showed that 50 per cent of the workers families were in debt and that the average debt per family was Rs 250, which was equivalent to 2.5 months' earnings for a worker (Roychaudhury, 1964: 409). A case study on the Bhowra Colliery (southwest of Jharia) in March 1962 points out that *adivasi* daughters were taking to wagon loading because their families were not able to survive on the scanty amounts their fathers could save from drinking (Ambasht, 1979: 14). Drinking and gambling constituted an enormous drain on resources.

Such situations played havoc with the vulnerable household economic security of the *adivasi*s and, by extension, the role of women as 'household managers'. The pressure was subtly and at times, violently transferred to the 'housekeepers'. 'Improper' household management was then linked to food shortages and other moral and emotional panics. Women were at the receiving end of the resultant frustrations: directly, through physical violence (wife-beating), or, indirectly, through other harrowing gestures like gambling, drunkenness or absenteeism. Cases of wife beating were particularly rampant where women were the primary bread winners (Kishwar, 1987: 99-100). Resistance would invite greater violence and abuse.

Some attempts were however made to make women's recruitment more remunerative after the nationalization of the coal mines in 1971. It is a different matter though that after the nationalization many *adivasi* families lost their jobs (Sengupta, 1982: 16). According to an estimate, the total number of families displaced by Coal India Limited and its subsidiary concerns were 32,751 while the number of jobs provided (to one member of the family) were only 11,901 (Areeparampil, 1989: 19). The rehabilitation procedures were also significantly influenced by patriarchal considerations. For after protracted agitation, if some vacancies came up in mining companies, they were generally offered to men, as according to *adivasi* systems of land inheritance, males are the owners of land (Das, 1992: 72). By enacting the Equal Remuneration Rule, 1976, the government

---

[9] The home-brewed *handia* (rice-beer) was being replaced by high-alcohol liquor.

perforce made the private mine owners and industrialists provide women labourers with equal remuneration. Yet in many cases, a politically correct wage structure contributed to their victimization.

In the industrial areas, the rules enacted by government for the safety and security of the women labourers acted as a major deterrent to women's recruitment. According to these rules, women labourers were prohibited from working in evening and night shifts and in underground mines. Women labourers were not taken in employment (generally in mines operated by private concerns) to avoid maternity benefits, equal remunerations, rest shelters and crèches. Progressive mechanization and gender discrimination (they were considered unskilled and less strong as compared to their male counterparts) also worked to their disadvantage. Attractive remuneration regimes, on the contrary, opened up new, disguised structures of oppression and exploitation. It is widely held that the Government of Bihar, in alliance with the so-called coal mafia connived to expel thousands of Jharkhandis from the pits and replace them with 'client workers' from North Bihar (Corbridge, 1999: 137). The much talked about 'voluntary' retirement schemes are often cited as a specific case in this context.

Falling prey to vicious manipulations of the unholy alliance between factory management, union leaders and job solicitors, many *adivasi* women had to make way for prospective job seekers under the garb of 'voluntary' retirement schemes. According to the provisions of the scheme, a woman labourer could voluntarily retire once she had reached the age of thirty-six and (if she thought appropriate) offer her job, and also her Provident Fund Contribution Number to her husband, son, brother-in-law, son-in-law, or any other person recognized by her as her kin. Studies conducted in and around Dhanbad and Jharia reveal that during the post-nationalization period, sums ranging from Rs 10,000 to 30,000 were taken as bribes from interested persons and later divided among the sons or husbands of the 'voluntarily' retired women, management, union leaders and the middlemen; all those who were involved in cajoling the women to retire. In this way a large number of outsiders from Arrah, Chapra (parts of north Bihar), Ballia (Uttar Pradesh) and other regions as well, appropriated the jobs of tribal and other low caste women by producing fake marriage certificates and false affidavits claiming to be their husbands. This is why one finds in the official records of those coal mines that the surnames of the 'fathers-in-laws' of *Jhas, Singhs*, *Dubeys* or *Mishras* are *Manjhis, Mundas, Oraons, Bhumiyans* or *Rabidas*. In some cases, genuine marriages indeed took place, but only temporarily; the outsiders estranged their wives immediately after getting the appointment letter or after making them pregnant, which compelled women to knock on the doors of the management or the union leaders (Prasad, 1988: 5). Women miners have suffered and continue to suffer from many similar disabilities in their day-to-day working conditions.

Non-payment or irregularity in payments has been one of the very common problems. Field studies conducted in Giridih and other places indicate that in many cases, wages, other perks, travelling expenses and even accommodation promised by the labour contractors (in the villages) were never made available to the *reza*s when they reached the workplace (Ray, 1982: 30-33). Sexual harassment has in fact increased over a period of time. Some recent figures that give us an idea of the crime can be seen in the case of conservative rape statistics provided by the Dhanbad district police. Accordingly, on average, six rape cases are reported every month in the Dhanbad-Jharia and Hazaribag coalfields (*The Telegraph,* 2002). But an exceptionally large number of such incidents involving assaults on women's sexuality go unreported. The situation on women's health also continues to be precarious. Women and reproductive health care in mines have to be understood in the larger context of direct and indirect impacts: exposure of women and children to mine disasters and mine pollution as well as to the reduction in quality of life due to denial of access to food security, natural resources and livelihoods. Respiratory problems, silicosis, tuberculosis, leukemia, asbestosis, arthritis, etc., are reported to be very widespread among women working in the mines. Surveys conducted on the impact of radioactive emissions around the Jaduguda uranium mines operated by the Uranium Corporation of India Ltd. have also reported several cases of women having disrupted menstrual cycles, miscarriages or difficulty in conceiving. Cases of stillbirths or mothers giving birth to children with deformities were, moreover, common (Sydney Morning Herald Online, 1999).

Adivasi women have thus become ruthlessly exposed to the operation of the market and other commercial forces and of unscrupulous elements that have been flocking into the tribal heartland in the wave of exploitation of mines and industries (Singh, 1987: 8). It was hoped that the 'women's programme' of the *adivasi* movements would also address the problem of women miners. But guided by hegemonic patriarchal traditions, Jharkhand Mukti Morcha was, on the contrary, enforcing 'sexual morality drives' to prevent 'sexual and moral corruption of tribal women by non-tribal men'. Women, either divorced or separated, working in the collieries were first warned that they should marry. Upon refusal to do so, they were beaten and forcibly married to someone chosen by the cadres and sent away to distant areas where their 'husbands' resided. Women willing to pay Rs 1,000 were however spared (Kelkar and Nathan, 1991: 155).

Colonial land and forest legislation led to a severe diminution in women's traditional economic space in Chotanagpur. The growth of mines and capitalist enterprises in such a situation opened up an economic opportunity for cheaply available casual women labourers. But this opportunity was again garbed in harsh living conditions, discriminatory work regimes and untold hardships for women. In addition, the interactions between the *adivasi* and *diku* cultures at the mining settlements were to unfold larger avenues for women's psycho-cultural exploitation bringing about significant shifts in traditional gender identities. With the ban on women's underground work and progressive mechanization of mines, women were fast turning into a 'reserve labour force' – to be hired and fired at will.

References

Ambasht, N. K., 1979. 'Santhal Coal Miners: A Few Case Studies', in J. Troisi (ed.), *The Santhals: Readings in Tribal Life*, vol. 8, Indian Social Institute, New Delhi.

Archer, W. G., 1974. *Hill of Flutes, Life, Love and Poetry in Tribal India: A Portrait of the Santals*, G Allen and Unwin, London.

Areeparampil, Mathew, 1989. 'Industries, Mines and Dispossession', in Walter Fernandes and Enakshi Ganguly Thukral (eds), *Development, Displacement and Rehabilitation*, Indian Social Institute, New Delhi.

Banerjee, Sukumar, 1978. 'Tribal Women in Coalfield Area', in *Tribal Women in India*, Indian Anthropological Society, Calcutta.

Census of India, 1951, vol. V, Bihar, Part II-B Tables (Economic Tables), 1953, Government Printing Press, Patna.

Census of India, 1961, vol. IV, Bihar, Part V-A, Special Tables for Scheduled Castes and Scheduled Tribes, 1965. Government Printing Press, Delhi.

Committee of Jharkhand Matters (COJM), 1990. *Report*, New Delhi.

Corbridge, Stuart, 1982. 'Industrial Development in Tribal India: The Case of the Iron Ore Mining Industry in Singhbhum District c. 1900–1960', in Nirmal Sengupta (ed.), *Fourth World Dynamics: Jharkhand*, Authors' Guild Publications, Delhi.

Coupland, H., 1911. *Bengal District Gazetteer: Manbhum*, Bengal Secretariat Book Depot, Calcutta.

Curzel, Dagmar, 1923. *Women's Medical Service of India: Report of an Enquiry into the Conditions of Employment of Women Before and After Child Birth In Bengal Industries* (File no. 2R-20/1923), April, Government of Bengal, Dept. of Commerce.

Damodaran, Vinita, 2002. 'Gender, Forests and Famine in 19th Century Chotanagpur, India', in *Indian Journal of Gender Studies*, vol. 9, no.2.

Das, Victor, 1992. *Jharkhand: Castle over the Graves*, Inter-Alia Publications, New Delhi.

Das-Gupta, Pranab Kumar, 1978. 'Tribal Women in Industrial Context', in *Tribal Women in India*, Indian Anthropological Society, Calcutta.

Engels, Dagmar, 1999. 'The Myth of the Family Unit: *Adivasi* Women in Coal-mines and Tea Plantations in Early Twentieth century Bengal', in Peter Robb (ed.), *Dalit Movements and the Meanings of Labour in India*, Oxford University Press, Delhi.

Government of India, 1969. *Report of the National Commission on Labour*, Ministry of Labour and Employment and Rehabilitation, New Delhi, p. 380.

Imperial Gazetteer of India, 1984 (originally published in 1909). *Provincial Series Bengal*, vol. I. and II., Usha Publications, New Delhi.

Ivern, Father, 1969. *The Chotanagpur Survey: A Study of Socio Economic and Health Development*, Indian Social Institute, New Delhi.

Jha, J. C., 1987. *Tribal Revolt of Chotanagpur, 1831-32*, Kashi Prasad Jayaswal Research Institute, Patna.

Kelkar, Govind and Dev, Nathan, 1991. *Gender and Tribe: Women, Land and Forests in Jharkhand*, Kali for Women, New Delhi.

Kishwar, Madhu, 1987. 'Toiling without Rights: Ho Women of Singhbhum', *Economic and Political Weekly*, vol. 32, no. 3, Jan 17.

Kumar, Radha, 1982. 'Will Feminist Standards Survive in Jharkhand', in Nirmal Sengupta (ed.), *Fourth World Dynamics: Jharkhand*, Authors' Guild Publications, Delhi.

Mohapatra, Prabhu Prasad, 1985. 'Coolies and Colliers: A Study of the Agrarian Context of Labour Migration from Chotanagpur, 1880–1920', in *Studies in History*, vol. 1, no. 2.

Majumdar, D. N., 1937. *A Tribe in Transition*, Longman's Green and Co. Ltd., Calcutta.

Peterson, J. C. K., 1910. *Bengal District Gazetteer: Burdwan*, Bengal Secretariat Book Depot, Calcutta.

Prasad, Narmadeshwar and Sahay, Arun, 1961. *Impact of Industrialization on Bihar Tribes: A Report*, Bihar Tribal Welfare Research Institute, Ranchi.

Prasad, Sushma Sahay, 1988. *Tribal Women Labourers: Aspects of Economic and Physical Exploitation*, Gyan Publishing House, Delhi.

Ray, S. N., 1982. *Migrant Women Workers*, Bihar Tribal Welfare Research Institute, Ranchi.

Report of the Royal Commission on Labour in India, 1931. His Majesty's Stationery Office, London.

Rothermund, Dietmar and Wadhwa, D. C., 1978. *Zamindars, Mines and Peasants: Studies in the History of an Indian Coalfield and its Rural Hinterland*, Manohar, Delhi.

Roychaudhury, P. C., 1958. *Bihar District Gazetteer: Singbhum*, Patna, Superintendent, Secretariat Press (*Singbhum District Gazetteer*).

Roychaudhury, P. C, 1961. *Gazetteer of India: Bihar, Palamau*, Patna, Superintendent, Secretariat Press (*Palamau District Gazetteer*).

Roychaudhury, P. C., 1964. *Gazetteer of India: Bihar, Dhanbad*, Patna, Superintendent, Secretariat Press (*Dhanbad District Gazetteer*).

Sengupta, Nirmal, 1979. *Destitution and Development: A Study of Bauri Community in the Bokaro Region*, Concept Publishing Company, New Delhi.

Sengupta Nirmal, 1982. 'Background of the Jharkhand Question', in Nirmal Sengupta (ed.), *Fourth World Dynamics: Jharkhand*, Authors' Guild Publications, New Delhi.

Simeon, Dilip, 1982. 'Jharkhand: Community or Proletariat', in Nirmal Sengupta (ed.), *Fourth World Dynamics: Jharkhand*, Authors' Guild Publications, New Delhi.

Simeon, Dilip, 1995. *The Politics of Labour Under Late Colonialism: Workers, Unions and the State in Chota Nagpur, 1928–1939*, Manohar, New Delhi.

Simeon, Dilip, 1998. 'Coal and Colonialism: An Exploratory Essay on the History of Indian Coal', *Research-in-Progress Papers, History and Society*, Third Series, no. 33, Nehru Memorial Museum and Library, May.

Singh, K. S., 1987. 'Colonial Transformation of Tribal Society in Middle India', *Economic and Political Weekly*, Vol 13, no. 30, July 29.

Singh, K. S., 1988. 'Tribal Women: An Anthropological Perspective', in J. Singh, N. N. Vyas and R. S. Mann (eds), *Tribal Women in Development*, Rawat Publications, Jaipur.

Singh, R. L., 1990. *India: A Regional Geography*, National Geographical Society of India, Varanasi.

Sinha, S. S., 2005. *Restless Mothers and Turbulent Daughters: Situating Tribes in Gender Studies*, Kolkata, Stree.

Sydney Morning Herald Online, 1999. December 11.

The Telegraph, 2002. Calcutta, October 28.

Vidyarthi, L. P., 1978. *Socio-Economic Implications of Industrialization in India: A Case Study of Tribal Bihar*, Planning Commission, New Delhi.

PART II
Gender and Ethnic Identities in the Mines

Chapter 5

Digging through Layers of Class, Gender and Ethnicity: Korean Women Miners in Prewar Japan

W. Donald Smith

Introduction

Korean women coal miners in Japan were a small group – truly a minority within a minority – that has received very little scholarly attention. As in many other countries, few mining historians in Japan take either ethnicity or gender seriously, and few specialists in women's history take ethnicity into account. As anthropologist, Sonia Ryang, has pointed out, research on colonial labour in Japan has focused on coercive wartime labour mobilization, contributing to 'the silence and/or silencing of other, marginalized groups of the colonized, notably in this case, women' (Ryang, 1998)[1]. Korean women miners deserve the attention of specialists in all these subfields, but their story has broader implications, showing how the prewar Japanese mining work force was shaped by gender and ethnicity, and how both factors, along with class, are central to the analysis of any labour force.

Korean women were part of an interwar Japanese mine labour force that was similar to that of many other countries in terms of its structuring along ethnic or 'racial' lines. Miners from Korea – a Japanese colony from 1910 to 1945 – were the largest minority group, and played a role similar to that of many minorities in world mining, such as African-Americans in Appalachia. Both groups made up a disproportionately large percentage of hewers, had lower incomes than majority group miners[2], and faced discrimination in housing and other areas of daily life.

---

[1] Ryang 1998: 3.

[2] In West Virginia in 1932, for example, 76.8 per cent of African-American coal miners were hewers, compared to 87.6 per cent of immigrants and 52.7 per cent of native-born whites (Trotter, 1990: 87). In Japan in 1930, 78.9 per cent of Korean coal miners were hewers, compared to 50.9 per cent of Japanese. Naikaku Tōkeikyoku, *Shōwa 5-nen kokusei chōsa hōkoku, zenkoku no bu*, vol. 2, *shokugyō oyobi sangyō*, 1935, 41-43, 225 and 228.

Perhaps because of the later timing of its industrialization, Japan was unusual among twentieth century industrial powers, however, in having tens of thousands of women working in large-scale mines decades after they had been banned in Western Europe. Their numbers peaked in about 1920, when the census counted 66,396 women working in Japan's coal mines, making up 26.6 per cent of the underground work force, and another 28,474 working on the surface (30.6 per cent of surface workers), for a total of 94,870 women mine workers (27.7 per cent of the total) (Nōshōmushō Kōzankyoku, 1920)[3]. Many women worked in rarely inspected smaller mines, so the true number of women working underground was probably close to 100,000[4].

Certain generalizations are commonly accepted about the occupational distribution of Korean and Japanese men and women in the mines. Koreans (most of them men) were much more likely than Japanese to work underground, probably because Koreans cost employers less and had fewer employment opportunities. Women, meanwhile, (most of them Japanese) made up the vast majority of coal haulers underground as late as 1930. Nearly two out of three women then worked underground, most of them married. Most women who worked on the surface were coal sorters, and they tended to be younger and single.

These generalizations about women and Koreans break down, however, when we examine gender and ethnicity in conjunction. The predominance of women among pre-war coal haulers, for example, was true only of ethnic Japanese. While Japanese women outnumbered Japanese men almost six to one in this occupation in 1930, male coal haulers outnumbered women almost two to one among Koreans. Similarly, the overconcentration of Korean workers underground was true only of men. Unlike Japanese women before 1933, only a minority of Korean women at the mines worked in the pits, and most were married wherever they worked.

Most of the little information available on Korean women in the Japanese coal industry is from the 1930 census (Table 5.1) and a handful of government and company documents prepared at about the same time, so this chapter will focus on that period in examining Korean women's distinctive place in Japanese mining and what it means for people concerned with gender and ethnicity in the workplace today. This narrow chronological focus, it should be pointed out, risks creating a misleadingly static impression of an industry in the midst of sweeping change. From about 1928 through to 1933, mine operators, seeking to heighten productivity and increase their control over the labour process, introduced new machinery and

---

[3] Nōshōmushō Kōzankyoku, *Honpō kōgyō no sūsei* for 1920, cited in Nishinarita Yutaka, 'The Coal Mining Industry', in Nakamura Masanori, ed., *Technology Change and Female Labour in Japan* (Tokyo: United Nations University Press, 1994: 87). Due to the poor quality of this translation, those who read Japanese are advised to consult the original version, published as 'Sekitan kōgyō no gijutsu kakushin to joshi rōdō', in Nakamura, ed., *Gijutsu kakushin to joshi rōdō* (Tokyo: Kokuren Daigaku, 1985: 71-105).

[4] Sone Sachiko estimates that as much as a quarter of the work force was ignored because official data on smaller mines were rarely gathered (Sone, 2003: 36).

new methods of mining, phased out women underground workers, and moved (especially at mines run by the *zaibatsu* conglomerates) from a poorly paid, diverse work force with high turnover toward a somewhat better paid, sexually and ethnically homogenous, and relatively stable work force.

Table 5.1 Occupational distribution in the Japanese coal industry, 1930

Occupation	Korean men	Korean women	Japanese men	Japanese women
Owners and Operators	3 (0.03%)	0 -	923 (0.7%)	5 (0.01%)
Engineers and Managers	37 (0.4%)	0 -	9,143 (6.7%)	0 -
Hewers	7,681 (78.9%)	0 -	69,898 (50.9%)	0 -
Haulers	148 (1.5%)	79 (38.5%)	3,245 (2.4%)	19,165 (54.4%)
Other in-mine transport	132 (1.4%)	4 (1.9%)	4,984 (3.6%)	61 (0.2%)
Bracers	67 (0.69%)	0 -	6,713 (4.9%)	0 -
Other in-mine workers	780 (8.0%)	4 (1.9%)	16,183 (11.8%)	2,881 (8.2%)
Coal sorters	41 (0.4%)	97 (47.3%)	2,102 (1.5%)	9,674 (27.47%)
Others aboveground	844 (8.7%)	21 (10.2%)	24,133 (17.6%)	3,433 (9.7%)
Total	9,733	205	137,324	35,219

Note: percentages indicate the prevalence of each occupation within each ethnic/gender group

Source: Naikaku Tōkeikyoku, *Shōwa 5-nen kokusei chōsa hōkoku, zenkoku no bu*, vol. 2, *shokugyō oyobi sangyō*, 1935, 41-43, 225 and 228

This chapter is primarily concerned with the Chikuhō coal field, Japan's most important during most of the modern period. This 787-square-kilometre area in Fukuoka Prefecture on Japan's southwestern island of Kyūshū employed the majority of miners in pre-war Japan and even higher proportions of women (of both ethnicities) and Koreans (of both sexes). In 1930, for example, Chikuhō and other, less important, coal fields in Fukuoka Prefecture employed 51.5 per cent of all male and 59.7 per cent of all female coal miners in Japan, along with 56.6 per cent of Korean male and 69.3 per cent of Korean female mine workers (Naikaku Tōkeikyoku)[5].

Japanese women and Korean men in the mines

As Sone explains in more detail elsewhere in this volume, women played an important role in Chikuhō mining for centuries before the first modern mines were opened in the late 19th century. As they did in so many other parts of the world, men, women and children in densely populated Chikuhō had dug coal for centuries in small mines as a sideline to agriculture, so modern mines hired both men and women as a matter of course. In sparsely populated Hokkaidō, Japan's second most important coal mining region, by contrast, little mining was carried out until after the 1868 Meiji Restoration that put Japan on the path of capitalist industrialization. With no tradition of family-based mining, very few women ever became underground miners in Hokkaidō.

Since women had worked in Chikuhō mines from the beginning, it is not surprising that they became structurally indispensable to large-scale industrial mining in the late 19th century. With the Japanese industrial revolution picking up momentum, companies began installing winch engines to speed the hauling of coal out of their mines. These winches had to be operated continuously to maximize profits but the workers who dug the coal were not provided with new equipment to help them keep pace. To produce a steady supply of coal for the winches to haul up, the rapidly expanding mines needed large numbers of new full-time workers, and many of these new workers were women. There was more female surplus labour and women could be paid less, so for management it made sense to use men only in jobs for which they were seen as indispensable, such as hewing, and to use women for other tasks, such as hauling the coal to a central loading point (Nishinarita, 1994: 60-62). Women's work in the mines made economic sense to their families, too, since men's somewhat higher wages were not high enough to support their families on their own.

Workers from a wide area of western Japan, meanwhile, came to work in the Chikuhō mines from midway through the Meiji period (1868–1912), joining residents of Fukuoka Prefecture, many of them outcast *burakumin*[6]. From late in the 19th century, some mines widened their recruitment nets beyond Japan and

---

[5] Calculated from Naikaku Tōkeikyoku, 1935, *Shōwa 5-nen kokusei chōsa hōkoku, zenkoku no bu*, vol. 2, *shokugyō oyobi sangyō*: 41-43 and 225 and vol. 4, *fukenhen, Fukuoka-ken*: 118.

[6] *Burakumin*, literally 'hamlet (*buraku*) dwellers', are ethnically indistinguishable from other Japanese but were stigmatized as unclean and restricted to jobs such as tanning leather and caring for the dead for centuries before their legal emancipation in 1871. Despite their legal equality, they remain subject to discrimination in employment, housing, and marriage today.

staged abortive experiments in hiring male Korean miners, but it was not until World War I that the region's mines recruited large numbers of Korean workers. The war cut off British shipments of coal to Asia, creating new markets for Japanese coal, and drastically reduced shipments of other products from Britain as well, opening new markets for Japanese manufactured goods and, in turn, increasing domestic demand for coal. From late 1916, with little if any surplus labour left in the domestic rural sector and many Japanese miners leaving for more attractive factory jobs – which paid more for men if considerably less for women during most of the prewar period[7] – the booming Japanese coal industry turned to Korea as a new source of workers.

Unlike the subsequent period of coercive migration, Korean migration to Japan before 1939 was voluntary, albeit within the context of colonial power relations, and largely driven by the higher wages and better living conditions available in the metropole. Koreans, on average, were poorer than Japanese and became even more so during the colonial period. Landless tenants, for example, increased from 39.4 per cent of the Korean farming population in 1913–1917 to 55.7 per cent in 1939 (Suzuki, 1942)[8]. Rice consumption per person, meanwhile, a basic indicator of East Asian living standards, was lower in Korea than Japan at the beginning of the colonial period and continued to fall, hitting 36.4 per cent of Japanese consumption in the 1930s[9]. Despite increased use of other, less desirable, grains, total daily calories fell from 2,133 per capita in 1912–1915 to 1,812 in 1931–1935 before recovering to 2,033 in 1936–1940 (Suh, 1978)[10]. There were relatively few industrial jobs available in Korea, especially before Japan's development of its colony in the 1930s as an industrial support base for Japanese military operations in China, and the jobs that were available paid less than work in Japan. Accordingly, leaving the farm, for many in southern Korea, meant seeking work in Japanese manufacturing, construction, or mining.

In Chikuhō, easily reached by ferry from Korea followed by a short train ride, national conglomerate Mitsubishi and local company Kaijima were among the first to hire Korean mine workers. Koreans also went to work at factories and on construction projects, increasing the number of Korean workers in Fukuoka Prefecture from under 900 in late 1916 to over 5,000 in May 1919. Women

[7] Women hauling coal underground were paid at least 21 per cent more and in some years 151 per cent more than female factory workers, and about double the pay of coal sorters. Underground male mine workers, meanwhile, were paid less than factory workers from 1929 to 1937 (hitting bottom at 62.9 per cent of factory wages in 1932), before wartime demand for coal drove underground mine pay ahead of factory wages from 1938 through 1945 (peaking at 124.9 per cent of factory wages in 1941). Calculated from *Rōdō Undō Shiryō Iinkai*, 1959: 284-285.

[8] Suzuki Takeo, *Chōsen no keizai* (Tokyo, 1942), 246, cited in Suh, 1978: 81.

[9] Calculated from Korean average consumption for 1934–1938 and Japanese consumption in 1939 as cited in Grajdanzev, 1944: 118.

[10] Suh, 1978: 87.

comprised only nine of these workers at the end of 1916 and 175 in May 1919[11], and the overwhelmingly male composition of the workforce suggests that most Korean migrants in this period saw themselves as sojourners who would return home once they had saved a certain amount of money. Inevitably, however, many stayed on, and the Korean community in Japan, still the country's largest ethnic minority[12], traces its modern roots to this period.

With Koreans providing a steady supply of flexible, low-cost labour – and, from the mine owners' perspective, a way to keep an often restive workforce from uniting against management – their numbers grew in the Chikuhō mines even as many Japanese were laid off during the post-World War I recession. With Korean men also hired to replace many of the Japanese women fired in anticipation of the 1933 enforcement of a ban on women's underground employment, Korean mining employment reached a pre-war peak in about 1928, when 8,468 Korean men and 243 Korean women worked in major Kyūshū mines, according to a government survey (FKKK, 1929)[13].

Koreans remained important thereafter to smaller mines, but major collieries like Mitsubishi laid off most of their Koreans from the late 1920s as they created a more ethnically and sexually homogenous and somewhat better paid workforce that managers saw as more compatible with the increasingly capital-intensive technology major mines were adopting in this period. After Japan launched the Asia-Pacific War with its 1937 invasion of China, however, demand for coal soared just as military conscription began to cut into the work force. Following the onset of forced migration in 1939, Koreans became more important to Japanese mining than ever. Korean coal miners in Japan reached a peak, according to official figures, of 136,825 in February 1945, when they accounted for 32.1 per cent of all workers in the industry[14].

For some Japanese women, the war brought new (if temporary) opportunities to advance into jobs previously monopolized by men. In coal mining, women workers hit a wartime peak in March 1945 of 65,613, 15.6 per cent of all mine workers[15],

---

[11] *Fukuoka Nichinichi Shimbun*, 27 June 1919.

[12] Japan had 613,791 officially registered South and North Korean nationals as of the end of 2003, 32.1 per cent of Japan's foreign population, down from 50 per cent as recently as 1994, according to the Japanese Ministry of Justice [http://www.moj.go.jp/PRESS/040611-1/040611-1.html]. This figure does not include the estimated 30 per cent of ethnic Koreans who have taken Japanese citizenship. See Chung, 2000.

[13] Fukuoka Kōzan Kantoku Kyoku, 'Chōsenjin kōfu rōdō jijō', Kōgyō shiryō no. 28 (1929), reproduced in Hayashi, 1991: 46. The survey covered mines of 'semi-important' status or above in Yamaguchi, Fukuoka, Saga and Nagasaki prefectures.

[14] Sekitan Tōseikai Kinrōbu, 'Shōwa 19-nendo zenkoku tankō rōmusha idō jōkyō shirabe', in Nagasawa, 1992: 39.

[15] Shōkōshō Kōzan Kyoku, *Honpō kōgyō no sūsei*, various years, compiled in Tanaka Naoki, *Kindai Nihon tankō rōdōshi kenkyū* (Tokyo: Sōfūkan, 1984: 160-161).

and some became *sakiyama* in charge of supervising Korean male coal haulers[16]. The war also opened opportunities for some male *buraku* and long-term Korean resident workers, but it brought little change at best for the vast majority of Korean women.

There is no record of even a single Korean woman becoming a *sakiyama*, although they worked in small numbers at collieries throughout the war. Sin Ta'e-wŏn, for example, who went by the Japanese name Takeyama Yasuko, went to work at age 15 in 1938 in the Yasutake mine in Chikuhō. Her mother, apparently also a miner, was pregnant, so Sin went in with her father as his coal hauler (Nakayama, 1983)[17]. Much later in the war, official figures showed 369 Korean women (along with 140,419 Korean men) working in or at Japanese metal and coal mines as of June 30, 1944[18], including nurses, hospital interpreters, and cooks as well as mine workers (FNS, 1940; RKK, 1943)[19].

Korean women were also sent – against their will – to wartime 'comfort centres' set up near coal mines in Japan. The issue of Japan's wartime coercive mobilization of up to 200,000 'comfort women' from Korea, China, and other occupied areas under Japanese control has been most often discussed in terms of Japanese exploitation of other nations. The mine comfort centres show, however, that the problem at times involved colonized men's exploitation of women from their own countries within the framework of Japanese imperialism. In Chikuhō, for example, mine supervisors reportedly rewarded diligent Korean men – most or all of them sent to the mines by coercive means – with visits to 10 or more 'hospitality establishments', many of which accepted company scrip in payment[20]. The number of Korean women sent to Chikuhō for sexual servitude is unclear, but some 80 were sent to two Mitsubishi mines on the Takashima and Hashima islands off Nagasaki (Takeuchi, 2003)[21], and at least 100 were sent to 'comfort centres' in at least 18 coal and other mines in Hokkaidō (Nishida, 2003)[22].

Within months of Japan's August 1945 defeat, most Koreans left the mines, and Japan's postwar mining workforce was virtually monoethnic until the last mines closed at the turn of the 21st century. With demobilized soldiers returning from Asia and the Pacific asking for their old jobs back, most women quickly left the coal pits as well. The exclusion of women from underground work was formalized by the US authorities who oversaw the 1945–1952 Allied Occupation

---

16 Interview with local historian Idegawa Yasuko, 16 May 1995, Kurate-machi, Fukuoka Prefecture.

17 Nakayama, 1983: 97.

18 Labour Ministry statistics, cited in Kang, 1976: 27.

19 *Fukuoka Nichinichi Shimbun*, 20 June 1940; Rōdō kagaku kenkyūjo, 'Hantō rōmusha kinrō jōkyō ni kansuru chōsa hōkoku' (1943), reprinted in Pak, 1982, vol. 1: 74, 113, 155.

20 *Nishinippon Shimbun*, *Yomiuri Shimbun*, and *Asahi Shimbun*, 21 February 1992, evening editions. I am indebted to Yokogawa Teruo for information on this topic.

21 Takeuchi, 2003: 31-32. The mines were on Takashima and Hashima islands.

22 Nishida, 2003: 16-36.

of Japan, and the last officially authorized women were gone by March 1947[23]. Some women – apparently including Koreans – continued to work illegally in small, rarely inspected mines as late as 1961 (Chikuhō Sekitan, 1973),[24] but mining came to be seen for the first time in Japan as a 'naturally' male occupation. Women kept working as coal sorters, but they were phased out as mechanization transformed sorting from a labour-intensive, largely female occupation into a relatively capital intensive and almost entirely male occupation beginning in the 1950s.[25]

Korean women's distinct occupational distribution

Most Korean miners, as we have seen, were men, but Korean women worked in and on Japanese coal mines from at least 1920 through 1945[26]. Even at their peak in around 1928, they numbered only some 500[27] but had a distinct occupational distribution. According to the 1930 census (see Table 5.1), nearly half (47.3 per cent) of Korean women in the coal industry were coal sorters, compared with 27.5 per cent of Japanese women. Another 10.2 per cent were other aboveground workers, compared with 9.7 per cent of Japanese. We can see that well over half – 57.6 per cent – of Korean women held relatively safe but poorly paid aboveground jobs, compared with just 37.2 per cent of Japanese women. Another 38.5 per cent of Korean women worked underground as coal haulers (*atoyama*), compared with

[23] 'The Coal Fields of Kyushu', 21, Supreme Commander for the Allied Powers, General Headquarters, Record Group 331, Box 2494, Folder 1, 'Records of Allied Operational and Occupation Headquarters, World War II', Washington National Records Center, Suitland, Md.

[24] Chikuhō Sekitan Kōgyō-shi Nenpyō Hensan Iinkai, Chikuhō sekitan kōgyshi nenpyō (Fukuoka: Nishi Nihon Bunka Kyōkai, 1973): 589. For a reference to some Korean women who 'headed to the coal mines to work' to help finance construction of a Korean school in northern Kyushu in 1956, see Kyodo World Service, Feb. 7, 2005. Also see the account of Tachibana Shizue, a pre-war mine worker who went back to work in a small mine at the age of 54 or 55 in 1956 or 1957, in Nakayama, 55.

[25] For a study of this phenomenon at two Hokkaidō mines, see Hokkaidōritsu Rōdō Kagaku Kenkyūjo, *Tankō ni hataraku fujin no rōdō jittai* (Sapporo, 1961).

[26] The 1920 census counted 65 women among the 5,190 Koreans in the Japanese coal industry. Naikaku Tōkeikyoku, *Taishō 9-nen kokusei chōsa hōkoku, zenkoku no bu*, vol. 2, 'shokugyō': 249. For three Korean women coal sorters at Meiji's Akaike mine from June 1944 through to June 1945, see Meiji, 1991: 1153, 1173, 1177, 1192, 1204.

[27] A June 1928 survey found 320 Korean women miners in Fukuoka Prefecture. Fukuoka Chihō Shokugyō Shōkai Jimukyoku, 'Kannai zaijū Chōsenjin rōdō jijō' (Fukuoka, 1929), rpt. in Hayashi, 1991, vol. 2, Part 1: 125. Assuming that Fukuoka accounted for 69.3 per cent of Korean women miners in Japan, as it did in the 1930 census, this would produce a national total of about 462.

54.4 per cent of Japanese, while 3.9 per cent were other underground workers, compared with 8.3 per cent of Japanese[28].

While Korean women were much more likely to work aboveground than their Japanese counterparts, Korean men were more likely to be hewers or to do other jobs underground than Japanese. Over three out of four Korean male mine workers in 1930 were hewers, as loosely defined by the census. Most of the rest held other jobs underground, while less than 10 per cent worked aboveground. Just half of Japanese male coal miners, by contrast, were hewers, and nearly 20 per cent worked aboveground. Korean men were excluded from skilled jobs and management almost as entirely as were women of either ethnic group; just 0.4 per cent of Korean men were engineers or managers, compared with 6.6 per cent of Japanese men[29].

The census does not tell the whole story, excluding as it does many workers at small, unregulated mines. In addition, the 1930 census, as we saw above, was conducted in the midst of sweeping change in Japanese coal mining. This means that the census isn't necessarily representative of Korean women's situation in the mines during the entire interwar period, but it does clearly show that Korean and Japanese women led very different working lives.

One of the few other detailed sources on Korean women mine workers – a roster of workers at major local company Kaijima – brings us a little closer to being able to see the lives of individual workers[30]. The women on this roster were from villages scattered across southern Korea, suggesting that they were not recruited directly by the company but came to the mine with husbands or other family members.

The roster, compiled by the Kaijima Oonoura colliery's no. 7 mine, lists 1,388 Korean workers employed from 1928 through to 1943, of whom 26 are clearly identifiable as women. Kaijima hired the women at ages ranging from 16 to 39, and all 26 were hired in 1938 or later, although it is known that women had worked at the mine earlier. Occupations are specified for just three women: two coal sorters, one hired at 16 and the other at 25, and a 'hewer' (probably an *atoyama*)

---

[28] Calculated from Naikaku Tōkeikyoku, 1935, *Shōwa 5-nen kokusei chōsa hōkoku, zenkoku no bu*, vol. 2, *shokugyō oyobi sangyō*: 41-43, 225, 228.

[29] Calculated from Naikaku Tōkeikyoku, 1935, *Shōwa 5-nen kokusei* chōsa, *zenkoku no bu*, vol. 2, *shokugyō oyobi sangyō*: 41-43, 225, 228. 'Hewers' (*saitanfu*) and *atoyama* are treated as separate categories in census reports, but *atoyama* are included in a broader category of 'hewers' in some other documents, such as the Kaijima roster cited below.

[30] Kaijima tankō dai-nanakō Chōsenjin kōfu iroha meibo, collection of Arirang Bunko, Tagawa, Fukuoka Prefecture. The roster, covering workers hired from 31 August 1928 through 22 October 1942, does not specify the sex of workers, but relationship to the head of household is listed for all but 164 workers, making it possible to identify these 26 women. The first identifiable woman was hired 19 September 1938, but it is clear from other sources that Korean women worked at Oonoura mine no. 7 before then, so it is likely that many of the 164 workers for whom no relationship is listed were also women. I am indebted to Hayashi Eidai for providing me with a copy of this document.

hired at 18. It can be inferred that a fourth woman, aged 37, was also an *atoyama* or other underground worker because she was fired for smoking in the mine, a serious violation of safety regulations (not to mention Korean and Japanese gender norms). None stayed long: anywhere from six weeks to two and a half years before fleeing the mine, resigning (because of 'housework', in one case), transferring to another Kaijima mine, or being fired.

The Kaijima roster does not give us a clear picture of Korean women workers' marital status, but this does not appear to have been as important in determining where in the mine compound they worked as it was for Japanese women. Among Japanese women, coal sorters were younger and less likely to be married than were *atoyama* (Nishinarita, 1994: 69-70), possibly because coal sorters were paid so little that they could not support themselves on their wages alone, much like the young 'office ladies' who live with their parents in Japan today. Korean women in Japan as a whole married much younger than Japanese, so Korean coal sorters were much more likely to be married than were Japanese in the same occupation, and probably were not much different from Korean *atoyama* in this respect. In 1920, for example, 64 per cent of Korean women aged 15 to 19 in Fukuoka Prefecture were married, compared with just 18.3 per cent of Japanese women in that age group. The proportion increased to 92.7 per cent for Korean women aged 20-24, compared with 65.5 per cent of their Japanese counterparts in Fukuoka[31].

Partly because Korean women in Japan married so young, just 17.6 per cent were formally employed as of 1930, compared with 33.1 per cent of Japanese women. (The comparable figures for men were 79.5 per cent for Koreans and 58.5 per cent for Japanese[32].) Marriage patterns were in turn shaped by a pattern of migration like that seen in so many other cases around the world: men first went on their own, sending for their wives and children after getting settled in. The first male migrants depended on the women in their home villages to subsidize their low wages, part of which they in turn remitted to Korea to help sustain the village economy. Women staying in Korea made clothes for men going to Japan as labourers (Sorensen, 1988)[33] and replaced them in the fields[34], continuing, of course, to cook and clean for children and elderly family members. As men became settled in Japan, however, they began to send for their families, and toward the end of the 1920s families began to migrate together, so the male-female ratio

---

[31] Calculated from Naikaku Tōkeikyoku, *Taishō 9-nen kokusei chōsa hōkoku, fuken no bu*, vol. 40, Fukuoka-ken: 286-287.

[32] Calculated from Naikaku Tōkeikyoku, 1935, *Shōwa 5-nen kokusei chōsa hōkoku, zenkoku no bu*, vol. 2, *shokugyō oyobi sangyō*: 224.

[33] Sorensen, 1988: 140.

[34] In one hamlet in South Chŏlla, for example, so many men had left by the late 1920s that women did all the farm work. 'Katei fujin manpyō', *Higashi*, 25 January 1928, quoted in Kim Chŏngmi, 1980: 115.

among Koreans in Japan narrowed from 7.6 to 1 in 1920 to 2.4 to 1 in 1930 and 1.5 to 1 in 1940[35].

Among those in the formal workforce, mining (coal and metal combined) employed 6.7 per cent of Korean men and 1.6 per cent of Korean women in Japan in 1930 (versus 1 per cent of Japanese men and 0.4 per cent of Japanese women). Many other Koreans worked in construction, but the greatest number had jobs in manufacturing – 51.7 per cent of employed Korean men and 67.5 per cent of employed Korean women – four to five times the proportions of Japanese in that sector [36]. Korean factory workers enjoyed less of an advantage in conditions over their compatriots in the mines than Japanese did. As historian Michael Weiner pointed out:

> Relatively few Koreans were employed as *shokkō* (permanent factory operatives) and those that were tended to be excluded from large and medium-sized firms. This meant that Korean workers were disproportionately represented in factories where working conditions were the poorest, and where housing and other facilities were either minimal or non-existent.
>
> Weiner, 1994: 49

Korean women's lesser participation in the formal sector meant they had fewer contacts with Japanese people than did Korean men and, perhaps as a result, were less 'Japanized'. Partly because so many of them were already married when they arrived in Japan, Korean women were much less likely than Korean men to marry Japanese; from 1938 to 1942, for example, 96 per cent of 'mixed marriages' involved a Korean man and a Japanese woman[37]. Women were also less likely to learn to read or speak Japanese, and they held on fiercely to gendered cultural mores in the face of official attempts to 'improve' the Korean lifestyle through imposition of Japanese norms of how to sit, walk, eat, and dress. While Japanese and Western clothing was nearly universal among Korean men in Japan by World War II, for instance, efforts to get Korean women to replace their traditional *ch'ima* (skirts) and *chŏgori* (blouses) with kimono achieved limited results at most. For example, the state-run Kyōwakai, an organization devoted to the control and rapid assimilation of Koreans, reported in 1942 that a three-year-long campaign in Kanagawa Prefecture near Tokyo had led to a 'gradual' increase in Korean women wearing Japanese clothing – but only during the group's meetings (Chūō Kyōwakai, 1942)[38].

---

[35] Calculated from census figures in Morita 1996: 41.

[36] Calculated from Naikaku Tōkeikyoku, 1935, *Shōwa 5-nen kokusei chōsa hōkoku, zenkoku no bu*, vol. 2, *shokugyō oyobi sangyō*: 224. For the stories of ten Korean women who worked in factories and dormitories, raised pigs, and carried out black market commerce, among other occupations, in pre-war Japan, see Jackie J. Kim, 2004.

[37] Calculated from Chōsen Sōtokufu, Chōsen jinkō dōtai tōkei, various years, cited in Morita, 1996: 76.

[38] Chūō, 1942, reprinted in Hyōronsha, 1990: 196.

Heavy burden of housework

As we saw above, the 1930 census demonstrates clear differences in the occupational distribution of Japanese and Korean women in the Japanese mines. Most notably, Korean women mineworkers were much less likely than Japanese women to work underground. How can we explain this difference?

One important factor can be found outside the mines. Most Korean women in pre-war Japan were there primarily to perform reproductive labour, so they faced a heavy burden of housework. Men far outnumbered women among Korean residents[39], as we saw above, but men expected women to do all the housework. Women also cooked and cleaned for boarders and took part in the underground economy, making everything from *kimch'i* to liquor.

When Korean women somehow found the time to work for wages, many of them probably found coal sorting – while tedious, exhausting work – easier than underground coal hauling to balance with the demands on their time and energy at home. Like Koreans, outcast *buraku* women were overrepresented among coal sorters in Chikuhō, for similar reasons. Sakamoto Yūichi argues persuasively that it was not because of discriminatory corporate policies that so many *buraku* women became coal sorters but because coal sorting, while it paid much less than underground work, was less exhausting. This left *buraku* women, many of whom commuted to the mines from nearby farming communities, at least minimal energy to do farm work with their families before or after their jobs at the mines (Sakamoto, 1997)[40].

Even if Korean women wanted higher-paying jobs underground, however, there may have been few available to them. In the room and pillar system employed until the 1920s at most major mines and later at smaller mines, an experienced hewer (*sakiyama*) dug out 'rooms' of coal, leaving 'pillars' in place to support the roof, while an *atoyama* – often the *sakiyama*'s wife – hauled the coal to a central collection point. Relatively few Korean men were able to become *sakiyama*, whether because of discrimination or a lack of experience, so there were correspondingly few jobs open to Korean women as *atoyama*. When mines switched from room and pillar to long wall mining, *sakiyama-atoyama* pairs were replaced by larger, usually all-male groups. This created new jobs for Korean men but, along with the partial mechanization of coal transport, meant that the number of underground jobs available to women (of either ethnic group) began to decline even before the 1933 ban on their work.

Cultural factors, while less important than the demands of household labour and the lack of underground jobs available to Korean women, also likely discouraged Korean women from becoming coal haulers. Underground mining was an exclusively male occupation in colonial Korea, although a few women worked

---

[39] The male-female ratio among Koreans in Japan was 7.6 to 1 in 1920, 2.4 to 1 in 1930, and 1.5 to 1 in 1940. Calculated from census figures in Morita, 1996: 41.

[40] Sakamoto, 1997: 101.

aboveground, so some Korean women (or their husbands) may have rejected underground work as improper for women. Unlike Japan, where men and women casually shared public baths, women in Korea were reluctant to even speak with men other than family members, although lower-class women were slower to embrace neo-Confucian sexual separation (and less able to put it into practice) than their elite counterparts. All of this must have made it difficult for Korean women to work in Japan's mines, where they would have had to work in the dark with men who were not only strangers but foreigners.

Women, regardless of ethnicity, had reason to fear everything from embarrassment to sexual harassment and assault[41] when they went down into the mines, especially considering the style of dress, or lack of dress, common in pre-war Chikuhō mines. It was hot in the mines, and exposed skin made it easier for miners to notice the small particles that often provided early warning of an impending roof collapse[42], so men wore only loincloths and women just short skirts, leaving their breasts exposed. Because women haulers usually worked with their husbands or other family members, nudity was not a particular problem at the working face. When women hauling coal alone through dark mine shafts encountered other men, however, 'mistakes could easily occur', as a government report coyly put it (Osaka, 1926)[43].

Workers in the coal sorting plants, by contrast, were fully dressed and almost all of them were women. While coal sorting, like other mine work, was generally off limits to women in the United States, the few males in Japanese sorting plants were either supervisors or boys too young to go into the mines[44].

It is often assumed that, like coal sorters, almost all the *atoyama* who hauled coal before the 1933 ban on most female underground work were women, but this is a modern oversimplification of a complex situation[45]. In Hokkaidō, almost all underground workers, including *atoyama*, were men, as were the majority of Korean *atoyama* nationwide. The 1930 census counted 19,165 Japanese women and 3,245 Japanese men, along with 148 Korean men and 79 Korean women, working as *atoyama* (Naikaku Tōkeikyoku)[46]. There were important differences,

---

[41] Mining was not the only industry nor Japan the only country, of course, in which women had (and have) to endure sexual harassment and assault to earn a living. For examples from the Japanese textile industry, see Tsurumi, 1990: 89, 119, 144, 165-167, 196.

[42] Interview with Idegawa.

[43] Osaka Chihō Shokugyō Shōkai Jimukyoku, *Chikuhō tanzan rōdō jijō* (Osaka, 1926: 76).

[44] Retired miner Nagaoka Iwao, for example, was a coal picker as a boy at Mitsubishi Hōjō mine, while his father was a coal sorting supervisor, the highest job available, according to Nagaoka, to *buraku* workers. Interview, Hōjō-machi, Tagawa-gun, Fukuoka Prefecture, 25 May 1995.

[45] This is a common misconception, shared by Regine Mathias, for example, who writes that: '*atoyama* … became a synonym for women miners in general' (Mathias, 1993: 101).

[46] Naikaku Tōkeikyoku, 1935, *Shōwa 5-nen kokusei chōsa, zenkoku no bu*, vol. 2, *shokugyō oyobi sangyō*: 41-43, 225, 228.

however, in the career paths of male and female coal haulers. Most male *atoyama* were young, inexperienced miners. They hoped to eventually advance out of the job, not because they saw hauling coal as a job for women, but because of the higher pay and increased respect accorded to *sakiyama*, according to retired miners interviewed by historian Ichihara Hiroshi[47]. Most men could in fact expect to advance to *sakiyama* status, but women only rarely became *sakiyama*, no matter how skilled or diligent they were.

Wages for women mine workers as a whole, meanwhile, were well below those of men, and the evidence suggests that Koreans as a whole were paid less than Japanese[48], but where did Korean women fit into the wage hierarchy at the mines? Specific wage data for Korean women miners are lacking, but anecdotal evidence suggests that they were paid less than either Korean men or Japanese women. Officials at one Hokkaidō coal mine, for example, said during World War II that they hoped to hire Korean women as coal sorters because of their low wages[49].

Korean women, while exploited in the home and on the job, were by no means passive victims of patriarchal, corporate and colonial domination. Korean women miners, unlike textile workers[50], may never have walked off the job in Japan, but Korean miners' wives played a major supporting role in the 1932 Asō strike, the biggest in Chikuhō that year and the largest strike ever by Korean miners in Japan. The some 400 strikers, all men, showed courage and determination in taking on capital and the state for 20 days in the harsh anti-union, anti-Korean atmosphere of the times, but their struggle would have been much harder without the support of Korean women[51].

Locally owned Asō, a major employer of Koreans of both sexes in the 1920s, had apparently laid off all its underground women workers by the summer of 1932[52]. That is when it sparked the strike by laying off or transferring 300 male

---

[47] The division of miners into *atoyama* and *sakiyama* was not, in general, a matter of company rankings, but something left up to the miners themselves. Personal communication from Ichihara Hiroshi, September 4, 1997.

[48] According to a secret Home Ministry report, for example, Japanese workers in mining and metallurgy in Fukuoka Prefecture had 'maximum' daily wages in May 1924 of 3.4 yen, compared with 2.9 yen for Koreans, 'minimum' wages of 0.9 yen, compared with 0.8 yen for Koreans, and 'normal' wages of 2.0 yen, compared with 1.5 yen for Koreans. Naimushō Shakaikyoku Dai-ichibu, *Chōsenjin rōdōsha ni kansuru jōkyō*, unpublished report, 1924, reproduced in Pak 1975-1976) vol. 1: 492.

[49] Rōdō kagaku kenkyūjo, 'Hantō rōmusha kinrō jōkyō ni kansuru chōsa hōkoku' (1943), rpt. in Pak, *Chōsen mondai shiryō sōsho*, vol. 1, 39. The mine is not identified.

[50] Korean and Japanese women joined forces in spring 1930, for example, in a strike against the Kishiwada Cotton Spinning Co. See Weiner, 1994: 171-174.

[51] For a detailed account of this strike, see W. Donald Smith, 'The 1932 Asō Coal Strike: Korean-Japanese Solidarity and Conflict', *Korean Studies*, no. 20 (1996): 94-122.

[52] The five Asō mines involved in the strike stopped using women underground sometime between February 1930 and March 1932. Asō's sixth mine still employed women underground as of March 1932 and it is unclear when they were dismissed. Moji

miners, the majority of them Koreans, in the face of slumping demand for coal. Workers demanded severance pay for laid-off miners, along with better working and housing conditions and the reversal of earlier wage cuts for remaining workers[53].

The strikers were determined to persist despite the arrest of at least 271 workers[54], their tactics – centred on the distribution of handbills and strike newspapers – won them little leverage with the company. Asō, already trying to cut output in the face of a depressed coal market, was hardly worried about lost production, and it could count on the police to keep strikers from blocking mine entrances or attacking company property.

The actions of miners' wives, many of them probably laid-off former miners themselves, were another matter, however. The women staged sit-ins and, as Korean miner Hwang Haksŏng recalled, barged into company offices, where they would spring at company officials, scratching and clawing, while yelling out their grievances. They also attended demonstrations, often, according to Hwang, with each woman carrying a baby on her back and holding another child by the hand (Kim Ch'anjŏng, 1980)[55]. Company officials were clearly unsure how to deal with this 'unladylike' behavior, while the police hesitated to beat women and children.

Two weeks into the strike, with the company still refusing to negotiate, over 50 women arrived at company headquarters to find their way blocked by police. Some 15 of the women managed to rush past the police into the office, but the company refused to meet with them (Kyōchōkai, 1932: 19-20). With the dispute becoming more heated, prefectural authorities on the same day sent the head of the Fukuoka Special Higher Police to urge company officials to work toward a settlement (Kyōchōkai, 1932: 22), probably fearing that the unrest would spread to the 7,000 Koreans working at other Chikuhō mines. In response, negotiations began that led to a settlement providing for severance payments for dismissed workers, although it did not address most of the miners' demands for higher pay or improved conditions (Kyōchōkai, 1932: 32-35). There is no evidence that women''s actions led directly to prefectural intervention, but government officials probably saw continued clashes between women and the authorities as a greater threat to public order (or at least to public perceptions of the police) than the male strikers' petitioning and leafleting.

Tetsudōkyoku Unyuka, *Ensen tankō yōran*, March 1930 and March 1932, cited in Ogino, 1993, no. 1, 401.

[53] None of the demands were specifically tailored to women's interests.

[54] Arrest figure compiled from incidents cited in Kyōchōkai Fukuoka Shutchōjo, 'Kabushiki Kaisha Asō Shōten Chōsenjin kōfu rōdō sōgi gaikyō', 1932: 16, 18, 19, 34; and Chikuhō Sekitan Kōgyōshi Nenpyō Hensan Iinkai: 367. The number arrested in these incidents could be anywhere between 271 and 285 due to the ambiguity in one Kyōchōkai account.

[55] Kim Ch'anjŏng, *Hi no dōkoku: Zainichi Chōsenjin kōfu no seikatsushi* (Tabata Shoten, 1980: 62).

Korean women also stood up for themselves in countless daily encounters. Early one morning in October 1929, for example, a supervisor at Asō berated several Korean women coal sorters for their lacklustre work. It was 3 a.m., and the workers were no doubt exhausted, but they began working with all the enthusiasm they could muster, all except for one 19-year-old who worked under the Japanese name Nakajima Masako. She was not just tired, she was tired of being bossed around, so she stopped working altogether. The Japanese supervisor, by his account, 'pushed' her shoulder twice. Nakajima, dodging her supervisor's blows, wrapped herself around his leg. He tried to push her away, and they both fell over. She went back to work but, later, with the help of a male Korean mine worker active in leftist politics, filed an assault charge against the supervisor. The company refused to negotiate, however, and the police, after conferring with the company, persuaded her to drop the case[56].

The role of Japanese women in mine disputes, meanwhile, deserves more detailed treatment than is possible here, but it is clear that even the limited gains of the pre-war labour movement would have been impossible without women's participation. Just as they did at Asō and have in many countries around the world, women played an important supporting role during many Japanese-led mine strikes, banging pots and pans, taking over company offices and feeding strikers. Because many women in Japanese mining communities were also miners themselves, however, they played a direct role as well, taking part from time to time in male-led strikes and occasionally staging strikes of their own accord (Hayashi Eidai, 1990)[57]. All indications are that women would have played an even more active role if not for resistance from male activists. During a major strike in 1931, for example, 76 women miners joined 197 men in occupying a mine until strike leaders decided the takeover should be an all-male affair (Ogino, 1993)[58].

Even after their exclusion from the mines, women played a key role in several post-World War II strikes that linked the workplace and community. During a nationwide mine and electrical strike in 1952, for example, women blocked coal shipments at one mine and played a key role in opposing the formation of a pro-management second union at another (Mitsui Kozan, 1963)[59]. In a defensive strike at the Mitsui mines in 1953, the housewives' militancy frightened company officials so much, by some accounts, that they insisted on holding negotiations at sea and then agreed to reinstate miners who had refused to accept early retirement (Price, 1993; Isayama Hiroshi, 1960; Tanro, 1992)[60].

---

[56] Asō rōmu gakari nisshi, 19 and 26 October 1929, reproduced as Doi Taira, ed., 'Chikuhō ni okeru rōdō undō shiryō: Taishō 14-nen-Shōwa 12-nen', *Chikushi Jogakuen Tanki Daigaku kiyō*, no. 6 (March 1971: 7, 9-10).

[57] Hayashi, 1990: 194, 200.

[58] Ogino, 1993: 389.

[59] Mitsui Kozan, 1963: 123. For an English-language account of the 63-day strike, see Price, 1997: 114-115.

[60] Price, 1997: 115-117; Isayama Hiroshi, 1960: 186; Tanrō, 1992: 285.

Women were also key players in the 1960 strike at the Mitsui Miike colliery, a walkout that began as a battle over layoffs and control of the workplace but escalated into a contest over the future of Japanese labour. Workers from around Japan supported the 16,000 strikers, while big business interests lined up behind Mitsui in its attempt to destroy the union. The walkout ended in defeat for labour, but women provided the base of support that kept it going for nearly a year. They took part in strike strategy sessions, underground sit-ins, and demonstrations, helping to bring attendance to 100,000 at one pithead rally[61].

Conclusion

Korean women's experience in and at the Japanese mines shows that it is not enough to analyze work and life in mining communities from the standpoint of either gender or ethnicity alone. We must bring both ethnicity and gender – along with class – to bear simultaneously to bring the true picture into focus, whether we are examining occupational distribution, daily life in coal communities, or the labour movement's struggles with capital and the state. This is true even when we examine an all-male, monoethnic work force such as that in most postwar Japanese mines. When there is no gender and ethnic diversity, we should not accept this as 'natural' but ask why. Rather than accepting categories such as 'Japaneseness' (or 'whiteness' in the United States or Australia[62]) and 'manhood' as givens, we must examine how these culturally constructed identities came into being and whose interests they serve.

References

Asahi Shimbun, various issues.

Asō rōmu gakari nisshi, 19 and 26 October 1929, reproduced as 'Chikuhō ni okeru rōdō undō shiryō: Taishō 14-nen-Shōwa 12-nen', in Doi Taira (ed) *Chikushi Jogakuen Tanki Daigaku kiyō*, nos. 2, 4-8 (March 1967, March 1969–March 1973).

Chikuhō Sekitan Kōgyō-shi Nenpyō Hensan Iinkai, 1973. *Chikuhō sekitan kōgyōshi nenpyō*. Nishi Nihon Bunka Kyōkai, Fukuoka.

Chōsen Sōtokufu. 'Chōsen jinkō dōtai tōkei', various years.

Chung, Erin Aeran, 2000. 'Korean Voluntary Associations in Japanese Civil Society', *JPRI Working Paper*, no. 69, July.

Chūō Kyōwakai, 1990. *Kyōwa jigyō nenkan* (Shōwa 16-nenban), 1942; Reprint. Shakai Hyōronsha, Tokyo.

---

[61] Miike comprises a separate coal field southwest of Chikuhō. For an overview in English of the 1960 Miike strike, see Price, 1997: 191-218.

[62] For the construction of whiteness in the United States, see Roediger, 1991.

Fukuoka Chihō Shokugyō Shōkai Jimukyoku, 1929. *Kannai zaijū Chōsenjin rōdō jijō*, Fukuoka.

Fukuoka Kōzan Kantoku Kyoku, 1929. 'Chōsenjin kōfu rōdō jijō', *Kōgyō shiryō* no. 28.

Fukuoka Nichinichi Shimbun, various issues.

Grajdanzev, Andrew J., 1944. *Modern Korea*, John Day Co., New York.

Hayashi Eidai (ed.), 1990. *Yami o horu onnatachi*, Akashi Shoten, Tokyo.

Hayashi Eidai (ed.), 1991. *Senji gaikokujin kyōsei renkō kankei shiryōshū*, Akashi Shoten, Tokyo.

Hokkaidōritsu Rōdō Kagaku Kenkyūjo, 1961. *Tankō ni hataraku fujin no rōdō jittai*, Hokkaidōritsu Rōdō Kagaku Kenkyūjo, Sapporo.

Idegawa Yasuko, 1995. Interview by author, 16 May, Kurate-machi, Fukuoka Prefecture.

Isayama Hiroshi, 1960. *Mitsui Miike*, San'ichi Shobō, Tokyo.

'Kaijima tankō dai-nanakō Chōsenjin kōfu iroha meibo', Collection of Arirang Bunko, Tagawa, Fukuoka Prefecture.

Kang Chaeŏn, 1976. 'Zainichi Chōsenjin no rokujūgonen', *Kikan sanzenri*, no. 8 (Winter), pp. 22-37.

'Katei fujin manpyō', 1928. *Higashi* (25 January).

Kim Ch'anjŏng, 1980. *Hi no dōkoku: Zainichi Chōsenjin kōfu no seikatsushi*, Tabata Shoten, Tokyo.

Kim Chŏngmi, 1980. 'Chōsen nōson josei ni taisuru Nittei no seisaku', *Chōsen shisō*, no. 3, June: 115.

Kim, Jackie J., 2004. *Hidden Treasures: Lives of First-Generation Korean Women in Japan*, Rowman and Littlefield, Lanham, MD.

Kyōchōkai Fukuoka Shutchōjo, 1932. 'Kabushiki Kaisha Asō Shōten Chōsenjin kōfu rōdō sōgi gaikyō', unpublished.

Mathias, Regine, 1993. 'Female Labour in the Japanese Coal Mining Industry', in Janet Hunter (ed), *Japanese Women Working*, Routledge, London and New York.

Meiji Kōgyōsho, 1991. 'Akaike tankō, Rōmu geppō, June 1944–October 1945', in Hayashi Eidai (ed.), *Senji gaikokujin kyōsei renkō kankei shiryōshū*, vol. 2, Part 2, Akashi Shoten, Tokyo, pp. 1153-1204.

Mitsui Kozan K. K., 1963. *Shiryō Miike Sōgi*, Tokyo.

Moji Tetsudōkyoku Unyuka. *Ensen tankō yōran*, various issues.

Morita Yoshio, 1996. *Sūji ga kataru Zainichi Kankoku Chōsenjin no rekishi*, Akashi Shoten, Tokyo.

Nagaoka Iwao, 1995. Interview by author, 25 May, Hōjō-machi, Tagawa-gun, Fukuoka Prefecture.

Nagasawa Shigeru, (ed.), 1992. *Senjika Chōsenjin Chūgokujin Rengōkoku horyo kyōsei renkō shiryōshū*, Ryokuin Shobō, Tokyo.

Naikaku Tōkeikyoku. *Shōwa 5-nen kokusei chōsa hōkoku. zenkoku no bu*, vol. 2, *shokugyō oyobi sangyō*, pp. 41-43, 225 and 228.

Naikaku Tōkeikyoku. *Taishō 9-nen kokusei chōsa hōkoku.*

Naimushō Shakaikyoku Dai-ichibu, 1924. *Chōsenjin rōdōsha ni kansuru jōkyō.*

Nakayama Akira, 1983. *Chikuhō Zanzō: onna kōfu wa ima*, Nihon Shashin Kikaku, Tokyo.

Nishida Hideko, 2003. 'Senjika Hokkaidō ni okeru Chōsenjin "rōmu ianfu" no seiritsu to jittai: kyōsei renkō to no kankeisei ni oite', *Joseishi kenkyū Hokkaidō*, no. 1 (August), pp. 16-36.

Nishinarita Yutaka, 1985. 'Sekitan kōgyō no gijutsu kakushin to joshi rōdō', in Nakamura Masanori (ed.) *Gijutsu kakushin to joshi rōdō*, Kokuren Daigaku, Tokyo.

Nishinarita Yutaka, 1994. 'The Coal mining industry', in Nakamura Masanori (ed.), *Technology Change and Female Labour in Japan*, United Nations University Press, Tokyo.

Nishinippon Shimbun, various issues.

Nōshōmushō Kōzankyoku. *Honpō kōgyō no sūsei*, various years.

Ogino Yoshihiro, 1993. *Chikuhō tankō rōshi kankeishi*, Kyūshū Daigaku Shuppankai, Fukuoka.

Osaka Chihō Shokugyō Shōkai Jimukyoku, 1926. *Chikuhō tanzan rōdō jijō*, Osaka.

Pak Kyŏngsik (ed.), 1982. *Chōsen mondai shiryō sōsho.*San'ichi Shobō, Tokyo.

Pak Kyŏngsik (ed.), 1975–1976, *Zainichi Chōsenjin kankei shiryō shūsei*, San'ichi Shobō, Tokyo.

Price, John, 1997. *Japan Works: Power and Paradox in Postwar Industrial Relations*, ILR/ Cornell University Press, Ithaca.

Rōdō kagaku kenkyūjo, 1943. *Hantō rōmusha kinrō jōkyō ni kansuru chōsa hōkoku.*

Rōdō Undō Shiryō Kankō Iinkai (ed.), 1959. *Nihon rōdō undō shiryō*, vol. 10, Rōdō Undō Shiryō Kankō Iinkai, Tokyo.

Roediger, David R., 1991. *The Wages of Whiteness: Race and the Making of the American Working Class*, Verso, London and New York.

Ryang, Sonia, 1998. 'Inscribed (Men's) Bodies, Silent (Women's) Words: Rethinking Colonial Displacement of Koreans in Japan', *Bulletin of Concerned Asian Scholars*, vol. 30, no. 4, pp. 3-15.

Sakamoto Yūichi, 1997. 'Chikuhō sekitan kōgyō to hisabetsu buraku: Nihon shihonshugi to 'hisabetsu rōdō o megutte', *Buraku mondai kenkyū*, no. 140 (August), pp. 82-116.

Sekitan Tōseikai Kinrōbu, 1944. *Shōwa 19-nendo zenkoku tankō rōmusha idō jōkyō shirabe*, unpublished.

Shōkōshō Kōzan Kyoku. *Honpō kōgyō no sūsei*, Various years.

Smith, W. Donald, 1996. 'The 1932 Asō Coal Strike: Korean-Japanese Solidarity and Conflict', *Korean Studies*, no. 20, pp. 94-122.

Smith, W. Donald, 2005. 'Gender and Ethnicity in Japan's Chikuhō Coal Field', in Stefan Berger, Andy Croll and Norman LaPorte (eds), *Towards a Comparative History of Coalfield Societies*, Ashgate Press, Aldershot, Hampshire.

Sone Sachiko, 2003. 'Exploitation or Expectation? Child Labour in Japan's Coal Mines before World War II', *Critical Asian Studies*, vol. 35, no. 1 (March), pp. 33-57.

Sorensen, Clark W., 1988. *Over the Mountains Are Mountains: Korean Peasant Households and their Adaptations to Rapid Industrialization*, University of Washington Press, Seattle and London.

Suh Sang-Chul, 1978. *Growth and Structural Changes in the Korean Economy, 1910–1940*, Harvard University Press, Cambridge.

Suzuki Take, 1942. *Chōsen no keizai*, Nihon Hyōronsha, Tokyo.

Takeuchi Yasuto, 2003. 'Mitsubishi Takajima tankō e no Chōsenjin kyōsei renkō', *Zainichi Chōsenjinshi kenkyū*, no. 33 (October), pp. 21-50.

Tanaka Naoki, 1984. *Kindai Nihon tankō rōdōshi kenkyū*, Sōfūkan, Tokyo.

Tanrō (ed.), 1992. *Tanrō: gekitō ano hi ano toki*, Nihon Tankō Rōdō Kumiai, Tokyo.

'The Coal Fields of Kyushu', Supreme Commander for the Allied Powers, General Headquarters, Record Group 331, Box 2494, Folder 1, *Records of Allied Operational and Occupation Headquarters, World War II*, Washington National Records Center, Suitland, Md.

Trotter, Joe, 1990. *Coal, Class and Color: Blacks in Southern West Virginia, 1915–32*, University of Illinois Press, Urbana.

Tsurumi, E. Patricia, 1990. *Factory Girls: Women in the Thread Mills of Meiji Japan*, Princeton University Press, Princeton.

Weiner, Michael, 1994. *Race and Migration in Imperial Japan*, Routledge, London and New York.

Yomiuri Shimbun, various issues.

Chapter 6

Women Working in the Mining Industry in Papua New Guinea: A Case Study from Lihir

Martha Macintyre

Introduction

The legacies of Australia's colonial relationship with Papua New Guinea (PNG) are highly visible in the ways that Papua New Guinean women participate in the workforce. A recent study of the gender profile of citizen employees by the National Statistics Office reveals that in 2000 while 67 per cent of the total female population participated in the labour force, only 5 per cent of women were in a waged job (Papua New Guinea National Statistics Office, 2002: 28-44; 2004: 55-68). There are no female dominated areas of employment and the only sector where women make up a substantial percentage is in community and social services where they are 50 per cent of the workforce. In Mining and Quarrying their numbers are so small as to not register on the graphs. In (commercial) agriculture, forestry and fishing women comprise about 10 per cent and in the wholesale and retail trade, about 20 per cent. The reasons for this are historical and cultural – all industries involving waged employment were introduced to Papua New Guinea by white people, usually Australian, who brought with them extremely conservative ideologies about female participation in the labour force. Moreover female subsistence production is essential in all Papua New Guinea communities and both men and women tend to be conservative about female employment outside the house and garden.

During the early period of colonization, women were legally excluded from working on plantations and the ideals of the maternal and domestic roles that permeated mission and secular education were drawn from those current in late 19th and early 20th century Britain and Australia. The relegation of women to the domestic sphere was seen as part of the 'civilizing process' by missionaries, colonial administrators and plantation owners (Jolly and Macintyre, 1989). They deplored the fact that Melanesian women routinely performed heavy labour, preparing gardens, planting, tending and harvesting food in the hot sun and then carrying it back to their villages, heavily laden with root vegetables, firewood and water. The pejorative image of the Papua New Guinean woman as a 'beast of burden' dates from this period and the problems relating to women's status in the transforming economy owe much to misconceptions about traditional and continuing women's roles as breadwinners – providers of food for their families.

Nursing and teaching had been encouraged as feminine careers by missionary schools and the expatriate role models that women encountered in missions and in the colonial administration reinforced the view that such jobs were feminine. They were also consonant with the prevailing ideals of women in traditional settings – where women were the carers of the sick and the educators of children. The hard physical labour that women performed every day providing food for their families, while highly valued in their own cultures, was not held in the same regard by many white people. Not surprisingly too local women were rarely eager to go way and labour in plantations, although they often performed similar heavy work on their own land.

Historically, the introduction of forms of waged labour into Papua New Guinea rarely involved employment of women – during the early colonial period they were not even employed as domestic servants. The forms of economic activity introduced by whites, then as now, were in primary industries – fishing and pearling, coconut plantations, mining and logging – and both Europeans and Melanesians were in agreement that such work was 'masculine'.

Papua New Guinean women have always played a major part in the subsistence sector: in agriculture, animal husbandry, harvesting of forest and marine resources and supplying water and fuel to their villages. In terms of hours of labour invested, I would argue that women dominate this sector. But when these industries involve paid employment, men dominate overwhelmingly.

So, an analysis of women working in the PNG mining industry has to acknowledge the long history of exclusion of women from the paid workforce and the strong ideologies that perpetuate this. Historically the prevailing ideas of women's roles, both Melanesian and British or Australian, were marked by ideas of distinct domains of work. The campaigns against employment of women and children in British mines during the 19th century, were so successful in constructing the view of mining as 'men's work' that by the 20th century most miners maintain that women have never been allowed in mines and that it is 'bad luck' for women to go down mines or work on rock faces. The everyday myths and superstitions of British working miners persist today – and I have been informed on both PNG and Australian mine sites that women *never* go into the pits because it brings bad luck. The PNG Mining Safety Act of 1977 prohibits women from working underground.

Mining in Papua New Guinea involves large multi-national corporations. While there is some artisanal mining, the major forms of mining are large scale, open-cast mines where the shareholders and management are predominately foreigners. Small scale alluvial gold mining has been a small part of the industry – in 2001 it accounted for a mere 1,779 kilograms out of a total of 67,083 kilograms of gold produced in the country (PNG Chamber Of Mining, 2003: 22). With the rising price of gold on world markets there has been a marked increase in small scale mining recently and for 2003 SSM production was estimated at about 4.5 tonnes (Trevor Neale, personal communication). There are currently four large goldmines operating in PNG: Ok Tedi Mining Ltd., Lihir, Tolukuma and Porgera. Kainantu is under construction and three others are likely to begin production in the next decade. Mining is estimated to contribute 15 per cent of real GDP and employs more than 20 per cent of the rural paid workforce – the vast majority are men.

At all the large mine sites, even Porgera which is owned by a Canadian company, the majority of expatriate employees are Australian. They bring with them ideas about the roles of men and women drawn from the work cultures in Australia. Recent research by Marianne Yrke at the University of Western Australia that described and analyzed these gender ideologies in the context of commuter mining based in that state, found that a large majority of miners (and their wives) held very strong views about women's roles being biologically determined and believed that women were 'naturally' mothers and homemakers who should not work if their husbands are able to provide for them (Yrke, 2004). On the basis of her evidence it is clear that within the mining industry prevailing attitudes to women's employment tend to be conservative, even reactionary. This is reflected also in the numbers of women employed in mining within Australia. A study undertaken for the Minerals Council of Australia (Pattenden, 1996) found that women constituted only 11 per cent of mine employees at a time when Australian women were 46 per cent of the workforce. While the percentage has increased since then the overall numbers are probably similar. Mining employers were generally reluctant to change conditions of employment to accommodate women's specific needs and many women surveyed reported sexism and hostility to their presence on a mine site – especially when working in positions that their colleagues considered 'men's work'. Many Australian managers carry these beliefs and attitudes with them to Papua New Guinea where they tend to be reinforced by prevailing ideas among local people.

Since independence there has been a very small increase in the percentage of women who work in the cash economy. Figures are far from exact, but according to the Pacific Human Development 1999 Report (UNDP, 2000), women constitute 18 per cent of the paid workforce, the lowest of any Pacific nation. At the same time there has been a very small gradual increase in female education at all levels (Brouwer et al., 1998: 29). There is a lack of tertiary educated women in Papua New Guinea which is reflected in the gender composition of the public service and in the professions. Women occupy 15 per cent of senior positions in the areas of government services, legal services and administration. Salary levels for people in the mining sector are higher than for any other industry group: the average salary/wage for a Papua New Guinean man working in a mine is K16,775; for a woman K15,223. The national averages are K6,524 (male) and K7,237 (women) (PNG Chamber of Mining, personal communication). The reasons why the *average* female wage is higher are complex and reflect the fact that employment as such is viewed primarily as a male activity so that although females are a minority, they are concentrated in jobs that are skilled (teaching, nursing, clerical work) and not in the menial jobs that are extremely low-paid. This situation is changing gradually in towns, but the location of several large mining projects in rural areas skews the trend. The average wages for expatriate workers in the mining industry are K199,743 (male) and K179,000 (female).

In this chapter I shall describe and discuss the lives of Papua New Guinean women working in mining on Lihir. I want particularly to draw attention to that small group of women with post-secondary qualifications who work in the mining industry as employees of mining companies. While they constitute a minute fraction of the female population, their situation warrants analysis as the experiences of women in mining illustrate most clearly the problems faced by working women generally in Papua New Guinea. The majority of women are employed by contractors, usually in menial jobs such as cleaning, doing laundry and working in the kitchens. I have been unable to get comparative figures for female employment in these areas, but for Lihir, where my research is based, the catering and cleaning company accounts for about 90 per cent of female employment. Quite a few of these women have four years of secondary education and are relatively over-qualified for the tasks they perform, reflecting the lack of opportunities for paid employment in rural areas.

This chapter draws on unpublished material from a much broader research project into the changing patterns of women's work in Papua New Guinea that I have been engaged in since 1998. On Lihir I have collected data through surveys, held focus group discussions with women in paid employment and village women whose work is primarily subsistence agriculture. In Port Moresby I have concentrated on women working in government jobs and for NGOs. In West New Britain I am collecting material from women who work in the oil palm and timber industries there and in Goroka I have focused on the same groups as in Port Moresby. My interest lies in women's experience of work, their ideas about careers and the ways that their roles as wage-earners affect their roles as members of families, as wives, mothers and sisters. In the conclusion I shall draw a few tentative comparisons between Lihirian women working in mining with women workers elsewhere in Papua New Guinea.

Lihir

The Lihir group of islands lies off the coast of New Ireland, northeast of the mainland. The largest island in the group, Niolam, is the site of a large gold mining project which began in 1995. When exploration began on Lihir in the mid-1980s there was virtually no paid employment available for Lihirian women on the island. A few women had been to secondary school and had left the island to find work in towns – mostly as nurses, teachers, welfare workers and office workers. Several educated Lihirian women had joined religious orders. The types of careers favoured by Lihirian women were those that are typically feminine – performing and providing services for others.

Since production began in 1996 the gender ratio of employees on Lihir has also remained substantially the same, with women constituting just under 12 per cent of the workforce. This is considerably higher than on other PNG sites – at Porgera it is 5 per cent and at Tolukuma 2 per cent. At the international conference in August 2003 in Madang, 'Women and Mining: Voices for Change' sponsored by The World Bank, The Department of Mining and Oxfam-CAA, it was noted that initiatives for gender equity in employment in mining were common in advanced industrial countries but almost non-existent in developing nations. Lihir Management Company, which manages production on Lihir, is a subsidiary of the British-based multi-national Rio Tinto. As such, its employment policies conform to those operating in countries such as Australia and the United States, incorporating equal opportunity provisions. No advertisements specify gender and candidates are chosen on merit – their qualifications and experience being the main criteria for selection. As there is a PNG law that requires foreign companies to have a localization policy that includes a plan for decreasing the number of expatriates employed, this influences appointments. At present 91 per cent of all employees are Papua New Guinean. The majority of expatriates are in senior management. In 2003 there were two white women employed – one as a mining engineer, the other a media/public relations officer.

On Lihir, the number of women employees with tertiary qualifications or professional training is marginally higher than on other PNG sites, but low internationally – it could hardly be otherwise given how few women attend university in Papua New Guinea. While a few come from Lihir, the majority of highly qualified women come from other parts of Papua New Guinea and are fly-in/fly-out employees. This means that they are rostered on for a period of about 28 days and then have a break of 14 days where they fly back to their home base. Most of these women have completed post-secondary training in business or administrative studies (including one with an MBA) while there are several who are trained nursing sisters. Those with tertiary qualifications include an archivist/librarian, a chemist, a hydro-geologist and an agricultural officer. There are a couple who have had teacher training and who are now working as office staff. Given the findings by an AusAID study (2002) on the decline in domestic violence in places within PNG where female employment or income earning was relatively high (notably in West New Britain), there are strong social reasons for increasing the proportion of female employees. The high levels of unemployment in Papua New Guinea prevail against any affirmative action for women however as there are few women in politically powerful positions and even fewer who support such strategies. It is difficult to ascertain precisely why women resist ideas of developing policies that would encourage and support female employment, but in interviews many women appealed to principles of merit derived from the education system as the fairest form of recruitment. A few suggested that any system that appeared to favour women would generate hostility from men and make the workplace even more difficult for women. Moreover, several women mentioned that they were concerned that there were so many young educated men who had finished high school yet were unable to find work and that in this situation, it would be unjust to employ people on the basis on sex rather than educational qualification.

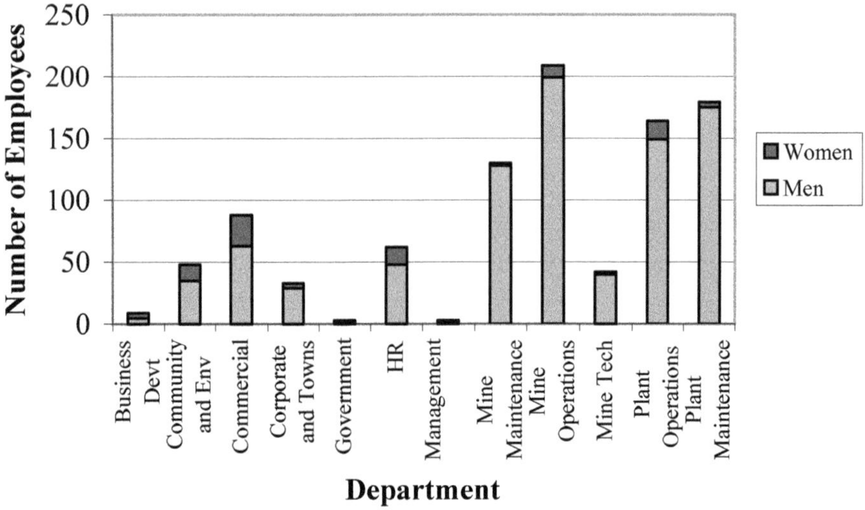

Figure 6.1 Breakdown of employment by gender

The changing economic role of women?

First let me set women's work in context. Lihir is a group of islands in New Ireland Province off the NE coast of mainland PNG. Until the discovery of gold in the 1980s, it was a small rural community with a total population of about 4,000, where all people engaged in subsistence gardening, only occasionally producing copra for cash to pay for school fees, clothing and a limited range of household items. It was poor and neglected by the national government. Women were the major producers of staple foods. For the majority of women this role has changed little, although now most have other ways of gaining some monetary income. Much of my interest in women workers derives from the fact that on Lihir we can observe the dramatic and rapid changes that industrialization brings in a developing country.

The continued role of Lihirian women as primary providers of food is evident from two surveys relating to gardening activity on Lihir undertaken in 2002 and 2003. 60 women who had no formal employment were asked for information relating to diet, the numbers of gardens they had and sources of cash income. While all had ways of obtaining cash for store-bought food (rice, flour, tinned fish being the primary purchases) almost all women had two gardens and half sold some of their produce to other villagers at some stage during the year. Of the 60, 11 said that they regularly worked in the gardens of employed relatives and were paid in cash or given some of the produce for their labour. These sorts of activities are the main ways that Lihirian women participate in the cash economy. The amounts of money they gain are small and the work they perform is much the same as it was before the mine began.

Lihirian working women

The lack of change in notions of women's role as food providers is reflected also in the fact that even women who are in full-time employment grow substantial gardens to feed their families. A total of 57 Lihirian women who work full time were interviewed for the purposes of a survey of the extent to which they continue to be dependent on garden food[1]. Of these, 88 per cent had at least one garden, with 48 per cent having 2 or more gardens. All those with gardens said they had a wide range of foods in their gardens, including staple foods (yam, mami, sweet potato, tapioca, taro, etc.) and fruits and vegetables (bananas, pumpkin, greens, pineapples, watermelon, etc.). Those women who did not have gardens said that they worked in their mother's or father's gardens and so 'earned' their share of produce.

The time women spent gardening depended upon their roster at work. No woman worked in their gardens on a Sunday, though they might visit them to collect some food. Lihir Management Company employees and employees of small businesses tended to have Saturdays off, and this was the time that they spent in the gardens. Women working for Niolam Catering Services mostly worked 6 days a week, their rostered day off changing every month. When this day fell on a Sunday they could not work in their gardens for a month because of their strict adherence to keeping the Sabbath. Otherwise NCS workers used their day off to go to the gardens. Nurses at the hospital worked a 4 week on-2 week off roster, and worked in their gardens while on break. Those women who lived close to the township of Londolovit where the mining camp is located sometimes went to the gardens in the afternoon as they would reach home while it was still light.

Women working for NCS said that they could ask for a day off if there was work that needed to be done in the gardens. Women from the islands said that they

[1] This survey was undertaken in the context of a social impact study for Lihir Gold and I thank them for their permission to draw on this work.

took the opportunity of bad weather, when the boats could not bring them to work, to work in their gardens.

All women said they relied on the help of others to maintain their gardens. Women listed their mothers, sisters, husbands and children as the most common helpers in the gardens. Their help is repaid with food from the garden, store bought food and goods, and money. Women said that they gave relatives between K5 and K50 for help in the garden each fortnight (K20 to K30 was the most common amount given).

Women also hired help in the gardens when major work was needed, such as harvesting or clearing new gardens. On Masahet island, another of the islands in the Lihir Group, people who do not have paid employment use this as a method to gain money. These people are generally paid between K50 and K200 for the work they perform, depending upon how much needs to be done and how long it takes. Thus gardening is a way of spreading money from employed women to their relatives and other non-employed Lihirians.

Almost all of the women interviewed said that if they did not hold paid employment their gardens would be different. Many said their existing gardens would be bigger, or they would have more gardens than they have. Others said that their gardens would be weeded much more regularly. Some attributed a decline in yield to the reduced time available for tasks such as weeding and aerating the soil around the roots to encourage growth. There was a common complaint among the women about the neglect of their gardens or inability to properly care for them, suggesting that for Lihirian women their status as food producers is still primary.

The number of gardens and confidence in their ability to provide food for their families varied by area. Women from the three outer islands had the most gardens, and were most confident of their ability to provide, with 62 per cent having 2 or more gardens, and none with no gardens. 81 per cent reported that they grew as much food now as they did prior to their taking a job. Of women from the affected area, those villages close to the mine, 48 per cent had 2 or more gardens, but 14 per cent had no gardens. 62 per cent said their gardens would provide enough food to feed their families completely. Villages in the south of Niolam (Matakues to Samo) had fewer gardens, with 29 per cent having no garden, and only 29 per cent had 2 or more gardens. Less than half of these women (47 per cent) said that they would have enough food if the mine closed. The distance these women travel to work was a factor in their gardening activity – several commented that they did not get home in time to go to gardens. Four said they occasionally worked in the dark by the light of lanterns.

Overall it is clear that employment on Lihir has not resulted in major economic change in terms of the dependence of the population on women's labour as subsistence gardeners. Income (from both men and women) is being spent on food, but not to the extent that there has been a transformation from dependence on subsistence to immersion in the cash economy, even in those households where a husband and wife are both employed. This is very telling, as it shows that many women are working much more – they are employed (usually working nine hours a day) and they remain subsistence producers. This is working the 'double shift' with a vengeance.

But the continued dependence on subsistence production has another important dimension for Lihirian women. The mothers, daughters and sisters of women who are employed often work in the employed woman's garden, planting, keeping weeds down and harvesting food – and for this they gain access to money. The relationship is not conceptualized as one in which the relative actually employs her relative, rather it is a way of redistributing money within family groups.

Working women with post-secondary qualifications

The majority of Papua New Guinean women with professional qualifications working at the mine come from places other than Lihir. They are employed on fly-in/fly-out (FIFO) contracts which mean that they live in camp while at work and fly home to their point of hire (usually their home village) during their breaks. Rosters vary, but most stay on the island for a minimum of two weeks. This system was chosen to obviate problems of accommodating large numbers of people and their families on Lihir, where there is very little land available on which housing could be built. At the beginning of the project the local people were adamant that they did not want 'migrants' from other parts of the country living on Lihir. While some Lihirians now see that this means that they cannot benefit from businesses catering for wage-earners and so want to change the agreement, the majority of Lihirians remain hostile to the idea of many migrant workers living on the island. Workers themselves are divided. Many would prefer to have a normal existence with their family and so would like to reside on Lihir. Others suggested that the FIFO arrangement enables them to save more money and to avoid the 'double shift' of gardening and wage-labour that Lihirian women have. Those who are based in towns such as Port Moresby where rents are high were more likely to favour the idea of renting a house on Lihir.

Women's living conditions while on site are not luxurious. The 'dongas' comprise rows of prefabricated buildings in which women have single rooms with shared bathrooms. Many have bought themselves television sets and have decorated their rooms in ways that make them less institutional. Some dongas have a shared recreational room, others have tables and chairs in the covered open space between the rows of rooms so that women can sit outside and chat in the cool of the tropical evening. Women grow pot plants to make the shared space more attractive. As many have worked on the mine for several years, the creation of a home on site is part of the adjustment to long periods away from family.

The majority of women are married or divorced and are supporting their own children as well as members of their extended family. Some are single mothers who have never married. Even young women who have never married often take responsibility for their brothers' or sisters' children. Some have adopted children while others simply pay their school fees and support them financially while they are studying. In short, all women are supporting other members of their extended families and all are paying for the education of at least two children. This pattern of income expenditure by working women shows the ways that education has become a means of social mobility and how cash income is used in ways that remain embedded in customary relations of reciprocity and intergenerational responsibility. The commitment of women to the education of their own children and their nephews and nieces indicates the way that this select 'elite' group of women aims to reproduce the following generation of educated people.

In Papua New Guinea education is not free. It is also selective, so that even if parents can pay fees, the child has to achieve a certain standard to progress. The women whom I have interviewed all stress their gratitude to parents who struggled to pay school fees, encouraged and supported them throughout their education. While a significant proportion (about 25 per cent) were the daughters of educated parents, most were among the first generation to receive secondary education. Overwhelmingly women reported that their parents, particularly their mothers, inspired them to work diligently and to pursue their careers. In focus group interviews, women described their mothers as the people who were dismissive of some of the ideas of gender stereotyping. One woman who was working as the only female in a technical area of the mine suggested that she had gone into that type of work because:

> …my mother didn't really know anything about mining work and she wanted me to be a miner. I think that she thought lots of women would be working with me. She's a humble and illiterate woman and was so happy when I got good marks at school. I went into this job knowing I'd be an only woman though. I learned that at university, but it didn't discourage me.
>
> Anna, interview June 2002

Women working in jobs that are viewed as 'masculine', such as geology and technical jobs, were proud of their strength of character and their resistance to forms of sexism or discrimination. One women, Lonatandrea, observed that while she sometimes experienced discrimination, 'I'm of a character [where] I put men where they belong' (June 2004). While several answered the question, 'What do you think are the biggest problems facing Papua New Guinean working women today?' by describing entrenched views of Papua New Guinean men about female inferiority and subordination ('Men think women have an inferior mentality'; 'The attitude of the menfolk in undermining the capabilities of a woman'; 'Women being degraded as second class to men'), almost all of them indicated that they felt able to resist and counter such sexism at a personal level – although several believed that these attitudes were blocking them from promotion.

But most women choose to work in areas that are considered appropriate for females – by themselves, society at large and most of the men with whom they work. Those who mentioned working women who inspired them to pursue a career wrote down jobs such as nursing, typing or office administration as their ideals. One woman explained her career choice in detail:

> I wanted to work in an office because it is clean work and not so heavy and tiring as being a village woman. I saw my mother come back from the garden, carrying big loads of taro and food for us, dirty and tired. Even when I was a small girl I thought I would get a job where I am clean and wear nice clothes and buy our food. I want to be able to make my mother's life easier and to have a better life for myself.
>
> Dorothea, 2001

The overwhelming majority of qualified women expressed high levels of job satisfaction – in marked contrast to those who have no formal qualifications or work in menial positions. Although I have not interviewed male employees, conversations with many men suggest that they do not experience the same levels of satisfaction and are much more resentful of expatriates in authority over them. Women seemed to view their work more autonomously, taking pride in working competently and efficiently. They said that if they did their work conscientiously their bosses showed that they appreciated it, which made them feel valued. Women also appeared to me to have much more realistic expectations of the range of experience required to assume managerial positions than men do. Four women commented directly that they aspired to work as senior managers eventually, but observed that their (expatriate) bosses had skills and knowledge that they had not yet attained. All believed that they had the capacity to do so and were no less ambitious than male counterparts – but their understanding of the complexity of managerial work was clearly expressed, whereas my conversations with dissatisfied male employees suggests that they often fail to see that they are relatively inexperienced. Men tended to be much more conscious of status difference and to perceive it as a reflection of racism. Their resentment of expatriates in senior and supervisory positions appears more often to be expressed in terms of race and unjust privileges derived from the colonial past, whereas women constantly emphasized their desire to learn more in order to gain promotion.

The work environment for the women employed at Lihir is one that is undeniably modern and globalized. This in many ways accounts for the ways that women on large mining sites enjoy better wages and work conditions than those who work in government or smaller businesses. The computers work, the phones work, all offices have electricity – people have free transport to and from work, they get paid on time and they have benefits as employees that are not available in many other sectors – such as special savings and loans schemes. For the women I interviewed, these factors made the job very desirable and in my opinion explain why they tolerate some of the anti-social elements of their jobs. Young, unmarried women tend not to even see the long hours and separation from family as stressful. People who have resigned or been dismissed often reapply for work, not only from

Lihir (where the desire to live in one's own village might be sufficient to make one swallow one's pride) but from places that are very distant.

Expatriates who work in Papua New Guinea are paid handsomely – in part because they have high qualifications and skills in the industry, but also because of the disruption to their domestic lives. Fly-in/fly-out mining is common in Australia, parts of Indonesia and in Papua New Guinea. While the expatriates thus engaged tend to see their employment as relatively short-term, Papua New Guinean women working in Lihir saw themselves as long-term employees. They too earn higher than average wages, but the stresses on marriages and the separation from children are acknowledged as hardships that are borne in the interest of financial security. In a country lacking social welfare provisions, working women see their earnings as a way of securing a better future for themselves through the education of children who will then care for them in old age.

The research that I am doing concentrates on the ways that women experience the conflicts between participating in a modern, globalized world through employment and those imposed by custom and familial obligations within Melanesian society. The speed with which women have changed and adapted is extraordinary, and has few historical counterparts. The shift from rural life to working in factories in the industrial revolution of 19th century Europe is comparable – but with major and significant differences. In Papua New Guinea people are not peasants – working the land and producing food means working within an autonomous kin group and owning the means of production. Papua New Guinea has few of the landless labourers that were factory fodder for European industrial development. The transformation of subsistence gardeners into career women has occurred within one or two generations. There are many aspects of working women's lives that are unique.

In conclusion, I shall outline a few findings. First, is the degree of financial independence that women have and their control over their earnings. All women interviewed have their own bank accounts and only 3 per cent also have joint accounts into which they pay a proportion of their wages, keeping the rest under their own control. Joint bank accounts were just as likely to be with brothers and sisters as with husbands – illustrating the continued strength of lineage kin as an economic unit and the tension between the notion of a nuclear family as the unit of consumption and traditional ideas of the lineage as fundamental kin group. 76 per cent of women maintain that they make all decisions about expenditure while the others consulted with their husbands or other members of their households. Apart from expenditure on school fees for relatives, women spent almost all their money on their households. Few are able to save, although all expressed aspirations about saving for a house. For those who are FIFO employees, their household in the village or in town almost always includes parents or sisters (along with their children) who take care of the woman's children during her absence and grow food gardens. The woman's wage thus supports a household of extended kin. In one instance a woman was helping to support 15 children as well as her parents – thus the economic unit was effectively the same as for the Lihirian women workers.

Most women were paying school fees for at least three children and supporting at least three other adults.

Second, is the extent to which women in the study are highly critical of customs such as brideprice, even those who come from places where it is not customary practice. I asked the question, 'What are the customs/traditions relating to women's lives that you think are most important today?' expecting that most people would respond with things that were positive. While most did, several wrote responses that described the ways that they believed 'traditions' or 'custom' restricted women's lives and made them appear inferior: 'As for me, where I come from we do not pay brideprice, I don't see anything good about it. It is like a woman is being sold or traded for a man's own needs or ends' (Kinavause, aged 42). The majority believed that it was one of the main institutions that oppressed women in Papua New Guinea because payment is perceived to confer a husband's rights over a woman. A minority objected to it simply because as it has become monetized, the amounts have inflated and so families are forced into debt to pay brideprice. Those who saw positive sides to brideprice stressed the ways that it cemented relationships between groups of people and so incorporated the couple into their new families. Women who perceived it as creating social bonds would contribute money to kin as part of brideprice payments, others avoided doing so, but gave freely in other contexts, especially the exchanges associated with death that are important in the economies of almost all Papua New Guineans communities.

Third, and perhaps the most significant change in terms of different conceptualizations of the women as independent persons, were the numbers of women who had left unsatisfactory marriages since they had been working. The main reasons given for marital problems included drinking by the husband, the husband's adultery and domestic violence. A few mentioned that their husbands became very jealous and suspicious of them because they were working away, but in the group interview women suggested that this would not be sufficient reason to leave in itself, but was often a factor precipitating violence. Women who had left husbands maintained that they would not have done so if they did not have a job. Several noted too that their children were subsequently better off because they did not have to give money to their husbands for beer. Male alcohol consumption was considered one of the major problems affecting Papua New Guinean women by over 90 per cent of respondents.

At present in Papua New Guinea the currency is weak and the cost of living is increasing. Mining is an important industry for the national economy and one that has very mixed effects on women's lives. Much has been written about the social impacts on women in communities affected by mining (Macintyre, 1993; 2002; 2003; Macdonald and Rowland, 2002), in this chapter I have attempted to shift the focus onto the women employees. There, too, the effects are mixed. For historical and cultural reasons relatively few women are employed in mining, but for those who are, it provides opportunities for career development and higher wages than other sectors. Studies of women's participation in the industrial workforce in advanced capitalist countries have consistently shown that women's work remains undervalued and that women work a 'double shift' – as mothers, housekeepers and carers as well as wage-earners. My study is revealing that while access to employment in the industrial sector is even harder for Papua New Guinean women, once they are employed there is greater equity in terms of salaries and conditions in the workplace. But the double shift for these women is often more arduous and the number of people who have to be supported much greater than in countries such as Australia.

References

Brouwer, Elizabeth C., Harris, Bruce M., and Tanaka, Sonomi, 1988. *Gender Analysis in Papua New Guinea*, The World Bank, Washington.

Jolly, Margaret and Macintyre. M. (eds), 1989. *Family and Gender in the Pacific: Domestic Contradictions and the Colonial Impact*, Cambridge University Press, Cambridge.

Macdonald, Ingrid and Rowland, Claire (eds), 2002. *Tunnel Vision*, Oxfam CAA, Melbourne.

Macintyre, Martha, 1993. 'Women and Mining', in *Papua New Guinea and Australia – Towards 2000*, Community Aid Abroad/Freedom from Hunger, Fitzroy, pp. 43-49.

Macintyre, Martha, 2002. 'Women and Mining Projects in Papua New Guinea: Problems of Consultation, Representation and Women's Rights', in Ingrid Macdonald and Claire Rowland (eds) *Tunnel Vision: Women Mining and Communities*, Oxfam-Community Aid Abroad, Melbourne.

Macintyre, Martha, 2003. 'Petztorme Women: Responding to Change in Lihir, Papua New Guinea', *Oceania*, vol. 74, nos. 1 and 2, September/December, pp. 120-134.

Pattenden, Catherine, 1996. *Women in the Australian Mining Industry*, Report of the Minerals Council of Australia, Canberra.

PNG Chamber of Mining, 2003. *PRIME*, vol. 4. May-July.

PNG National Statistics Office, 2002. *1998 Annual Business Census*, Port Moresby.

PNG National Statistics Office, 2004. *Papua New Guinea 2000 Census Report*, Chapter 5, 'Economic Activity', Port Moresby.

The World Bank, 2003. *Women in Mining Conference: Voices for Change*, CD of Proceedings, Madang.

Yrke, Marianne, 2004. *'What You Do is Who You Are': Gender Identity in the Resources Industry*, Thesis submitted for the degree Master of Arts, The University of Western Australia.

Chapter 7

Traditional Small Scale Miners: Women Miners of the Philippines

Evelyn J. Caballero

Introduction

Small scale mining in the Philippines is the major leading mineral industry, operating in more than 30 provinces involving 300,000 miners. Based on 1999 data of the Mines and Geosciences Bureau of the Philippine Department of the Environmental and Natural Resources, small scale miners produced about 62.38 per cent of the country's gold production. In Benguet Province alone, where I did my research among the Kankana-ey (Caballero, 1991a, 1996). Bayle (1998) documents 52 small scale miner's associations with 27,000 members. This number does not include those associations that are not registered.

Who are traditional small scale miners? Where did the term traditional small scale miners originate? When I started my research among the Kankana-ey of Itogon Municipality in Benguet Province in 1979, I went into the field thinking that they were primarily dependent on their swidden fields, some wage-labour from Benguet Corporation mines, and last a little gold panning on the side. I was wrong. By the end of my research, I had discovered and documented a culture of small scale miners whose subsistence base was predominantly mining with some shifting cultivation and a few other occupations (Caballero, 1991a, 1991b, 1992, 1996). The implementing of Rules and Regulations of the People's Small Scale Mining Act of 1991 or Republic Act 7076 legally incorporates and defines traditional small scale miners as 'Filipino citizens who have a distinctive socioeconomic cultural tradition with a subsistence base focused on small scale mining. They live in stable sedentary communities and employ a mining technology that is labour-intensive and simple; and employ physical separation methods for the extraction of mineral/s and/or metal/s from the ore' (RA 7076, Chapter II, Section 4.29).

Women play a very important role in traditional small scale mining. An understanding of traditional small scale miners and, in this case, women traditional small scale miners, is best appreciated by placing them in their historical context and by a description of their culture.

Historical and archaeological data

Gold is one of the earliest forms of metal found in the Philippines. The first appearance dates back to about 300 to 500 B.C. and is from the island of Palawan at the southern end of the Philippine archipelago. The limited Chinese records translated by historians, such as Chau Ju-Kua's *Chu Fan Chi* which was written between 1209 to 1214, indicate that the Philippines was one of the islands drawn into a large gold trade network established by the Sung Dynasty. Historical evidence also argues that the present day Kankana-ey are part of a 450+ year old small scale mining tradition which predates the Spanish arrival. Later, during the Spanish period from the 1500s to 1898, there are meticulous records regarding gold.

Spanish records of gold used in the Philippines

Spanish records regarding gold begin during the early 1500s and extend to the early 1900s. Earliest accounts by Pigafetta (1525) and de Saavedra (1527) indicate the bountifulness of gold by everyone wearing gold jewellery, the quantity of which depended on the person's rank and the presence of household items of gold. Spanish records indicate that gold was used as tributes given either as gifts to the Spaniards or given out of fear of them; as a trade item for metal, for food items such as rice, pigs, carabaos, or for cloth and other things they needed (de la Vega, 1609; Morga, 1609). Gold was also used as jewellery, which for women included gold pendants, rings, bracelets, earrings, rosaries, gold buttons on chemises, and anklets. Colin (1663) recorded gold being used for teeth ornamentation and he observed that both women and men richly adorned themselves with gold jewellery. For rituals, gold was given as dowry by the man to his bride or as a wedding gift by a chief to his betrothed daughter (Blair and Robertson, 1903–1909); in burials as grave goods (Loarca, 1582) or like those from Zambales as payment to a priest (*bayoc*) for performing a death ritual or in performance of a baptism, and as offerings to appease ancestors as in the *an-anito* ritual (Perez, 1680). Gold is also noted in the historical records as payment for crimes committed like murder, arson, and theft. Abusive language against a village leader was also punishable by either a fine in gold or slavery. Gold was also used for the buying of slaves; if a slave was set free, gold was one of the items given to the slave's master. 'Gold' was also used as a personal name. For example, *Vray*, which means fine gold, was given to a woman because of the nobility of her lineage (Bobadilla, 1640). One having the most gold was considered to be wealthy.

Expedition of Quirante

One of the most meticulous records documents an expedition in February of 1624 to locate the gold mines in Luzon headed by Captain Alonso Martin Quirante (Quirante, 1624). He went inland from Aringay, north of Agoo, to what is now the province of La Union. His documentation gives detailed account of the expedition's day-to-day progress, the confrontations between the Spaniards and the Igorots, and the success of the latter in repulsing the attacks and preventing the Spaniards from obtaining the mines. He noted various characteristics of the lode mines and their miners. He describes them as going to the river during the rainy season with their wives and children to placer mine. Quirante also describes wooden pickaxes tipped with iron which were used by the miners to chip the ore in the tunnels. The ore was then crushed by a 'stout rock' and 'other smaller stones' by hand until the ore was reduced to powder. It was then washed in streams where the gold dust or grains were recognized by their gleam in the sunlight. The large grains of ore were milled and washed several times until very little of the metal was left. I recorded similar tools and mining practices among the Kankana-ey (Caballero, 1991, 1996).

He describes in much detail the tunnels of the 'Ygolotes' who were mining around Benguet Province. He mentions 5 'hills' within a distance of '5 or 6 leguas' (leagues) being mined during the dry season. The first was Galan which is now the modern gold-bearing Tublay area. This was used as the main base for the Spanish. Quirante (1624, 282) observed that the tunnels:

> ...extended northwest and southeast, for it is flooded with sunlight most of the day. When I reached that place the Ygolotes were working the said mines through many mouths or passages that they had opened, following the metal of one large vein, from which they were taking out the ore that was the softest and easiest to dig, although it contained blue iron pyrites that contain antimony.

The second location was the mines of Arisey and Bugayona. He describes these mines as very steep with deep ravines. The tunnels are also described as being flooded with sunlight during the major part of the day. He writes:

> ...a slope ... where the natives washed (gold) and gathered certain small stones known to them, which they crushed for their profit; for in no other way is there any known or constant source from which to obtain the metal – but only loose dirt with certain ores, and those of the said red metal, which traverse the soil without digging down to the bottom.
>
> Quirante, 1624, 282

Based on my ethnographic research this act of collecting ore along the streams and rivers is called *sen-eng* (Caballero, 1992). The third location was the Baranaban mine. Keesing (1962: 16) identifies Arisey, Bugayona and Baranaban as the 'gold-bearing Ambuklao of the upper-middle Agno river'. Quirante describes the narrow tunnels containing loose dirt and flooded with sunlight all day long. He noted that the tunnels had caved in.

The fourth location is Antamog (modern Antamoc mining area of Benguet Corporation). He describes Antamog as having several tunnels and that it looked like it had been mined the longest. He observed that stakes and boards were used by the Ygolotes for tunnel support. He notes that these mines were more neglected than the others. However, it was from Antamog that he got the best ore. The fifth location is Conog, the modern Balatoc-Itogon area were Dalicno is located. Conog was documented as being one-half league from Antamog. He described Conog tunnels as being flooded with sunlight all day. The mines were abandoned and contained the same high quality of ore as Antamog.

In terms of general statements about the Kankana-ey, Keesing (1962: 103) writing about a Spanish document notes that 'many of the present-day settlement areas of these ethnic groups are recognized by name in this seventeenth-century list. This great stability in residential patterns stands in marked contrast to the more ephemeral nature of settlements in most of Benguet.' This statement applies to the Dalicno area because as just discussed, Quirante made a specific reference to the Conog area as a place where gold was mined. Thus it would appear that settlement in this area has been stable for at least 400 years. In another reference to historic gold mining, Keesing (1962) notes that in Buzeta's survey of the Philippines (1850), gold was being panned from the tributaries in Benguet Province, particularly that of the Agno River.

In other historical records, when the Spanish in 1856 established a concession to mine copper in the Mankayan area of northern Benguet, there are records of Igorot women helping in the 'washing' of the metal (Keesing, 1962). Archaeological evidence supports the argument of a stable small scale mining settlement pattern in the Balatoc-Itogon area. In 1991, Reynolds and I conducted an archaeological survey of the ridge where the mining tunnels of Dalicno are situated. We recorded two habitation sites and four burial caves (Reynolds and Caballero, 1993). The conclusion at that time was that one of the habitation sites represents a gold processing site which pre-dates the occupation of Dalicno. This is based upon the fact that while the elders of Dalicno were aware of the archaeological site, none of their oral traditions or their personal memories noted people living at the site. The site thus supports the observation of Quirante about the Conog area.

The following is a brief description of one of the habitation sites where Reynolds and I conducted test excavations in 1994. The postulated living area of the site does not conform to the topography of the ridge because it has been built up and leveled through the construction of rock walls which function as terraces. Our test excavations confirmed that the flat areas are a direct result of the terrace wall construction.

The living areas of people presently residing in Dalicno are constructed in a similar fashion. The Dalicno-Kankana-ey build rock wall terraces to provide a level area on which their houses are built and other activity areas are located. These terraces are built against the side of the mountain with each family having spatially separate terraces.

The artefacts recovered at the site included many pieces of *naba* which are the rock bearing gold ore in its pre-ground state and a number of pieces of pottery resembling the *gangi*, which is the pottery container which present day small scale miners use to roast the gold to form gold beads. A number of *gaidan*s, or grinding stones, were also found. A test pit around one of the *gaidan*s produced many pieces of naba and more pieces of *gangi* to a depth of 40 centimetres. The size ranges of both the *naba* and *gangi* fragments were the same in both the test trench and the test pit. The presence of these kinds of artefacts fits into the gold processing technology that I documented for the Dalicno Kankana-ey.

The conclusion at this time is that the archaeological site represents a gold processing site which pre-dates the occupation of Dalicno. The types of artefacts found, especially the *gaidan*s, are mentioned by the Spanish when they described the manner in which gold was processed. The grinding rocks are still being used by the present day Kankana-ey. Last, the construction of terraces for living areas is also a practice that the present day Kankana-ey of Dalicno engage in. Therefore, based on this evidence, there is a high probability that the site represents a Spanish period, or earlier, gold processing site.

Ethnographic data

In the analysis of ethnographic data of traditional small scale miners it is not always easy to isolate the role of women. There are many activities in their culture that are considered to be joint endeavors by both genders and are perceived as such by both males and females. There are, however, distinctive roles for women. The following ethnographic description of the Kankana-ey traditional small scale miners sheds more light on these roles (Caballero, 1996).

Ownership

Among Kankana-ey traditional small scale miners, the founding members of a community become the nucleus of a corporate descent group. During their lifetimes they are the elder leaders of the community and are called *panglakayan* or *lallakay*, both male terms of reference, and are held in the highest esteem. However, in one community I studied, the female founder was referred to by her name in their historical accounts and in their ritual prayers. When the elders die they become *anitos* who are spirits who join the ranks of the ancestors. Traditional small scale miners believe that ultimately their god (Kabunian) and *anitos* own the resources and the elders manage these resources for the community. Consultation and the resolution of conflicts regarding mining are through the elders. They determine the party at fault and impose on them the necessary penalties (for example, the party at fault is punished by not being allowed to continue mining in his/her tunnel).

It is the *panglakayan* who own and manage the primary resources of the community, and these traditionally included mineral, agricultural, and water resources. The descendents of this corporate group are entitled to the use of the primary resources (for example, subsurface claims for mining or land for swiddening) and can own the gold which they mine and the crops which they grow. Other individuals may acquire temporary-use rights (rights of usufruct) by affiliating themselves with the corporate kin group. For example, miners may have use rights to a tunnel and only own the gold extracted. A miner's use rights cease once he or she abandons a tunnel. Other miners may be given permission to mine on lands where rights to surface or subsurface areas are owned by another ethnic group or a commercial mine.

There are also cases where a woman is a founding member of a community and such was the case in the area of my research, Sitio Dalicno. In the 1920s two American prospectors in Benguet Province by the name of Icard and Dugan married Kankana-ey women. Icard and Dugan purchased claim rights and operated what is now known as the Itogon-Suyoc Mine. Icard encouraged his wife, Kid-ing, to establish legal claim rights to a mining area. Because she was a woman she could not own a residence certificate, and so her claim was registered under a male relative's name, Baukan who in turn solicited three brothers-in-law and another male relative who was called Balio to operate the mines. By 1925 their families established residence near the mines which Kid-ing had delineated to be their mining claim. In January of 1931, with their combined capital, Kid-ing and her male relatives surveyed a mineral claim which they staked called Kating Fraction. Today the Kankana-ey from Dalicno claim that Kid-ing is the founder of the community of Dalicno with the first group of elders as the co-founders – they are the nucleus of the corporate descent group. From a core group of 6 families in the mid-1920s, Dalicno today has a population of 3,000.

Among the Kankana-ey, a husband, wife and their extended kin work together in both placer and lode mining. While both men and women participate in mining, some tasks are performed more by women. The processing of gold, for example, is a woman's job, as is the performance of household tasks like cooking, caring for the children, washing laundry, etc. However, it is not uncommon to see husbands doing these chores if their wives are busy at other tasks. Women are involved in the mining and processing of gold-ore or gold-bearing sand, and in the decisions regarding the health, education or the general welfare of the community members.

There have emerged in communities women leaders. In Dalicno, for example, there are key women leaders whose words of advice are valued by elders and local government officials. One, a former principal of the elementary school, won in the Philippine elections in May of 2004 and is the only female of the 8 Municipal Councillors of Itogon Municipality. Another is a mid-wife who has been successful in the extension of health services and infrastructure to Dalicno. Both play key roles in the health, education and general welfare of the community. For the former principal, where the community local government leaders in the past were predominantly male, they would turn to her for resolutions that needed to be brought forward to the local government unit.

Lode miners

The Kankana-ey refer to lode mining as *agabukay* and to their tunnels as *abukay*. Lode mining involves the digging of tunnels following the lode. The extraction of gold is generally a family activity where both women and men participate in mining. New tunnels are located and opened through a combination of ritual, prospecting and/or visual. Old tunnels may be re-mined. Decisions as to which tunnels to mine are made after consultations with the previous owners and with the *panglakayan* (elders).

Once a site is located, an offering of pig and prayers is given to their gods and *anitos*. The miner then assesses the location of his/her prospective tunnel in relation to other tunnels, making sure that their tunnel does not intersect the tunnel(s) of others.

The man or woman who owns a tunnel is called a supplier, for they are the ones who finance the expenditures needed for the tools and food of their workers. The number of miners in a tunnel can range from three to thirteen or more. The number of workers is dependent on the number of shafts mined and the money the supplier can provide to finance the venture. When a woman's husband dies or is physically handicapped, she recruits male relatives from her bilateral kin group to help her mine, while she herself continues to actively participate in the mining and supervision of her mines.

Mining tools are simple, consisting of iron chisels, a double-sided iron hammer with a wooden handle, iron shovels, iron crow bars, battery operated lamps or a battery operated mining helmet lamp, fabricated iron carts made of wood or iron with wooden or rubber wheels, a sack made from burlap or any other material used for containing mullock (which is waste ore) and gold ore (*naba*), ladders, and pulleys.

Placer mining

A placer deposit is a loose sedimentary deposit (gravel, sand, silt, etc.) containing valuable minerals. These deposits are commonly treated by sluicing to concentrate the valuable material. The work force in placer mining consists of family units divided into pairs, for example, a husband and wife or a mother and daughter; however, mining can sometimes be done alone. In placer mining rock wall channels with floodgates called *kaangs* are built in the river bed. These channels can range from 25 to 59 meters in length and function as launders which are constructed from rocks dug while mining the river. Towards the end of the channel is a long screened rectangular iron box called a sluice box. A portion of the river is directed through the channel in order to carry away lighter rock and waste material. The sediment inside the channel is sifted several times through the sluice box until *linang*, which is substantial amounts of tiny gold particles, along with minute amounts of sand with pyrite, nonmetallic rock grains and metal shavings are in the sluice box. At the end of each workday, the gold-bearing sand in the sluice box is brought home and placed in containers.

Rights to where to locate a channel are based on a first-come-first serve basis and are temporary. A channel may be worked for as long as a miner is willing to work it, and as long as it is not destroyed by the rains. A miner will reoccupy a channel that has been recently flooded and covered by river sediments only if she or he has a good gold sample from it. Abandoned channels can be mined again by another miner as over time the river washes gold bearing sand in to the channel making it again productive. Ownership can also be transferred to another miner.

Processing

It is in the processing of the ore and concentrate that the Kankana-ey women are known to excel. Ore is collected until such time when the miners decide that it is time to process. This decision is dependent on the quantity of the ore, its gold content, or the need for immediate cash.

When large quantities of ore are processed, the miners use small mechanized, fabricated crushers. The crushed ore is then loaded in rod mills or ball mills for grinding. Much of the processing of the ore is done by the women. After the ore is crushed in the rodmill or ballmill and becomes slurry, women do most of the processing and smelting. It is not uncommon to see men process the gold ore, but in general, the women are better at processing, especially with the handling of the separator (*sabak)*. Figure 7.1 is a flow chart of the gold recovery process.

The Kankana-ey women use physical separation techniques to separate gold from ore. There is a marked contrast between traditional small scale miners and gold rush miners. This distinction focuses on one of the main environmental issues confronting the Philippines today: mercury pollution. In the gold rush areas, mercury is used in the recovery of gold (see Figure 7.2). By contrast, the traditional small scale miners employ milling, gravitation and panning methods for gold recovery. Unlike the gold-rush areas, traditional small scale miners use painstaking measures and great care to recover gold from virtually all the solids and also recycle the water used in the process. The miners scrape the surface of the soil around the work and roasting area, collect this soil in sacks and process it for its gold content (*kidkid*); crush and regrind the used crucible (*gangi*); and recycle the middlings and panning tails (*kibo*) in the holding tank (*dayasan*). Miners in gold-rush areas showed far less concern with spillage and overflow resulting in extensive mercury pollution and substantial amounts of sediment buildup.

The tool complex for milling gold-bearing sand is made of basalt rocks derived from the river. After the ore is crushed, it is milled using a tool complex of basalt grinders and mortars. The crushed ore is milled with a large lower grinder called an *alintegan* and an upper stone grinder called an *alinteg* until the ore is ground into the fine texture of what they call *linang*, resembling very much the gold bearing-sand mined from the river.

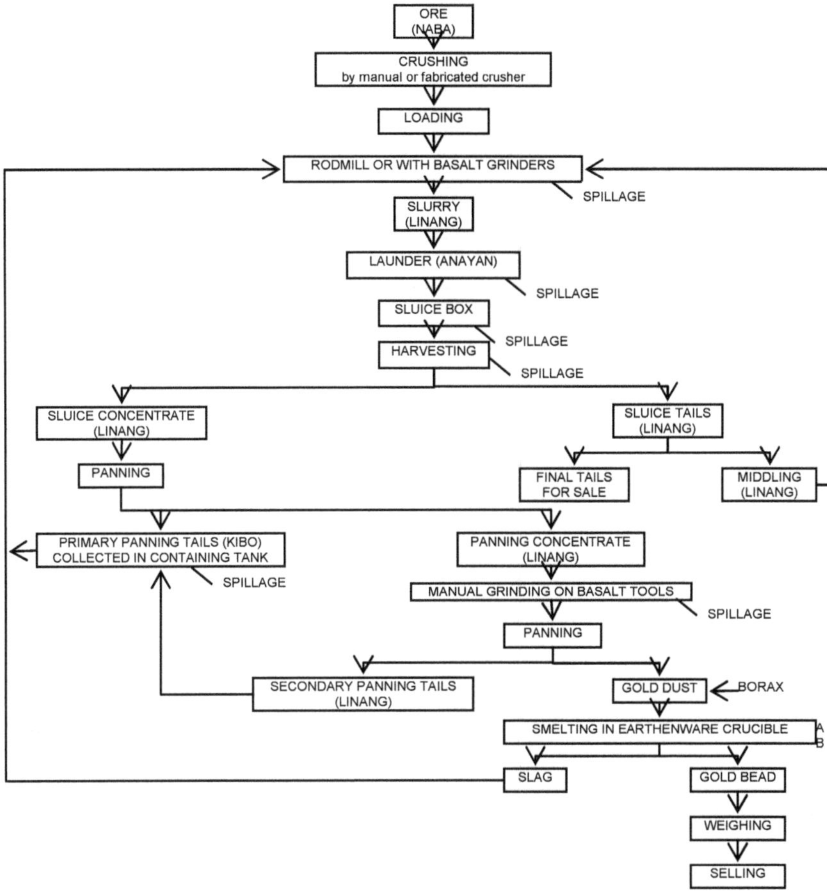

A - The crucible (gangi) after being used for several times is later collected and pulverized for processing of its gold content.

B - The spattering, which is a result of heating the material to high temperatures, is collected for further processing.

Spillage - The spillage is collected and recycled for its gold content (kidkid).

Figure 7.1 Flow chart of traditional small scale gold recovery process

After milling they pan the concentrate in a galvanized iron separator (*sabak*) over a small holding tank (*dayasan*), where the gold is separated from the unwanted materials. While some men have been observed panning with a separator, the women are known to be more adept at separating the gold during this procedure.

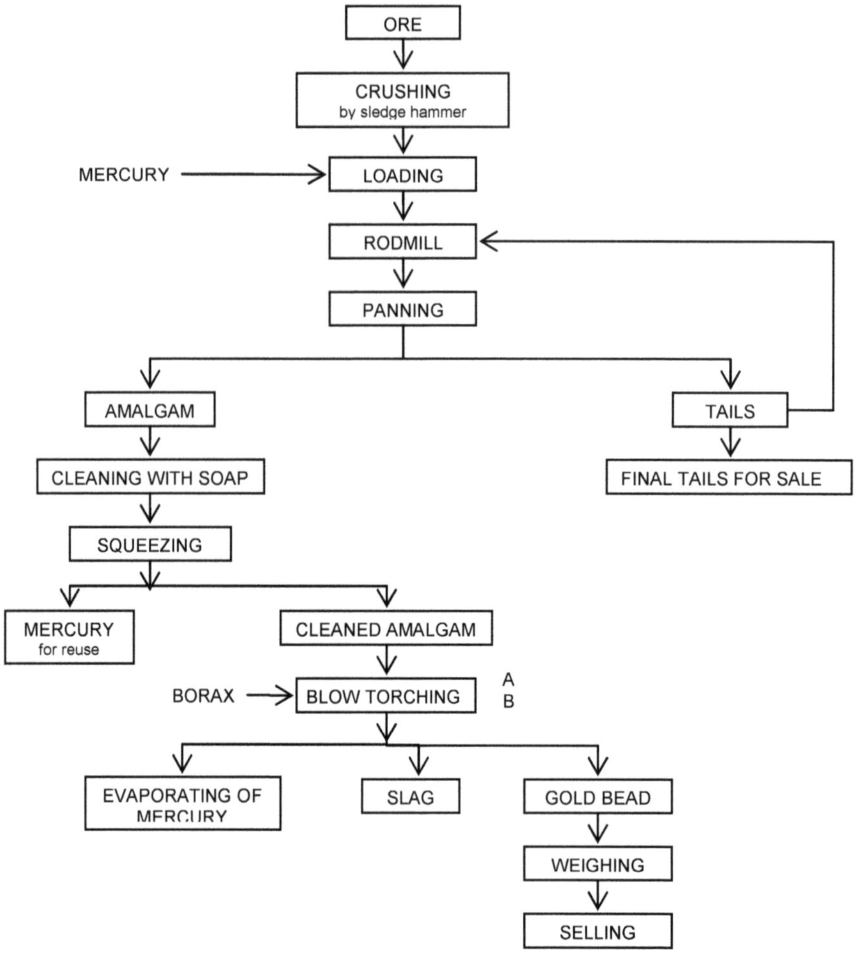

A - The crucible after being used for several times is later collected and pulverized for processing of its gold content.

B - The spattering, which is a result of heating the material to high temperatures, is collected for further processing.

Figure 7.2 Recovery process flow chart: Amalgamation gold rush areas

After panning, the sluice concentrate is further milled with other grinding rocks. The panning concentrate is ground several times until a powder like consistency is achieved. During this process most of what is milled is gold grains (*ngamngamoy*) with a few minute non-gold particles. There is still perhaps 5 per cent or less of unwanted materials mixed with the gold. Sometimes it is not uncommon to see the processors squeezing the juice from sunflower leaves over the concentrate. This is to prevent the very fine gold from floating. Other plant juices used for the same function come from the leaves of the sayote, tobacco or citrus mitis (calamansi). After this step the panning concentrate is placed back into the separator for further panning. What remains after this procedure is almost pure gold.

Smelting the gold

After this final step of milling, the gold grains are wrapped in plastic and sprinkled with about 1/4 teaspoon of borax for flux. This is placed in an earthenware crucible and brought to a smelting area, where the crucible is placed in an open furnace and covered with charcoal.

When a manual blower is used two workers are needed for smelting the ore; one for pumping air through the blower, keeping the heat constant during the melting process, and the other for stirring the coals with a stick around the gold. When smelting is complete, the remaining mass is picked up with tongs and placed in a cup containing water, cooled and the slag (which is the black particles around the crucible) is knocked off exposing a gold bead. The gold bead is weighed on a balance scale and sold to dealers located either within the community or the nearest city.

Sharing the ore and concentrate

One of the striking and unique features of the Kankana-ey traditional small scale miners are the social mechanisms which allow them to share the ore which they mine and the concentrate acquired during the processing stage. The most common ways of sharing are through *sagaok* and *makilinang*.

Sharing the ore (sagaok)

Sagaok is the process of going from one tunnel to another requesting ore with gold. It is not uncommon to see a line of men and women of varying ages waiting outside a tunnel with their knapsacks waiting for their turn to request gold-ore (*makisagaok*). It is important that the miners share a portion of the ore that they are mining. Ore is always given first to the elder women and men, although miners refer to elder women more than they do to men when they talk about *sagaok*.

In Dalicno, when a miner finds high-grade ore by the kilogram, ore with gold nuggets have to be given to the elder men (*panglakayan*), the elder women, teachers, barangay officials, and the midwife. Those held in the highest esteem – the elders – are expected to be given more of the share of the gold. In addition to sharing the gold with the community, the miners have to hold a ritual, where several pigs are killed as an offering to their deities and *anitos*. The ritual killing of pigs is a community affair, as are the taboos that follow after each ritual.

Sharing the concentrate (makilinang)

Another way of sharing gold is through *makilinang*, a process of distributing gold through the sharing of the concentrate. Men and women of varying ages can request concentrate (*linang)*, although this is mostly the women's and young adult's activity. Elder women need not ask for *linang*. They are given concentrate and need not be present during the gold extraction process. Because of this, the processor always sets aside some concentrate (*linang*) to be shared with the elder women within the community. Many will help with the work while waiting for *linang*. Like the sharing of ore (*sagaok*), only those who are consanguines and affines can participate in this sharing activity. Strangers to the community are always accompanied by a kin.

It should be stressed that the shared concentrate is not only a product of the processing of the ore, but is also from the recycling and the processing of minerals that takes place as a result of the gold recovery process. This cultural practice of sharing the concentrate reinforces the concepts of recycling and the processing of other materials during the recovery process. It encourages and continues the practice of the physical separation methods in gold extraction. This is a dimension which makes the Kankana-ey unique and is one of the major factors which contributes to the continued success of their technology through time.

The processing areas are by their homes and facilitate the work of both men and women when it comes to household chores like cooking, care of the children and their elders. In addition, family-centred processing areas are also venues where affairs and decisions of the community are discussed. Locating the processing areas by their homes has been a practice that has been continued from one generation to another.

In both practices of *sagaok* and *makilinang*, elder women always had first priority to the shared ore or concentrate. Sharing gold with the older women of the village was interpreted by Torres (1993) as one way for the Kankana-ey to properly provide for them without marginalizing their existence.

One of the most striking features of the traditional miners are those aspects of their technology that go as far back as the Hispanic period as described earlier. The Kankana-ey, in spite of their proximity to large commercial mines, have continued through generations to use their simple and inexpensive tools. Today there are still many traditional small scale miners who remain small scale, but their technology is changing with the use of more ball mills. However, much of their mining and the gold recovery process is still predominantly done with traditional tools. Others have become medium scale miners extracting gold through carbon-in-pulp and heap-leaching.

Rituals

> There will always be gold. We will never run out of gold, but you have to mine it the right way.
>
> Elder from Sitio Dalicno, Benguet Province

Mining the right way involves not just their technological/productive system but also involves their social and ritual systems. Mining the right way means management of the resources by the elders who are present among the living and in the spirit world. The social and the ritual systems are interlinked. The deities and the spirits or *anitos* give the gold and other natural resources to the community. In their five-level hierarchy of deities and spirits there are women and men. For example, in the third-level, the spirits are composed of ten female and ten male *anitos*. The elders, both women and men continue to play a predominant role even while in the spirit world as the fifth level of *anitos*. As *anitos* they are always called upon for the proper guidance and management of the community and are invoked and appeased so that good fortune will always abound in all the endeavors of the community. As the elders before shared their gold and other natural resources with the living and made the lives of these prosperous and healthy, so too must the living share it with the rest of the community.

The values of traditions and customs which have been put into place for the proper management of the resources and for the continued unity of the community are emphasized during all rituals. The rituals also function in part to regulate the over-exploitation of gold by having periods of abstinence (*ngilin*) from mining and other activities. Examples of abstinence from mining are for those who are newly weds, during a death ritual or when certain taboos are broken (for example, the eating of cow, dog, goat, ducks or eel). Like the gold they share when they *sagaok* and *makilinang*, rituals, and other community projects are a collective effort involving everyone's participation. During rituals the women cook the rice and other lowland dishes while the men butcher, cook, and distribute the pigs to the women, men, and children. Today, the priests (*mambunongs*) who perform rituals are predominantly male. However, historical records and folktales indicate the presence of female priests.

Events that affected the Kankana-ey

I would now like to discuss how the Kankana-ey were affected by the policies of the Philippine government (Caballero, 2001). Again, the role of the women will be the focus of the discussion. In summary these were:

1. In 1991, the introduction of RA 7076 also known as the People's Small Scale Mining Act.

2. In 1990, the granting to Benguet Corporation of an Environmental Compliance Certificate to conduct Reverse Circulating Drilling and Trenching in Dalicno.
3. In 1990, the discovery of archaeological sites in the immediate area of Dalicno with material culture similar to that used by the traditional small scale miners.
4. In 1993, the start of the Itogon-Suyoc Mines Load and Haul Operations.
5. In 1993, the collapse of the tailing dam of Itogon-Suyoc Mines.

The Small Scale Mining Law (RA 7076)

In the autumn of 1991 the Mines and Geosciences Bureau worked with representatives from the Philippine Social Science Council (PSSC) in an Inter-Agency Committee which was tasked to formulate the rules and regulations for the implementation of The Small Scale Mining Law or RA 7076. The PSSC representatives found the law to negatively impact traditional small scale miners. The law ignored the existence of traditional small scale miners as all the statute's provisions were premised on the misconception that all such miners are of the type exemplified by gold rush miners. Utilizing anthropological data from my research, several reasons for the inapplicability of the law to traditional small scale miners were cited. Based on this information, we recommended for repeal of the statute or at the very least, that substantial necessary amendments be made. With the realization that this process would entail considerable time, we worked with the Inter-Agency Committee in drafting the Implementing Rules and Regulations for the law. Through our efforts, the present Implementing Rules and Regulations for the Small Scale Mining Law make a distinction between gold rush small scale miners and traditional small scale miners.

A signature sheet of endorsement by individuals from the Philippine social science community was attached to PSSC's position paper. More importantly, there was overwhelming support of PSSC's recommended provisions by the traditional small scale miners in Benguet Province, several of whom also wrote their own position papers. In the public hearings in Baguio City men and women came from their communities carrying placards opposing the Small Scale Mining Law and actively voicing their opposition to the law. They said that when confronted with the new law their customary sharing practices and values will be destroyed. To meet the fiscal provisions stated in the implementing rules and regulations, one miner from Mankayan-Suyoc area simply stated, 'You will force us to be selfish. We will not be able to give *linang* to our elder women'.

The term traditional small scale mining or miners is a term that has been institutionalized and is used by the Mines and Geosciences Bureau of the DENR and other sectors in the Philippines.

Benguet Corporation's drilling operations, archaeological sites

At about the time the Kankana-ey miners were responding to the potential negative impacts of The Small Scale Mining Law, the community of Dalicno increasingly felt that it was going to be displaced by the Drilling and Trenching Operations of Benguet Corporation which started in 1990 and later in 1993 by the Load and Haul Operations of Itogon Suyoc Mines. Their response to Benguet Corporation's operations was to barricade the road and as a result BC stopped part of their explorations. The archaeological sites were discovered in 1990. The negative impact of Benguet Corporation's operations to the archaeological sites was brought to the attention of the national government. After a multisectoral investigation of the area in 1991, Benguet Operations were suspended.

Itogon-Suyoc mines' load and haul operations and tailings dam

In 1993, Itogon Suyoc Mines started its Load and Haul (LHD) operations. The Dalicnoans saw the LHD operations as negatively impacting their homes, water sources and mining tunnels. During the LHD operations the tailing dam of Itogon Suyoc Mines collapsed and endangered the municipality of Itogon. The predominantly Ibaloi residents of Itogon, along with local government representatives and the Dalicno community joined hands to oppose the operations of Itogon Suyoc Mines and presented their concerns to the national government. During their intense resistance – there are stories of men and women barricading the area where the company's machinery was operating – mention is made of how the elder women (*babbaket*) from Luneta picketed and barricaded the area. In addition, key individuals, who composed the local leadership structure and also participated in the community discussions in Dalicno regarding the impact of RA 7076 and the Environmental Impact System, used information from these discussions plus anthropological documents to help two Local Government leaders to successfully uphold their rights in court as occupants of the area prior to the operations of Itogon Suyoc Mines.

All through these major events, it is difficult to say that during the meetings in Dalicno that it was just the men or the women who attended. There was equal interest and participation by both men and women. In our public meetings with the community it was observed that it was the school principal who would facilitate the meetings. It is difficult to separate the roles of mobilization and resistance by gender as both men and women informants perceive each other as acting together to help each other (*nagtinulungan kami*, literal translation 'we helped each other'). This is in consonance with their communal values of sharing resources. Because there is the value of shared resources and shared management of resources, it is possible that the conflict-resolution in response to outside threats to their own resource base is a strategy of collective organized resistance reinforced by indigenous social and political structures. This is supplemented by information, education and dissemination of existing policies and events related to the relevant critical issues.

Conclusion

Based on the findings in the archaeological site, it is most likely that women miners in the past processed and milled their ore in much the same way as the Kankana-ey women of today. The miner's work areas and management of mining claims indicates reciprocity as an organizing factor in the form of the sharing of resources. The women traditional small scale miners play a key role in the technology, and in the social and economic exchange. While the Kankana-ey refer too many of their activities as a community collectively, in their conversations they highlight the women irrespective of age as better processors of gold ore. They revere their elders but give particular attention to the elder women (*babbaket)* in the social and economic distribution of gold through the social mechanisms of sharing of the ore (*makisagaok)* and the concentrate (*makilinang).* While it seems more common to hear the elders referred to as the male elders or the *panglakayan* they mention their female founder in their folktales and in their prayers to their deities and *anitos*, where women are equally represented at various levels. In the areas of advocacy and organized resistance, the women again play a predominant role. While gender relations in the communities are generally more equal, the women traditional small scale miners play a central role in solidifying the interpersonal relationships that results in the cohesiveness of the group.

References

Bayle, Romeo, 1998. 'Small-scale Gold Mining', *Community Habitat*, Issue 1. Philippine Rural Reconstruction Movement.

Blair, Emma H. and Robertson, James H. (eds), 1903–1909. *Philippine Islands, 1493–1803*, 54 vols. Cleveland.

Bobadilla, Diego de, 1640. 'Relation of the Filipinas Islands', in Blair, Emma H. and Robertson, James H. (eds), *The Philippine Islands, 1493–1803*, vol. 29, Cleveland.

Buzeta, Manuel, 1850-51. *Diccionario Geografico, Estadistico, Historico, De las Islas Filipinas*, 2 vols, Madrid.

Caballero, Evelyn J., 1991(a). 'Mining for Gold: The Niche Concept and the Survival of Traditional Small Scale Miners', *Philippine Sociological Review*, January-December, vol. 39, Number 1-4, pp. 17-23.

Caballero, Evelyn J., 1991(b). 'Formulation of Policy and Standards for the Disposal of Mine Tailings from Mining Operations in the Philippines', Report submitted to Dames and Moore under contract to USAID/Philippines.

Caballero, Evelyn J., 1992. *Gold from the Gods: Traditional Small-scale Miners from Benguet Province, Philippines*, U.M.I: Dissertation Services, 300 N. Zeet Road, Ann Arbor, Michigan.

Caballero, Evelyn J., 1996. *Gold from the Gods: Traditional Small-scale Miners in the Philippines*, Giraffe Books, Quezon City.

Caballero, Evelyn J., 2001. 'Strategies of Survival for a Community of Traditional Small-scale Miners', in *Towards Understanding Peoples of the Cordillera*, vol. 1, Cordillera Studies Centre, Baguio.

Colin, Francisco, 1663. 'Native Races and their Customs', in Blair, Emma H. and Robertson, James H. (eds), *The Philippine Islands, 1493–1803*, vol. 35. Cleveland.

de la Vega, Juan Manuel, 1609. 'Expeditions to the Province of Tuy', in Blair, Emma H. and Robertson, James H. (eds), *The Philippine Islands, 1493–1803*, Cleveland.

de Saavedra, Alvaro, 1527. 'Voyage of Alvaro de Saavedra', in Blair, Emma H. and Robertson, James H. (eds), *The Philippine Islands, 1493–1803*, vol. 2, Cleveland.

Keesing, Felix, 1962. *The Ethnohistory of Northern Luzon*, Stanford University Press, California.

Loarca, Miguel de, 1582. 'Relation of the Filipinas Islands', in Blair, Emma H. and Robertson, James H. (eds), *The Philippine Islands, 1493–1803*, vol. 5, Cleveland.

Morga, Antonio de, 1609. 'Events in the Filipinas Islands', in Blair, Emma H. and Robertson, James H. (eds), *The Philippine Islands, 1493–1803*, vol. 14, Cleveland.

Perez, Fray Domingo, 1680. 'Relation of the Zambales', in Blair, Emma H. and Robertson, James H. (eds), *The Philippine Islands, 1493–1803*, vol. 34, Cleveland.

Pigafetta, Antonio, 1525. 'First Voyage around the World', in Blair, Emma H. and Robertson, James H. (eds), *The Philippine Islands, 1493–1803*, vol. 33, Cleveland.

Quirante, Alonso Martin, 1624. 'Expedition to the Mines of the Igorrotes', in Blair, Emma H. and Robertson, James H. (eds), *The Philippine Islands, 1493–1803*, vol. 51, Cleveland.

Reynolds, William E. and Caballero, Evelyn J., 1993. 'A Preliminary Survey in Sitio Dalicno, Barrio Ampucao, Itogon Municipality, Benguet Province', *National Museum Papers*, vol. 4, no.1, pp. 47-58.

Torres, Amaryllis T., 1993. 'Environment, Society and Development: Philippine Report on Social Sciences Perspectives and Sustainable Development', in *Empowerment and Accountability for Sustainable Development: Towards Theory Building in Social Sciences*, Proceedings of the Third National Social Science Congress, Philippine Social Science Centre, Quezon City.

Chapter 8

Mining Gender[1] at Work in the Indian Collieries: Identity Construction by *Kamins*[2]

Kuntala Lahiri-Dutt

Introduction: Identity work

Mining's notoriety as a difficult job for women has lead to a profusion of myths surrounding the roles of women in the mines[3]. These beliefs arise from the portrayal of mining jobs as unsuitable for women, creating an isolation and hence solidarity among men, forming working class spirits and building up foundations of male resistances, and finally, leading to the need for 'protecting' women from these unsafe and excessively demanding working practices that are deemed essential in mining.

---

[1] Gender as used here is not just men and women, male and female, but a social and linguistic construction, a non-stable social meaning ascribed to the male and female. Women, femininity and female experiences are understood as constituted and open rather than as robust unities. Thus, experiences are not associated with the whole, coherent person, but are constituted through the discursive invocation of a person as a 'man' or as a 'woman'. Here, I follow Hollway et al. (1998, p. 252) 'Gender-differentiated practices depend on the circulation between subjectivities and discourses which are available'.

[2] As explained before in this book, '*kamin*' is the common name of Indian women mineworkers working at manual jobs in the mines. The term is a feminine opposite of 'coolie', the physical labourer in almost all contexts in India. *Kamin*s can be found also in the quarries such as stone. Writing in 1940, B. R. Seth in his book *Labour in the Indian Coal Industry* (p. 63) describes: 'The word *kamin* refers to female worker(s) employed on the surface and underground, and the word "khalasi" to one who cleans and handles the pump, haulage and winding engines during their operations…'.

[3] Sometimes these myths involve the propagation not only of simple macho superstitions but also pervade the psyche of researchers, mining technocrats, trade unions and policy-makers. It is beyond the scope here to go into these myths although I have discussed them in the introduction. For a recent publication portraying the mines as masculine places see Kideckel, 2004.

All work is socially constructed; mining as work too has been socially constructed historically as dirty, heavy and risky causing it to be considered as not fit for women. This social construction is based on older images of mining work and becomes invalid if one contemplates mining as it is practiced today in many large and formal mines. Modern mines in many countries, including the state-owned collieries of India, usually have extremely complex organizational structures where the proportion of demanding physical work in the actual pit face is relatively small in relation to the large range of other jobs even in formal mines. Moreover, with increased mechanization and much improved safety standards, the manual part of mining nowadays comprises only a very small segment of the entire cycle of operations, management, administration and marketing. However, all mineworkers tend to concede to the notions of mining as being a rather heavy, dirty and risky job; mining over the years of industrial civilization has become 'the' work attributing a special identity to its workers and creating a working class identity in turn. This is because when people construct their work practices, their identity is at stake and this is more so in cases of a 'dirty'[4] and 'difficult' work such as mining. Clearly, it is not possible to give a single and objective description of life in the mines. Many alternative narratives are constructed on mining work with available discursive[5] material. However, these images of mining have been challenged by a number of feminist researchers showing that this portrayal is simply an assertion of masculinity and not a reflection of reality. Empirical studies (see, for example, Furumura, 2003; Kaur, 2001; Yoshida and Miyauchi, 2003; Espley et al., 2004) have provided evidence of active participation in the mines, large and small, past and present, in the developing as well as in developed countries. How an individual perceives the world and communicates that knowledge to others conveys a potential wealth of information about who that person is. This is the essence of 'identity work'[6]: an active attempt to construct identity through various constructions of reality.

Regarding life and work in the pits, identity work in my view can raise certain important issues. If mining work is not always as demanding as it is commonly portrayed, then the practices based on this portrayal operating to exclude some women need to be examined. The question then is: do Indian women miners in

---

[4] Everett Hughes (1951) first invoked the term 'dirty work' to refer to tasks and occupations that are likely to be perceived as disgusting or degrading. Consequently, workers are seen to personify the dirty work.

[5] Following Henriques et al. (1984, p. 105), I would define the term a regulated system of statements, locked in an intricate web of practices, leading to the production of knowledge.

[6] This term has recently come to be used in ways loaded with meaning, for example, as the construction of lives through the unveiling of life histories (see Stewart and Strathern, 2000). In fact many books and papers identify mining itself as a special kind of work that attributes a special kind of identity to its workers. That is not the meaning I wish to use here but rather as 'The process by which the individual creates her/his own identity' (see Helve, 1998).

Indian collieries differ from men in their construction of pit life? If not, then understanding why women miners accept a version of the mining life that compromises their interests becomes a critical task. In this chapter, I intend to address the question of women's portrayal of working life in the *khadan*s[7] of eastern Indian collieries, their constructions of identities, and establish that it might be impossible to produce an accurate portrayal of what actual pit life involves. In my view, mining as a work is constructed in ways that enable men to assert a specific form of cultural masculinity. To understand why women mineworkers in India do not construct their roles in positive ways – say following the example of the woman gem miner in Andamooka in South Australia[8] – brings a special understanding of class, caste and gender interface as experienced in the Indian collieries.

I focus closely on examining how the work in the collieries is portrayed during interactions involving identity construction. The research interview is indeed a situation in which identity work is performed, due to the power relations existing not only in the mines but also between me as a researcher and the participant in the research, the *kamin*s in the Indian collieries, and due to the nature of research itself. In this chapter, I have attempted to identify the discourses produced through relations of power in the broader social context of coal mining, as well as in the more immediate context of the research interviews. Since, 1993–1994 I have examined the roles of women in the collieries and stone quarries, and the interviews forming the basis of this chapter were taken during several trips to the Raniganj coal fields. Initially I went into the field looking at gendered impacts of mining, but could not help examining questions of participation and status of women in the mining communities in the industry. A strong influence leading to this re-interrogation of my research material was the recent work of Dick and Cassell (2004) on the subjective position of the researcher, current strands of feminist thoughts and practices, and also my own, physically distanced, view of the Indian collieries, the location of this research.

The chapter is arranged as follows: after a brief discussion of the theoretical position on occupational identity that set the context of my research, I go on to the extracts of the conversations (with brief descriptions of the contexts in which the meetings took place) and then analyze them. The reconstructed history of work (from available historical material) of *kamin*s (and *malkatta*s or coolies) in the

---

[7] *Khadan* is a Hindi word meaning the mine pits producing a large range of mineral products from deep and shallow, and even open cut mines and quarries.

[8] See Heritage, 2001. This is just one small example of the increasing trend of re-discovering women's histories in the mines. In the International Mining History Conference in Japan in 2003, I met a passionate American woman who has been collecting the oral histories of women miners in the Appalachians. Many of these histories are easily accessed on the internet (see for example http://cass.etsu.edu/archives/afindaid/a355_06.html or http://www.coalminerstamp.com.pdfs/oth_women.pdf or www.mishmash.com/marat, all accessed on 1st July 2004).

Raniganj collieries, West Bengal, builds up the context of the particular woman miner's conversation. Through this analysis, I wish to show how, in constructing an identity, an individual woman consents to the idea that a job like mining entails working practices that are incompatible with traditional notions of womanhood in India. I explain this as her resistance towards being positioned by the researcher in broader ideological discourses that compromise her perceived notion as 'woman'. Thus, the chapter tries to bring out the subjectivities[9] inherent in my position as a researcher in relation to those of my research participants.

Theorizing gender identities in mining

To look into examinations of occupational identity, I turned to research material from organizational behaviour studies. Dick and Cassell's (2004) recent work on the occupational identity of policewomen provides the inspiration of this research. Traditional social-psychological theories of self viewing the individual/ organizational relationship, the tendency to see oneself in positive terms, originating from the self-enhancement motive, have been subject to intense critique in the last ten years or so (see, for example, Henriques, 1998) for their neglect of the social aspects and the contexts. Ashforth and Kreiner begin their chapter on social construction of meaning from the experiences of dirty workers (1999, p. 413) by quoting a gravedigger:

> I have this question lots of times: 'How can I take it?' They ask if I'm calm when I bury people. If you stop and think, a funeral is one of the natural things in the world… I enjoy it very much, especially in summer.

Ashforth and Kreiner (1999) argue that the stigma of dirtiness often fosters relatively strong occupational and workgroup cultures – that is widely shared and deeply held systems of values, beliefs and norms – with attendant ideologies and social weighting practices, that is, selective social comparisons and differential weighting of outsiders' views. This is not uncommon in mining, as we have seen in Burke's chapter. Thus the occupational identity becomes a resource that members draw upon, enact and affirm through their daily actions to enhance the meaningfulness of their work.

An important aspect of identity at work is how it affects gender relations. All work has strong gender divisions of labour which are perpetrated by organizations. Alvesson (1999) in her ethnographical research of an advertising agency, a workplace requiring significant professional services, shows how efforts to secure

---

[9] Following Henriques et al. (1984, pp. 2-3), I will define the term as referring to individuality and self-awareness – the condition of being a subject. I understand that subjects are dynamic and multiple, always positioned in relation to particular discourses and practices and produced by these.

an occupational identity for males interact with gender divisions of labour and workplace level constructions of gender. He argues that masculinities and femininities do not form simple patterns of domination and subordination, but interact in more complex ways.

At the same time, in recent years, what we mean by the term 'identity' has been conceptualized in ways that are best described as 'social constructionist'. According to Dick and Cassell, much of the research has used Foucault's (1997, 1981) work on the relations between forms of knowledge and forms of power and subjectivity to examine gender relations and identity at work (see Alvesson, 1998; Ball and Wilson, 2000; Casey, 1995). These see identity (or subjectivity) as effect of the regulatory operation of discourse. Foucault saw discourse as knowledge: '...the underlying rules, the *a priori*, assumed parts, of knowledge which limits the forms of the sayable ... Discourse ... determines what is taken as known and how this is established' (quoted in Townley, 1994, p. 2). Discourse then is not unitary and neutral, and 'is the outcome of relations of power in society, the consequences of continual struggles between various 'authorities' to regulate the behaviour of the population in certain ways' (Dick and Cassell, 2004, pp. 55-56)[10]. Discourse also 'produces' identities as power can be a productive force creating objects in the effort to target them. Following this line of argument, recent work on labour processes has examined the way organizational identities are constructed through discourse and the way 'consent' and 'resistance' operate within organizational power relations (Jermier et al., 1994). Resistance frequently contains elements and consent incorporates aspects of resistance (Collinson, 1994, p. 29). Thus, consent (or resistance) is better understood as a 'subject position' which is taken up within discourse to achieve a specific identity, for a specific purpose, at a specific time (Dick and Cassell, 2004).

These ideas are used in my chapter to explain the construction of mining as dirty, risky, heavy and hazardous work by women miners in the Indian collieries towards building up their identities. I intend to show how consenting to the idea of mining as tough operates as 'identity work' within the specific context of the research interview. I explain this as a resistance towards the potential building up of 'undesirable' identities by me, as the researcher of upper caste teacher woman from the city. This assertion is based on my point that the knowledge arising from a research inquiry too becomes equipped with power. I agree with some recent feminist research that asserts that it is not possible to give a single privileged interpretation in describing 'women's experiences' since 'women' as a group with common characteristics and interests does not exist (Heyes, 2000). Thus, the research outcome reaffirms that just as it is an essentialist practice to construct identity in a monolithic and exclusionary way, so it must be remembered that our

---

[10] Dick and Cassell (2004) describe the argument of Foucault (1981, 1988, 1990) that discourses on sexuality 'exploded' during the so-called Victorian era as a direct consequence of the struggles between various authorities to regulate sexual proclivities of people.

explanations of 'women's experiences' too are partial, interpretative and contingent.

Power takes the visible forms of caste hierarchy in Indian society, so consequently, in any research it is hard to conceal although its presence is often overlooked. Pandian in his 2002 essay interprets this omission as a construction of the other by denial:

> The subtle act of transcoding caste and caste relations into something else – as though to talk about caste as caste would incarcerate one into a pre-modern realm – is a regular feature one finds in most upper caste autobiographies. Caste always belongs to someone else; it is somewhere else; it is of another time. The act of transcoding is an act of acknowledging and disavowing caste at once.
>
> Pandian, 2002: 6

In this chapter, I intend to present my subjective position as a researcher being a key to understanding the reluctance of women miners in highlighting their jobs as exceptional. Cultural identities such as ethnicity, caste and religion have powerful influence on social and gender relationships within the formal industrial labour force in India. Class is inextricably intertwined with identities such as gender and ethnicity, a situation which necessitates that the assumption of a singular, monolithic working class be rethought. The relationships between the politics of class and community have been noted by subaltern historians (Guha, 1982–97) who opposed the Marxist notion of a unitary working class having universal validity in India. Gender introduces a complexity in this relationship, as has been pointed out by feminist historians (see Sen's 1999 research on gendered labour history of the jute mills of eastern India). Sen in 2001 incorporated the caste angle to her analysis, since the political economy of industrial resource extraction in India cannot ignore caste realities. Consequently, the complex interplay of the three hierarchies – caste, class and gender – including my own subject position constitute the backdrop of this research report.

Women in Indian collieries

Coal mining began in 'modern' India in 1774[11] in the area between the rivers Ajoy and Damodar in Bengal, but it was not till the mid-19th century that coal began to play a major role in empire building in British India. Coal mining manifested the symptoms of colonial modernity that descended on feudal economic relations and

[11] On 11 August, 1774, two employees of the East India Company, John Sumner and S. G. Heatly made an application to Warren Hastings to produce and sell coal in Bengal and its dependencies. They offered to supply to the East India Company 'ten thousand maunds of pit coal every year for a period of five years at the price of two Arcot rupees and three-quarters per maund (of 80 Sicca Weight to the Seer)'. H. D. G. Humphreys, 'The Early History of Coal Mining in Bengal', MGMII Volume, p. 147 (quoted in Ghosh, 1990: 42).

production systems of eastern India and yet remained feudal in its organization and labour relations.

Rothermund and Wadhwa (1978) described how collieries grew up as the secondary 'enclaves' meant to serve the primary metropolitan enclaves located in Calcutta within a vast sea of subsistence agriculture. The coal economy was characterized by private investment and the involvement of indigenous capital facilitated by a low level of technology. The import of labour from other parts of the country to build up a captive reserve did not begin till the mine owners realized that the indigenous family unit preferred to go back to work in the farm in sowing and harvesting seasons[12]. The labour was extremely heterogeneous as a result. Deshpande wrote in 1946:

> The Bauris, a semi-aboriginal tribe, were the earliest miners … and are still noted for their skill and hard work. Since the opening of the East Indian Railway in 1890, the aboriginals from Chota Nagpur such as Santhals and Koras started coming into the field. In the course of time, other aboriginals such as Bhuiyas, Mundas, and semi-aboriginals such as Nunias, Beldars, Meahs, Dosadih, Chamars, Goala, Ghatwals, etc. entered the coalfields in increasing numbers. Since the early years of the development of the industry, certain castes and tribes have shown special aptitude for particular types of operations in the mines.

Thus, ethnic identity and caste were, from the very beginning, an undeniable part of labour identity in the eastern Indian collieries. In Raniganj collieries, according to Ghosh (1990: 130), 90 per cent of the coal cutters were Santhals and Bauris. Bhuiyas and Rajwars became mostly loading coolies and trammers, Beldars and Nunias were mainly earth cutters and surface workers, whereas migrant Bilaspuris took up the more skilled jobs. The historical work of Ray Chaudhury (1996) provides a closer view of the kamins amongst this caste-marked colliery labour. In 1901 women formed about 48 per cent of total mine workers in Raniganj. Of these women, 65 per cent worked in underground collieries. The proportion remained more or less the same till 1921 (61 and 60 per cent, respectively). The participation of women was significant till the 1930s, after which it decreased to about 20 per cent in post-colonial India and now has fallen to only about 6 per cent (Ray Chaudhury, 1996). Most substantial declines, however, have taken place in recent decades under state ownership of the mining industry. Wage rates of men and women in the mines have been the same since nationalization. The All India Labour Appellate Tribunal (LAT) took a strong objection to women receiving 75 per cent of the wage payable to a male worker (as per Mazumdar Award), and observed (1958, Part 2, p. 2118):

---

[12] As early as in 1896, the Report of the Labour Enquiry Commission (p. 36) pointed out that although the mine manager would like his man to work for 6 days a week, on an average he does not work for more than four or five days. Sen in her 1999 and 2001 works examined the jute mill labour during the colonial period in Bengal and the similarities with the collieries are not negligible.

> We ourselves have had occasion to observe female workers engaged in strenuous work, and it is monstrous to suggest that in work of that kind she does lesser work than the male, and yet we are aware of the lamentable fact that for that very work she has been given in the past a lesser wage as a matter of tradition; that tradition must go with the upsurge of social conscience …We have no hesitation in coming to the conclusion that the time has been reached when the female worker is entitled to receive the same wages as the male worker for the same class of work, unless it can be shown that her output is demonstrably lower than that of the males.

Times have changed since and with equal wage rates, working conditions have improved immensely making 'strenuous work' as described by either the LAT non-existent (Murty and Panda, 1988). The current organizational structure of a typical colliery is rather complex with many layers of intermediaries between the top and bottom ranks. Figure 8.1 describes the overall hierarchical structure of mining in a typical colliery. It shows the large range of non-manual workers involved in the mining organization in Indian collieries. The personnel departments differentiate further between 'staff functionaries' ands 'line functionaries' – the latter group being involved in production directly. If only the production line is taken, a non-mechanized operation has at least five jobs:

1. Coal preparation: involving the works of driller (making holes in the seam), explosive carrier, and blasting (usually performed by a mining *sardar*) – all these are done by waged men. A few women are still involved in preparing clay cartridges (*gulli*s) on a piece rate basis (at least 1000 pieces on an average work day).
2. Tramming: involving the work of trammer, usually waged men.
3. Loaders: always a piece rate job (depending on a fixed payment per tub or 40.5 cubic feet with a minimum of two tubs) done at one time by 'pick miners' or *malkatta*s.
4. Hauling: involving the work of haulage *khalasi* (meaning operator) – also men.
5. Direct loading: in the absence of a coal handling plant, reserved coal from depots is loaded manually by wagon loaders or mechanically (by pay loader). Many women work on the surface in this job as wagon loaders.

General Manager
↓
Additional General Manager
↓
Area Manager
↓
Project Officer
↓
Agent
↓
Various engineers – civil, industrial, medical officers, mechanical and personnel managers
↓
Mine Manager
↓
Deputy Mine Manager
↓
Assistant Mine Manager
↓
Upper Division Clerks
↓
Lower Division Clerks
↓
Foreman/Overman
↓
Chargeman or Mining *Sardar* (workpoint supervisor)
↓
Workmen – *kamins* and *coolie*s, i.e. operators, loaders and runners (depending on the level of mechanization of the operation)

Figure 8.1 Production-line organization: Typical Indian colliery

Only the lowest level of workers in this structure is engaged in blue-collar jobs in actual coal production and the rest are either white collar or supervisory in nature. Yet, in India, mining continues to be perceived as a dirty, risky and heavy work that is unfit for women. Consequently, *kamin*s are put at the bottom positions without any managerial titles or senior or supervisory positions. The question that my research explores in this chapter is why women mineworkers themselves do not object to the portrayal of their jobs as hard although such portrayal goes against their own interests in the coal mining industry.

Conversations with research participants

This chapter focuses on the collieries located in the eastern part of India, which till recently dominated Indian coal production. The collieries in which the research took place are mainly located in the Raniganj region although some are located further west in the Jharia-Bokaro regions. Accessing the research participants usually happened through a process of establishing contact through mine management, local social and political leaders/key resource persons or a trade union member, after which I explained my research objectives to her and made repeated visits if she agreed to talk to me. These meetings happened at home or at points of work such as dumps or wagon loading points. Often the contexts were not 'ideal' for a research interview in the conventional sense; it was always hard to get a woman mineworker alone with lots of time to spare.

In selecting my participants, there was no conscious effort on my part to do 'scientific sampling'; however, the resultant interviews can indeed be described as a 'sample' of women workers in the collieries who were interested in my research and were willing to participate in it. In this way, I believe, it was possible to get the richest data possible, because of their interest in my research[13]. Consequently, my examinations of identity construction take place through open-ended discussions with no fixed set of questions. Casual conversations initially focused on routine everyday chores at home and at work as well as perceptions of their work. Garcia-Dungo (1999: 7) elaborated on the question of getting access to research participants in her account of ethnographic study, and following her instruction that 'the researcher should consistently flow with the direction of the "talk" riding on what people are conscious of at any moment of time', the discussions were free-flowing in nature (see also Chaudhary, 2004, for an example of such work). Because the conversations were open-ended, it was not possible to analyse the data using line-by-line analytical techniques such as those used by sociologists. Instead, particularly for the purpose of putting this chapter together, I selected those parts of conversations that related specifically to identity work. Some of these related to the nature of their work as represented to be difficult, skill requirements and their balancing traditional 'feminine' jobs such as cooking, housework and mothering as against the jobs in the collieries.

---

[13] Such questions have been raised for a long time. Marshall and Rossman (1989: 145) discuss the technicalities of designing qualitative research, and particularly address the question of trustworthiness of the findings of a research in view of the 'natural subjectivity of the researcher'. 'How can we be sure that the findings are reflective of the subjects and the enquiry itself rather than the product of the researcher's biases or prejudices?' Questions of internal and external validity, reliability and objectivity emanate from the positivist paradigms, and Lincoln and Guba (1985) proposed four early alternative constructs to more accurately reflect the assumptions of the qualitative paradigm. These are credibility, transferability, dependability and confirmability.

Excerpts

Jaba Bauri

Jaba is a young woman who works in a mine workshop coiling the large fans required to ventilate the underground galleries. We spoke in the coiling workshop. Our conversation started with the description of her returning to the job following the birth of her child. From a lower caste local community that came early to work in the coal mines, Jaba's husband too is in the same colliery. They travel together in the local bus from their *dhaora*[14] every morning. Women like Jaba formed a large segment of the workforce till about the 1930s after which several laws were enacted to prevent women from working underground and during night shifts. At the same time, laws were enacted to make the provision of crèches and separate pithead toilets, catering to the 'practical'[15] gender needs of women workers. The impact of such measures was such as to reduce their numbers greatly, although women continued to work on various surface jobs that required special skills[16].

Technically, a law introduced in the late 1930s assures her of a crèche and separate pit-head bath in the colliery, but in reality none of these exist. We talked about it for some time:

> *Jaba*: Yes, I know that these facilities should be here but then see there are so few of us working here so I guess its not that important.
> *Kuntala*: Did you feel that you could take it up with the (mine) manager?
> *Jaba*: Look, I am a good married woman; I don't want to get into trouble here. My husband wouldn't have any of it. Also, I have my folks at home to take care of my child. I know my going out to work isn't easy ... but I am working here in the workshop, not in the *khadan*, am I?
> *Kuntala*: So folks at home – do they look after your child while you're away at work?
> *Jaba*: Look, it's a job ... they do, I think... I do my job properly ... I took up the training to coil up fans and here I am. I do not blow up the money like my husband and do the job well.
> *Kuntala*: But this issue of crèche – its really such a big necessity for other women with young children – do you not feel that it should be taken up to the managers?

---

[14] *Dhaora*s were early labour housing, the *coolie* quarters, around collieries. Provided by the owners, *dhaora*s often they had poor living conditions not much better than urban squatters.

[15] Feminists have identified two distinctly different forms of gender needs: practical (such as the provision of separate toilets) and strategic (such as the provision of means of livelihood itself). See Moser's 1989 work for more on the diverse needs.

[16] *Kamin*s in Indian collieries have specialized, over the years, in several job areas such as *guli* or pellet making, loading the wagons, cleaning the siding, and such other skilled jobs in less mechanized mining processes. One important job was shell picking and sampling, which meant picking up the non-coal rocks and dividing the coal according to size. One of my first memories of visiting a colliery is that of squatting *adivasi* women workers sorting coal by hand in open dumps.

> *Jaba*: Look, I am a good woman – I stay out of trouble and do my job. It is for the union babus to find issues to fight on.

In my view, Jaba is justifying the lack of attention by the mine management to her practical gender needs while at the same time implying that it is not fit for a married woman like her to 'get in to trouble' by expressing dissatisfaction with her working conditions. She is attempting to defend herself because she feels that the researcher might be suspecting her of not being feminine enough to be working in the mines. Again, an elaborate justification regarding the reason for leaving the child at home follows that statement, reinforcing her being 'within the fold of the family'. She differentiates again between her work and that of the work in the *khadan*, where her husband is. Thus, there is an inconsistency in her statements; probably because the researcher has asked her about her child-minding and she feels that she might appear to be 'less than motherly' if she agrees to her notions of motherhood and femininity. We can see that the relationship between the researcher and the participant has influenced the nature of justification. As the less powerful party in the interaction, it is incumbent on her to provide a convincing account of the reasons for her to work in the colliery; although she would like to be portrayed as a good worker she stresses her role as a mother with family support. Possibly she posited the researcher as one perceiving her as a lower class woman because she works in the mine and hence feels obligated to resist my notions of childcare provision by stressing the family support.

Buri Kora

Buri Kora, aged over 50, has known mining since she grew up as a child in a *dhaora* around the Selector colliery[17] in Raniganj. Both her parents had worked in 'Company' collieries as part of the family unit of labour, and she too started working in one before the coalmines were nationalized. Buri now works in a coal dump as a wagon loader. Buri is one of those few women who have continued in the collieries since before its phased nationalization in 1971–1973. Buri reminisced about the life of her mother, Ashari Kora, who worked long hours on the surface. Coal used to come to the surface from the *khadan* in *dulis*[18] of the *chanock*[19], and women picked them up with fork and shovel to carry them to the wagon car in baskets over their heads. Buri remembered joining her parents at the work when she was only about 10 years. This was in the late 1960s. At that time, Buri received Rs 12-18 (US 20 cents) per week. The work was piece rate; filling up one wagon was rewarded with Rs 12.50. If 5 people worked on it, the amount was shared by

[17] I tried to trace the history of this colliery, which does not exist any more. It was located near Raniganj town, close to Mahabir colliery. The owner, according to Buri, was Sitaram Bhaluk, but his family has long since left the area.

[18] The lift.

[19] The shaft is still called *chanock* in eastern India.

them. One day saw the group filling up from 2 to 4 or 5 wagons. She described her work as it was 40 years ago and what it is now:

> *Buri*: We used to work all the time, day and night, without break. We had no fixed time for food. We took a meal of cold rice in the morning to start work by 6am.
> *Kuntala*: So most of your waking hours were spent in the colliery?
> *Buri*: Yes, of course. It was a tough life.
> *Kuntala*: Do you feel so because there was no lunch break?
> *Buri*: Around mid-day, one of the group went back to cook and brought the meal back to the colliery. We all paid for the cost. We washed our hands to start working again; often till 10-11pm at night. Yes, we had worked very hard. It was a different life.
> *Kuntala*: Yes, it indeed was very long hours. Now you have fixed hours - is that better?
> *Buri*: Well, I can never say that. How can I like it, this damn job. I am also getting on in years.
> *Kuntala*: Is there any other reason for you to not like your present job?
> *Buri*: Look, I never had children – there young girls here laugh at me, they say I am like a man. They laugh at my wrapping the gamchha (towel) on my head. They laugh at my *biri* (locally made cigars) smoking. Look, I grew up carrying the coal basket over my head, I am not like these girls.

Here one can clearly see that pride in Buri is creating an identity for herself, a coal ancestry that many local communities are proud to declare. She is not 'like others' because she has a 'colliery culture', and she has also experienced the tough days of low wages and hard work. But she does not want to recognize the present working conditions as necessarily better than the past. In fact she clearly expressed her dissatisfaction with her current colleagues. In her view, the present day women do not have the same difficulties as faced by her in the past. That gives her a masculinity that she perceives as a subject of fun-making among them. It is no longer common among women workers in *khadan*s to wrap towels over the head or smoke *biri*. She is conceptualizing her position primarily as an *adivasi* ('we Santhals'), but also as a woman. Although she is expressing pride in her 'coal ancestry', she is tired of it as tainting her interactions with her present colleagues. The fact that women miners were at the bottom of the pits' gender hierarchy too becomes visible but only if one considers the fact of her sex as an identifier. It is likely that Buri's construction of the caste/ethnic hierarchy follows from her seeing me as an upper caste woman from the town. This becomes more evident from the conversation with Dhani.

Benki Ruidas

Benki, about 29 years old, from the caste of *chamars* (leather workers), has been working as a security guard in Mahabir colliery for nine years now. Initially she worked in Bansra colliery as a wagon loader where she joined after her mother's retirement from work as a pellet-maker. Before her, Benki's grandmother too had worked in the 'company' collieries. Thus, like Buri, Benki also descended from those family labour units which came to work in the mines with women working as transporters or loaders of coal cut by the *malkattas*[20]. Benki now works as a security guard of the colliery, posted at the gate of the officeloader. I met Benki during a lunch break and sat under a tree to talk with her.

Although she has worked in a colliery for several years, Benki has not been underground. She looks at underground mining work with a curiosity. We discussed the underground for a while. She said:

Benki: Never have I been under the ground – working here for so long – I never dared. Any living thing becomes white there. There are white cockroaches and rats running about and even snakes rustle past at times. A pile of coal dust here, some logs there – but all one can see is darkness. It must be terrible down there – it must be a very smelly place.
Kuntala: How is the temperature down there?
Benki: I have heard that it is fairly damp and warm there. Some areas are warm, some are cooler – you get a sudden breeze of air as you walk past.
Kuntala: In all these years – did you not want to go and check it out for yourself?
Benki: No, didi. I am very scared of such places. The places are not right for women.

We then started talking about the fact that both of us are working women, and not the first generation in that, too. I asked Benki about her job:

Kuntala: What is your opinion of your job?
Benki: The salary is not too bad, but it is not suitable for women.
Kuntala: Why do you feel that it is not suitable for women?
Benki: Well, this job makes me stay in only one place for the whole day.
Kuntala: Oh, so you mean only your current job as a security guard does not satisfy you because you are at the gate all day?
Benki: No, no. I mean overall jobs are hard for women in the mines. This is not a place for women.
Kuntala: But you said your mother too worked in the collieries?
Benki: Well, they had no other choice but to work. We are working because we are forced to work. Otherwise, if we could sit at home, why would we not do so?
Kuntala: Did you not take transfer from being a wagon loader to seek a job that's not as demanding. So why is it hard?
Benki: Didi, we are less educated village women not like you. Still, we too want to have a comfortable life – tell me, isn't this job difficult? Would you be able to do it?

Dulu Bauri

Dulu comes from a Bauri family – Bauris being the traditional coal cutters of Raniganj collieries as described by a colonial district administrator (Paterson, 1910). Dulu joined the mines after the death of her husband who worked

[20] Literally, the term means 'cutter of stuff', although '*coolie*' is a broader term indicating all sorts of manual work performed by lowly labourers.

underground. She was a newly married girl from another village, and to her the colliery worker husband seemed from a different world. She insisted that her husband took her to see where he works. She described the day her husband took her down the shaft, not very long after their wedding:

> *Dulu*: It was not easy walking on there – I had to lift my sari and wrap it tightly, wear a helmet and take a battery and light. I went down in the *duli*, far down, same length as two palm trees.
> *Kuntala*: What did you see in there?
> *Dulu*: They were cutting the coal – making a road inside the earth. One can go a long way under the surface – even up to Topse (a nearby village). So many men were working there; everyone busy with his job. Someone was cutting the coal, someone else was loading it. Oh, it was so dark, like being in mother's womb. Someone was spreading limewater on the black wall. In one place there was a very big hole, like a pond. Four pumps were lifting water out of that. Ever since I saw that, I knew how hard his life was.
> *Kuntala*: Seeing his place of work, did you change a little bit?
> *Dulu*: Yes, I did. Ever since that day, if we had a fight, I told myself that his life has no certainty. I stopped quarrelling with him.
> *Kuntala*: Did you feel that his work was risky?
> *Dulu*: Yes, too much. Think of it: if by any chance the water fills up, people die. If the roof falls, then too people die.
> *Kuntala*: Did he die in an accident, your husband?
> *Dulu*: No, he passed away at home – had a heart attack.
> …
> *Kuntala*: So your work on the surface must be so much better than what your husband had to do underground?
> *Dulu*: Not really. All work in the colliery is dirty – do you not see all the black dust around here? When I had to come in here to work after his death, I cried and cried. There are all coarse men and lower caste women around me – is this a place for a gentle woman? I am here only because my husband is dead – otherwise I will be at home, tending to my work at home.

Dhani Mejhen

Dhani is from the *adivasi* community Santhal. Her husband died in a mine accident in Mahabir colliery about 12-13 years ago. Previous to that, she was a housewife, looking after her home and two children without much formal education. She took up the job after her husband's death, under difficult circumstances, much against the wishes of her in-laws. The in-laws had wished her younger brother in law – brother of her deceased husband – to join the colliery, but Dhani put her foot down. Although her disappointed in-laws refused to keep any contact with her after her 'unwomanly' behaviour, she not only took up the job but kept at it sending her sons to school. This must have been a minor revolution, a feminist step in a common housewife's life.

During our conversation, we talked about her present work and future plans. The Voluntary Retirement Scheme or VRS in colloquial terms has been used as an effective instrument in India in the last decade or so of liberalization of the economy to reduce the number of employees in the public sector units. Nationalized collieries took advantage of this scheme as much as did the banks. However, the coal mining industry has been actively using this instrument to release women from its workforce. The colliery manager has proposed to Dhani that she takes up VRS and stays at home. We spoke about what she wants to do in this situation:

> *Dhani*: I have been contemplating of late about taking up the VRS.
> *Kuntala*: Why do you want to take up VRS and stay at home?
> *Dhani*: Look, I have now spent my life in the colliery, this is not a place for women to be in. But still I worked here, because I had to bring up my two sons.
> *Kuntala*: So did you often feel that you are working because you 'have to'?
> *Dhani*: Yes, if my husband was alive, he would've looked after me. I would not be working here in this place of men.
> *Kuntala*: So what do you want to do now?
> *Dhani*: I would like my elder son to get my job in the colliery and take up retirement.
> *Kuntala*: And why do you want that to happen?
> *Dhani*: I am now getting on in years – I have done my bit for my sons. Now it is their time to look after me.
> *Kuntala*: So will you not miss the life here, sitting at home?
> *Dhani*: Huh, is this a life? I will be happy to leave here. This is not a place for me.
> *Kuntala*: But, your son, he will not get the same salary that you are drawing…
> *Dhani*: His incomes would gradually improve, I am certain.
> *Kuntala*: But Dhani, tell me, what are your problems at work?
> *Dhani*: Na, didi, I do not like this work – I wish I was in an office job like yours.
> *Kuntala*: If you leave, will you not miss your friends at work?
> *Dhani*: See, didi, you are from the city. You are educated. Your ways are different from ours. For us, home is the best place to be in.

Conclusion

As we can see from the above conversations, it is clearly impossible to put forth a single and objective description of women's work in the mines. Many alternative images of mining as a gendered work jostle with each other, and women too construct their own versions. When I am in the field as a researcher, I am not necessarily without a certain identity, and my participants do not necessarily ignore that. The excerpts in this chapter provide the essence of 'identity work': a conscious attempt to construct identity through various representations of the reality. These individual perceptions of the work and the communication of that knowledge to others can carry rich information about the social location of that person. Regarding life and work in the pits, identity work in my view can raise certain important issues. In my view, mining as a work is constructed in ways that enable men to assert a specific form of cultural masculinity. It is widely known that mining work as it is performed these days is no longer as demanding as it used to be portrayed, yet rules exist to exclude some women from some work in the mines.

In this study of women miners in Indian collieries I show that their construction of pit life does not differ significantly from the commonly held beliefs. Women's portrayal of working life in the *khadan*s of eastern Indian collieries and their constructions of identities showed the difficulties of producing an accurate portrayal of pit life. Examining why women miners accept a version of the mining life that compromises their interests proved to be a critical task. Why women mineworkers do not construct their roles in positive ways brought me closer to an understanding of class, caste and gender interface as experienced in the Indian collieries. The negative construction of work in the collieries during our interactions essentially involves identity work. The research interview, material which we use as data, creates a situation in which identity work is performed, due to the power relations existing not only in the mines but also between the researcher and the participant in the research, and due to the nature of research itself. Identifying the discourses produced through relations of power in the broader social context of coal mining, as well as in the more immediate context of the research interviews, is a most critical issue in field studies in gender. The process of creating an identity involves the politics of gender and is situated in particular contexts. However, the way the interview material is used as a text too can be subject of debate, but that is beyond the purview of this chapter.

References

Alvesson, M., 1998. 'Gender Relations and Identity at Work: A Case Study of Masculinities in an Advertising Agency', *Human Relations*, vol. 51, no. 8, pp. 969-1006.

Ashforth, B. E. and Kreiner, G. E., 1999. 'How Can You Do It? Dirty Work and the Challenge of Constructing a Positive Identity', *Academy of Management Review*, vol. 24, no. 3, pp. 413-434.

Ball, K. and Wilson, D. C., 2000. 'Power, Control and Computer Based Performance Monitoring: Subjectivities, Repertoires and Resistance', *Organization Studies*, vol. 21, no. 3, pp. 539-565.

Beteille, Andre, 1996. *Caste, Class and Power: Changing Patterns of Stratification in a Tanjore Village*, Oxford University Press, Delhi.

Casey, C., 1995. *Work, Self and Society: After Industrialism*, Routledge, London.

Chakrabarty, Dipesh, 1989. *Rethinking Working-class History: Bengal 1890–1940*, Oxford University Press, Delhi.

Chaudhary, Nandita, 2004. *Listening to Culture: Constructing Reality from Everyday Talk*, Sage Publications, New Delhi/Thousand Oaks/London.

Collinson, D., 1994. 'Strategies of Resistance: Power, Knowledge and Subjectivity in the Workplace', in J. M. Jermier, D. Knights and W. R. Nord (eds), *Resistance and Power in Organizations*, Routledge, London.

Deshpande, S. R., 1946. *Report of an Enquiry into the Conditions of Labour in the Coal Mining Industry of India*, New Delhi.

Dick, Penny and Cassell, Catherine, 2004. 'The Position of Policewomen: A Discourse Analytic Study', *Work, Employment and Society*, vol. 18, no. 1, pp. 51-72.

Espley, Samantha, Francis, Helen and Castonguay, Marie-Josee, 2004. *Hardrock Mining and the Evolution of the Roles for Women*, Copper Cliff, Ontario, Inco Limited.

Eveline, Joan, 1995. *Stories of Heavy, Dirty and Limp: Protecting the Institution of Men's Work*, Administration, Compliance and Governability Program, Research School of Social Sciences, The Australian National University.

Foucault, M., 1977. *Discipline and Punish: The Birth of the Prison*, Penguin, London.

Foucault, M., 1981. *The History of Sexuality: Volume 1, The Will to Knowledge*, Penguin, London.

Foucault, M., 1988. *The History of Sexuality: Volume 2, The Uses of Pleasure*, Penguin, London.

Foucault, M., 1990. *The History of Sexuality: Volume 3, The Care of the Self*, Penguin, London.

Furumura, Eriko, 2003. 'The Heritage that the Organization of Miner's Wives Left in their Community: Especially about "Tanpukyou"', Paper presented to the *International Mining History Conference*, Akabira, Hokkaido, Japan, September.

Garcia-Dungo, Nanette, 1997. *The Social Construction of Subordination/ Domination among Women in Urban Poor Households*, Monograph of College of Social Sciences and Philosophy, University of the Philippines-Diliman, Quezon City.

Ghosh, A. B., 1990. *Coal Industry in India: An Historical Analytical Account*, Sultan Chand and Sons, New Delhi.

Helve, H. (ed.), 1998. *RYPE Interim Report 1 and 2*, University of Helsinki, Helsinki.

Henriques, J., 1998. 'Social Psychology and the Politics of Racism', in J. Henriques, W. Hollway, C. Urwin, C. Venn and V. Walkerdine (eds), *Changing the Subject: Psychology, Social Regulation and Subjectivity* (2nd edition), Routledge, London.

Henriques, J., Hollway, W., Urwin, C., Venn, C. and Walkerdine, V. (eds), 1998. *Changing the Subject: Psychology, Social Regulation And Subjectivity* (2nd edition), Routledge, London.

Heritage, M., 2000. *Chasing Rainbows in the Desert*, Frontier Press, Andamooka.

Heyes, Cressida, 2000. *Line Drawings: Defining Women through Feminist Practice*, Cornell University Press, Ithaca and London.

Hughes, E. C., 1951. 'Work and the Self', in J. H. Rohrer and M. Sherif (eds), *Social Psychology at the Crossroads*, New York, Harper Brothers, pp. 313-323.

Jermier, J. M., Knights, D and Nord, W. R., 1994. *Resistance and Power in Organizations*, Routledge, London.

Kaur, Amarjit, 2001. 'Labour Dynamics in Plantations and Mining: An Historical Perspective', in Rebecca Elmhirst and Ratna Saptari (eds), *Changing Labour Relations in South East Asia*, Curzon, London.

Kideckel, David, 2004. 'Miners and Wives in Romania's Jiu Valley: Perspectives on Postsocialist Class, Gender and Social Change', *Identities: Global Studies in Power and Culture*, Jan-March, vol. II, Issue 1, pp. 39-63.

Lincoln, Y. and Guba, E., 1985. *Naturalistic Enquiry*, Sage, Beverly Hills, CA.

Marshall, Catherine and Rossman, Gretchen B., 1989. *Qualitative Research*, Sage, New Delhi.

Morris, Meaghan and Patton, Paul, 1979. *Micheal Foucault: Power, Truth, Strategy*, Feral Publications, Sydney.

Moser, Caroline, 1989. 'Gender Planning in the Third World: Meeting Practical and Strategic Gender Needs', *World Development*, 17, p. 2.

Murty, B. S. and Panda, S. P., 1988. *Indian Coal Industry and the Coal Miners*, Discovery Publishing House, New Delhi.

Pandian, M. S. S., 2002. *One Step outside Modernity: Caste, Identity Politics and Public Sphere*, International Institute of Social History, Sephis-Codesria, Amsterdam and Dakar.

Paterson, J. C. K., 1910. *Bengal District Gazetteers: Burdwan*, Bengal Secretariat Book Depot, Calcutta.

Ray Chaudhury, Rakhi, 1996. *Gender and Labour in India: The Kamins of Eastern Coalmines, 1900–1940*, Minerva, Calcutta.

Rothermund, Dietmar and Wadhwa, D. C. (eds), 1978. *Zamindars, Mines and Peasants: Studies in the History of an Indian Coalfield and its Rural Hinterland*, Manohar Publications, New Delhi.

Sen, Samita, 1999. *Women and Labour in Late Colonial India: The Bengal Jute Industry*, Cambridge University Press, Cambridge.

Sen, Samita, 2001. *Gender and Class: Women in Indian Industry, 1920–1990*, Publication no 16, NOIDA, UP, Indian Association of Labour History Institutions.

Stewart, P. J. and Strathern, Andrew (eds), 2000. *Identity Work: Constructing Pacific Lives*, Association of Social Anthropology in Oceania Monograph Series, University of Pittsburgh Press, Pittsburgh.

Townley, B., 1994. *'Reframing Human Resource Management': Power, Ethics and the Subject at Work*, Sage, London.

Yoshida, Kayoko and Miyauchi, Reiko, 2003. 'Invisible Labour: Comparative Oral History of Women in Coal Mining Communities of Hokkaido, Japan, and Montana, USA, 1890–1940', Paper presented to the *International Mining History Conference*, Akabira, Hokkaido, Japan, September.

PART III
Gender in the Mining Economies

Chapter 9

The Place of Women in Mining in the Cordillera Region, Philippines[1]

Minerva Chaloping-March

Introduction

This chapter describes the nature and extent of women's work in mining and the various factors that explain why their participation in the sector expands or contracts. Mining in this chapter refers to two types of operations in terms of scale and organization: 'formal' corporate large scale and 'informal' small scale (Casambre et al., 1987: 143). The former, which began in the Cordillera during the early part of the 20th century, is the enterprise of big mining companies and requires substantially large capital investments and highly mechanized technology. The latter, a centuries-old livelihood in some parts of the region, is carried out by individuals or group operators who may be organized locally usually involving kinfolk and in some instances a number of non-kin members. Large scale mining is described here as 'formal' owing to its being recognized as having a legal status to operate. Small scale mining is often associated with the idea of being 'informal' because the operators, in many cases, do not have a law-sanctioned claim over the areas where they carry out their activities. The notion of its being informal is also linked to the situation that not all small scale miners' production are reported and officially accounted for hence they form part of the 'underground' economy.

The chapter first presents a brief overview of the mining sector in the Philippines to provide a context for the subsequent sections. The chapter then focuses on three themes: a) women in the mining profession; b) corporate large scale mines in the Cordillera region and women's work; and c) small scale mining and women's work, and d) the relationship between large scale corporations and

---

[1] I am indebted to several people who have generously provided important information and comments for this contibution: Mrs. Redempta P. Baluda, Engr. Leonardo P. Josef, Engr. Lomino Kaniteng, Engr. Artemio F. Disini, Mr. Donald Sotero, Mrs. Alicia Follosco, Engr. Florencio Felipe, Engr. Roger Awingan, Dr. Marietta de Leon, Mr. Bryan Yap, Engr. Eduardo Austria, Dr. Carolyn McCommon, and Engr. Felizardo Gacad Jr.

small scale operations. The chapter concludes with a discussion on the prospects for women in the sector.

The data for this chapter is drawn from the author's ongoing research which looks into the interrelationships of large scale mines, small scale miners, mining communities and other key players (such as civil society organizations and the Government) in dealing with social impacts associated with mining in the Philippines. The chapter looks at three mining companies one of which is the oldest in the Philippines while the two others are the country's only remaining large scale mines. All three companies are found in the Cordillera region, northern Philippines.

Mining in the Philippines

The rise of the mining sector

The Philippines is favoured with rich and diverse mineral resources. Limestone, marble, rock aggregates, dolomite guano, sand and gravel, silica sand, feldspar and industrial gemstones are among the non-metallic and industrial minerals, which abound in the country. Gold and copper are the key metallic products followed by nickel and chromite. Owing to an extensive and highly developed tradition of mining, the country had been among the world's prime producers of these four metallic minerals (see Cambalda et al., 2002: 1).

Modern mining commenced in the Philippines during the early part of the 20th century with the introduction of American capital and technology. As early as the 1930s, there were 40 mines operating, mostly for underground gold vein deposits, throughout the country (Domingo, 2003). By the 1950s, large scale open-pit mining for low-grade porphyry copper deposits was opened. With ten large copper mines in operation, the Philippines was propelled to become one of the world's top ten producers during the following two decades. From the 1960s to the 1980s, the Philippines had as many as 70 medium to large mines producing gold, silver, copper, chromite, nickel, zinc, manganese and other metallic and non-metallic minerals (Domingo, 2003). Expansive exploration and development of copper and gold mines took place from the 1960s to 1970s towards launching the summit of the mining industry such that during the period 1970 to 1981, mining had its highest contribution to total exports, ranging from 15 to 25 per cent (MGB, 2004a).

Decline and stagnation of the mining sector

From the late 1980s onwards, the mining sector has exhibited a considerable decline. Although gold and copper prices in the international market soared in 1980, they dropped drastically the following year. Rising costs of operations, the devaluation of the Philippine peso that escalated interest rates, and the marked-up taxes for mineral products all contributed to the sombre business climate leading to huge losses and ultimately closures of a significant number of major mines (MGB, 2003a). With closures and suspended operations taking place from the mid-1980s to the 1990s, corporate large scale mining, in particular, has shrunk significantly.

The industry somehow posited a consistent growth, albeit at a sluggish pace, from 1989 to 1995. The Philippine government has endeavoured to rejuvenate mining through the legislation of the Mining Act of 1995 whose key provisions include allowing up to 100 per cent ownership of mining operations by foreign companies. Before this law was promulgated, foreign-equity ownership was confined to a maximum of 40 per cent. With the goal of drawing the much-needed capital to charge up the industry, the government opened – in full – for the first time, the investment avenue to foreigners. The move attracted about 20 of the world's largest mining companies to establish offices in the country and file applications for various mining tenements the following year (Domingo, 2003). However, developments for many mining projects were slow due to a combination of policy, technical, financial and environment-related factors (MGB, 2003a). These include the long-drawn-out permitting process, policy difficulties arising from the challenge posed by mining critics to the constitutionality of certain provisions of the Mining Act, overlapping coverage of several laws, and mounting public concerns over the environmental and ecological impacts associated with mining.

For over a decade now, no new major mining project has been opened in the Philippines. Whilst there are currently four medium-sized nickel laterite mines, three medium scale chromite mines, and three medium scale gold mines, the sector is left with barely one large scale gold mine and a lone large scale copper mine (MGB, 2003). Both are located within the Cordillera region. At present, the most critical issue faced by the mining sector is the Supreme Court decision, issued on 27 January 2004, that invalidated the provisions of the Mining Act permitting foreign-controlled corporations to have more direct participation in exploring, developing, and utilizing mineral resources in the Philippines.

Share of small scale mining

Small scale mining feeds a substantial share to the country's overall mineral production. With an aggregate PhP33 billion[2] generated by the mining sector in 2002, 43 per cent is shared by small scale miners which is much larger than that generated by large scale metallic mining at 20 per cent or non-metallic mining at 36.67 per cent[3]. Whilst there are currently over 2,000 quarries and small scale mines of various commodities (MGB, 2003), the commodity that has attracted huge numbers of small scale miners, including women, is gold. Until the 1970s,

---

[2] Roughly US$590 million (at US$1= PhP56.08).

[3] Computed from MGB, 2004c. The Mines and Geosciences Bureau recorded gross production by source of contributions: a) metallic mining, b) small scale mining, and c) non-metallic mining.

small scale gold mining was confined to infrequent panning along riverbanks involving gold recovery with the use of a crude sluice box. In the early 1980s, the soaring price of gold at US$800/oz and the depressed socio-economic conditions in the Philippines drew many families to small scale mining (Cabalda et al., 2002). The increasing number of small scale miners was further boosted by the discovery of the Mt. Diwalwal gold in the southern province of Davao del Norte that soon engaged tens of thousands of people and panning activities in other provinces were invigorated. Several studies have estimated that small scale miners in the country range from a peak of 500,000 in the mid-1980s to about 200,000 at present (MGB, 2004b).

Women in the mining profession

Mining engineers

Among the professions, mining engineering is the most direct means of entry into mining (Espley et al., 2004). In the Philippines mining engineering has essentially been a career pursued almost exclusively by men. As can be gleaned from Table 9.1, the number of registered mining engineers from 1927 to 2003 indicates the extremely faint presence of women. In a span of 76 years, the profession – on the whole – has drawn only 40 women. There was barely one female mining engineer during the period 1941–50, three in 1961–70, and five in 1971–80. The highest number of new female registrants was in 1981–90. However, just as the field has attracted increasing numbers of women, the mining industry started its decline.

The aggregate number of mining engineers has been consistently rising from 1951 to 1990 although the new registrants each year are virtually all men. From 1991 to 2000, however, there was a drastic drop of new mining engineers even as the ratio of women in relation to men has been rising. By 2003, there were merely a dozen new mining engineers including two women. It is most probable that the mining industry would have seen more women pursuing mining engineering as a career, hence creating a different male-female ratio at present, had it not been for the closure of major metallic mines in the 1990s. Instead, many male mining engineers themselves were laid off and had to seek employment in other professions. As one newly inducted female engineer observed:

> It feels good having finished my course, passed the board exams, even making it to the top. I am now a licensed mining engineer. But this is just the start. I am not sure if there is any good opportunity for work. I hope the mining industry in the country peaks up soon. Or, would you have any advice about opportunities abroad?[4]

[4] Myrna (23 years old), 29 August 2003.

The bleak condition of the mining industry gives men or women, a reason to hesitate in pursuing a career in mining. With the poor prospect for jobs, there has been a significant reduction in the number of schools offering courses associated with mining due to very low number of enrollees (Shea, 2002).

Table 9.1 Number of registered mining engineers in the Philippines, 1927–2003[5]

Year	Female	%	Male	%	Total
1927–1940	0	*0*	228	*100*	228
1941–1950	1	*1.5*	66	*98.5*	67
1951–1960	0	*0*	170	*100*	170
1961–1970	3	*0.6*	459	*99.3*	462
1971–1980	5	*0.7*	665	*99.2*	670
1981–1990	16	*2.0*	774	*98.0*	790
1991–2000	13	*4.7*	266	*95.3*	279
2001–2003	2	*15.5*	11	*84.6*	13
TOTAL	40	*1.49*	2,639	*98.5*	2,679

Geologists

A close adjunct of mining engineering is geology. Similarly, it has been perceived as a 'masculine' discipline. The increase and decrease of women geologists reflects the ascent and waning of the mining sector. This can be shown by the number of geology graduates from the National Institute of Geological Sciences, University of the Philippines, one of the country's four remaining institutions offering mining-related courses (Table 9.2). The decades of the 1960s to late 1980s which marked the zenith of the mining industry was also the period when there has been a steady rise in the total number of geology graduates. The ratio of women geologists relative to men has been rising but the actual numbers have been falling. Geologists were once the envy of many job seekers especially shortly after the enactment of the Mining Act due to intensifying exploration activities. But with the lack of alternative opportunities at present, many geologists have found themselves seeking academic jobs, business ventures, and short-term work in consulting (Flores et al., 2002).

I now turn to the mines in the Cordillera, a mountainous mineral-rich region where mining has shaped the *modus vivendi* of the people. The following section first provides a description of the corporate large scale mines operating in the region before discussing the specific types of work undertaken by women within

[5] Summarised by the author from Philippine Society of Mining Engineers (PSEM). (2003). The list provided by PSEM does not include 2002 registered engineers pending approval of the Professional Regulatory Commission.

these mines. The section shows that the interplay of mining technology with tradition and culture helps account for the lingering 'macho' undertone of mining. What is perceived as a typically male role has implications for the scope of work in mining available for women.

Table 9.2 Geology graduates, National Institute of Geological Sciences, University of the Philippines, 1957–2003[6]

Year	Female	%	Male	%	Total
1957–1960	0	*0*	24	*100*	24
1961–1970	6	*5.2*	109	*94.8*	115
1971–1980	27	*13.1*	179	*86.9*	206
1981–1990	64	*22.3*	223	*77.7*	287
1991–2000	59	*35.8*	106	*64.4*	165
2001–2003	29	*50.9*	28	*49.1*	57
Total	185	*21.7*	669	*78.3*	854

Corporate large scale mines in the Cordillera and women's work

The Cordillera region is located in northern Luzon, one of the major island groups of the Philippines. It occupies an area of almost 2 million hectares, about 6 per cent of the total land area of the Philippines. The region is distributed into six provinces (Abra, Apayao, Benguet, Ifugao, Kalinga and Mountain Province) and one city (Baguio). The 2000 national census places the region's population at 1.36 million, a slim 1.78 per cent of the country's population (NSCB, 2004). Whilst mineral resources are found in many places throughout the Philippine archipelago, the Cordillera region has been the location of the country's pioneer gold and copper mines, all Filipino-owned, which have been among the country's leading producers for many decades (see Figure 9.1).

Benguet Corporation

At the heart of the Baguio mineral district is the town of Itogon where the pioneer mining company in the Philippines, Benguet Corporation, is located. The company was established in 1903, initiating modern mining of gold. It was considered 'the richest gold mine in the orient' as early as 1921 (Gomez, 2003). It had been the country's largest gold mining company and the 16th largest in the world until the late 1980s (Benguet, 2003). It had several operating underground mines namely Antamok, Acupan, Kelly, Atok (Benguet Atok Contract Operation in Ampucao) and Itogon-Suyoc from the mid-1980s until 1996, and a total employment of as

[6] Summarized by the author from the alumni and student records of the National Institute of Geological Sciences, University of the Philippines, June 2004.

many as 6,000 miners and engineers at the company's peak operation in 1986 (Benguet, 2004). Benguet Corporation's 100+-year existence clearly demonstrates not only its decisive presence in Itogon but also the enormous gold reserves in the district that allowed the company to persist throughout an extensive period. Following a re-structuring of its operations in view of the bleak business climate of the 1990s, the company suspended its Acupan mine in 1992 and its Antamok mine (Benguet Antamok Gold Operation) in 1999 and diversified into water projects, real estate development, forest management and ecotourism (Benguet, 2003). In 2001, the company reopened its Acupan mine and contracted out production activities to groups of small scale miners, many of whom were laid-off miners of the company. The reopening of the Acupan mine has made possible the retention and maintenance of about 100 security and support services personnel of the company (Benguet, 2003).

Philex Mining Corporation

Adjacent to Benguet Corporation is Philex Mining Corporation's Padcal mine in the town of Tuba. Philex was established in 1955 and is the first block cave operation in the Far East. Philex is a primary producer of copper with gold and silver as by-products. Padcal mine is the only ISO 14001-certified mining company in the Philippines and is presently the country's sole remaining large scale copper mine. It employs 2,944 people who, with their families, make up a community of about 17,000. The last couple of years have ushered the Padcal mine into extracting lower ore grades. With this and the mine's depleted ore reserves, Philex is now planning way ahead for the closure of Padcal mine. The company's strategic mine plan projects Padcal to operate until 2011 (PMC, 2004).

Lepanto Consolidated Mining Company

More than a hundred-kilometre north of Itogon is the Mankayan mineral district where Lepanto Consolidated Mining Company is located. The company began operation in 1936 and built the country's first copper plant. It has operated as a copper-silver mine, maintaining the country's highest mill capacity of 4,500 tons per day until 1996 when the company shifted to gold mining (LCM, 2002). With the company's 2,104 employees, the mining community has a population of 12,000. Lepanto's significant discovery of the Victoria high-grade gold deposit in September 1995 has given the company the prospect for a twenty year-operation (LCM, 2004). The company's Victoria Project is currently the sole large scale gold mine in the Philippines.

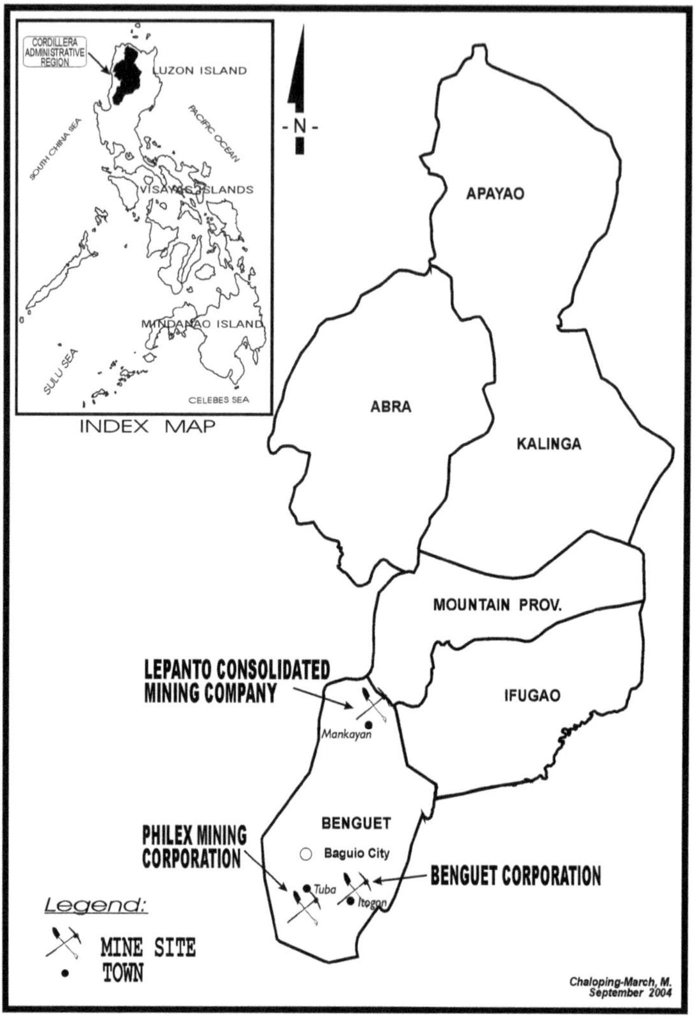

Figure 9.1 Large scale mines in the Cordillera

Employment ratio and types of work held by women

Using comparable data compiled by the Gender and Development Program of the Mines and Geosciences Bureau – Cordillera Administrative Region (MGB-CAR), Tables 9.3 and 9.4 show a snapshot of the employment profile of Philex and Lepanto in 2000[7]. In terms of aggregate manpower in the large scale mines, men

[7] MGB-CAR has no data on Benguet Corporation due to the company's suspended operations leaving only about 100 maintenance and security personnel.

occupy almost all jobs with 90.6 per cent in Philex and 96.1 per cent in Lepanto. Women employees in Philex make up 9.4 per cent, equivalent to slightly less than one tenth, of total employees. In Lepanto, women constitute a sparse 3.8 per cent of the total.

Table 9.3 Classification of work between male and female employees Philex-Padcal Mine, June 2000

Type of Work	Female	%	Male	%	Total	%
Managerial	12	*14.6*	70	*85.4*	82	*2.8*
Production	39	*2.0*	1,928	*98.1*	1,967	*66.8*
Administrative	225	*25.1*	670	*74.9*	895	*30.4*
TOTAL	276	*9.4*	2,668	*90.6*	2,94	*100*

Source: MGB, 2000

Table 9.4 Classification of work between male and female employees Lepanto Consolidated Mining Company, June 2000

Type of Work	Female	%	Male	%	Total	%
Managerial	4	*6.7*	56	*93.3*	60	*2.8*
Production	12	*0.7*	1,786	*99.3*	1,798	*85.5*
Administrative	65	*26.4*	181	*73.6*	246	*11.7*
TOTAL	81	*3.85*	2,023	*96.1*	2,104	*100*

Source: MGB, 2000

The largest number of jobs undertaken by women is essentially in administration with 26.42 per cent in Lepanto and 25.14 per cent in Philex. Production jobs were not specified in the classification of work provided by MGB-CAR. Nonetheless, data from Lepanto and Philex indicate that production roles involving women include strategic mine planning, processing-related functions in the mill, operation and maintenance jobs, and support tasks such as exploration research. The overall picture shows that the number of women filling these roles is very small, with barely 2 per cent in Philex and a negligible 0.67 per cent in Lepanto. Most of the production jobs involve excavation and are carried out by men, particularly miners and muckers.

The significance of women's work in large scale mining should be viewed not solely in terms of their number. It is important to look at their participation in terms of the specific echelon which they occupy within the organization. It is noteworthy that women are a visible part in the chain of command and responsibility within the corporate large scale mines as indicated by their holding of senior management positions. There are 12 women in the middle and top management of Philex, or one seventh of the company's management team. Such a fraction looks insignificant but is considered relatively sizable because no other company in the country's mining industry has ever had that segment of women occupying top positions[8]. Lepanto's women managers make up a modicum (6.67 per cent) of the company's management positions. The MGB-CAR does not include the employment data of Benguet Corporation, which currently maintains a heavily reduced manpower to oversee the company's Benguet District. However, it is notable that the current Vice President is female.

Whilst a large percentage of jobs held by women are in administration, the placement of women in strategic and important positions within the planning and processing spheres of large scale mining should not be overlooked considering that these functions used to be occupied solely by men. Table 9.5 provides information on the specific work areas where women are found as well as the level of roles they occupy. Out of 90 women employees in Lepanto in 2004, 67 (74.4 per cent) are 'staff', compared to 23 (25.6 per cent) who hold rank-and-file positions. Staff positions involve leadership and supervisory roles. The bulk of women are found in administrative services (24), and finance services (15). However, they also have a relatively strong presence in metallurgical services (9). A few are placed in mining (4), technical services (1), milling (1), and operations and maintenance services (1). The specific types of jobs in these various areas include women as chief chemist, analytical chemists, chief metallurgist, senior geologist, program (Gemcom) engineers, Gemcom geologists, and material test/instrumentation engineers[9]. 'Staff' roles involve, among others, duties such as overseeing other workers including men. A similar set of data was not available from Philex but information from the company indicate that key production and processing positions are filled by women with 20 mill operators and 12 chemists and research engineers.

Women's largest share in the mining sector is in performing the largely 'conventional' women jobs such as secretarial, clerical and administrative functions. However, their skills have also found deployment in the technical fields, which remain predominantly the sphere of men. Although very slow in pace, women are making gradual inroads into the core functions of mining operations. Whilst this is the case, there remain a few factors which continue to cast, to some degree, a masculine valance to the field of mining. One of these factors is the type of mining method employed by mining companies. The other is the taboo associated with women although this is true in only one mine. A short discussion of these factors in the following section is warranted.

---

[8] Redempta P. Baluda, Division Manager – Environment and Community Development, Philex-Padcal Mine. Personal communication (E-mail), 3 June 2004.

[9] From data provided by Lepanto Consolidated Mining Company, 'Job Titles of Female Employees Lepanto Consolidated Mining Company – Lepanto Mine Division', May 2004.

Table 9.5 Summary of manpower resources by functional groups Lepanto Consolidated Mining Company, May 2004[10]

Group	Women			Men			Total
	Staff	Other	Total	Staff	Other	Total	
General Management	4	0	4	27	21	48	52
Administrative Services	24	*23*	47	30	134	164	211
Finance Services	15	0	15	22	54	76	91
Environment & Social Develop.	6	0	6	7	5	12	18
Technical Services	3	0	3	34	173	207	210
Metallurgical Services	9	0	9	5	46	51	60
Operation Maintenance & Services	1	0	1	46	218	264	265
Milling Group	1	0	1	12	60	72	73
Mining	4	0	4	118	1,123	1,241	1,245
Total	67	23	90	301	1,834	2,135	2,225

Type of mining methods in large scale mines

Operating a mine involves both surface and underground methods. All mines in the Cordillera have employed underground methods, with the exception of the Antamok open pit that operated from 1992 to 1999, the Western Minolco Mines which shut down in the mid-1980s, and the Philex-Padcal Mine in 1958-59 (Baluda, 2003). Mining in the Cordillera has commonly been known to consist of tunnels and excavations. Since tunnels require heavy rock and earth works, women are seen as unsuitable to handle the activities thus women have never been included as workers. The nature of the tasks required is considered too strenuous for women. Letting women work in underground areas is deemed insensible because of the risks associated with drilling, blasting and shovelling. The following observation of a mining engineer who used to work at Benguet Corporation's

[10] Summarized by the author of 'Lepanto Consolidated Mining Company – Lepanto Mine Division Manpower Summary by Groups Ending May 2004', 28 June 2004.

Balatoc mine indicates how women are regarded in terms of involving them in underground work:

> We give very high respect to our women here in the Cordillera and in the Philippines in general. Women are to be protected and cared for. Why dare allow them to be exposed to the dangers underground? There's blasting, there's too much noise, there are fumes, there are accidents. Sure, women are as able as men. But that is not the issue. The issue here is about wantonly endangering women. The most difficult and dangerous jobs such as those performed by miners and muckers are meant to be for men, not at all for women[11].

Similarly, an engineer and recently retired senior officer of Philex remarked:

> I agree that mining is not a typical workplace for women, that is, if the workplace refers to the actual underground excavation site. This is due to the risks associated with it and also due to the physical exertion required of the job. However, mining cannot be done without the support services like geology, finance and accounting, metallurgy, medical, human resource, legal, environmental management, assaying, and others. These jobs, though not always, are known to have been ably handled by women[12].

If the mines in the Cordillera were open pit operations and required earth-moving methods such as truck-and-shovel excavation, jobs like truck driving could be conceivably available to women. However, there are scores of hundreds of experienced men, including laid-off miners, who are at hand to be recruited for such jobs. Thus, employing women for excavation jobs is not a move that mining companies would ever be enthusiastic to take. This was demonstrated in the seven-year open pit operations of Antamok where few women staff, although allowed to visit the pit, ever took part in the excavation work.

Taboo on women's entry into underground areas

The idea of mining being a man's territory interweaves with certain beliefs that women are not supposed to be entering underground work areas. In Lepanto, prohibiting women from entering underground workings has become an established rule. Among the Kankana-ey in Mankayan, stories have been told and re-told that the gold underground is owned by the *anito* (ancestor gods). People believe that specifically in Lepanto, the gold underneath is being protected by a lady *anito*. She becomes jealous if women are allowed into the underground workings and gets provoked to bring a curse in the form of an untoward incident. In 1989, the Resident Manager of Lepanto issued an official order prohibiting women from the tunnels following a couple of disasters which are believed to have been caused by the violation of the Kankana-ey custom.

---

[11] Interview with Engr. Roger Awingan, 19 June 2004.

[12] Engr. Leonardo P. Josef, personal communication (E-mail), 8 June 2004.

Miners and mining officers point to some incidents that they associate with the resentment of the spirit goddess. In 1989, for instance, a group of company guests including three women visited Lepanto and entered the Tubo underground shaft. After the visit by the guests, the Resident Manager – cognizant of the Kankana-ey belief – requested a *mambunong* (native priest) to perform the necessary ritual to appease the spirit goddess. The ritual was accordingly carried out. Shortly after the day the ritual was performed, the *mambunong* died in a vehicular accident. A few days after, three miners were killed in an unexplained accident in the Tubo underground shaft. People have speculated that the separate tragic accidents – the death of the *mambunong* and that of the three miners – could not be a plain coincidence. Also, people correlate the number of miners killed to the number of women who visited the underground workings earlier. In respect of the custom, the Resident Manager thus issued an official order that no women should ever go underground. That order is still enforced until the present[13]. Such a formal policy undoubtedly gives little chance for women to hold a job that involves entering the underground work areas.

Unlike Lepanto, Benguet Corporation and Philex do not have policies prohibiting women in the tunnels. A few engineers who used to work at Benguet Corporation's Balatoc and Acupan mines recall that women were not welcome to the underground areas simply because there are many risks. In addition, miners and muckers who perform most of the jobs within the underground workings usually do not wear clothes except underpants due to the extremely hot temperature. Under such circumstances, women entering the tunnels will undoubtedly cause distraction and commotion, if not accidents as a result. Thus far, the only women who used to enter the tunnels consisted of one or two engineers conducting safety inspections. In Philex, women visitors frequent the tunnels for mine tours.

Women's work in small scale mining

This section focuses on women's work in small scale mining. Generally carried out as a family or kin-based enterprise, small scale mining involves women working together with men and sometimes assisted by young children. Through small scale mining, non-formally employed women find opportunity to earn cash and contribute towards the household income. As the section will show, women are ubiquitous in all phases and activities.

A centuries-long livelihood

Long before the establishment of corporate large scale companies, mining – although in incomparably smaller scale – has always been an important livelihood

[13] Disini, Artemio F., 'Why Women Are Not Allowed Underground at Lepanto Mines: Reply to the Query of Minerva Chaloping-March', mimeo, 2004.

in many parts of Benguet, frequently serving as a complementary activity to farming (Chaloping, 1991). Prior to the arrival of the Spanish colonizers in the 1500s, the Igorots – particularly the Kankana-ey and the Ibaloy – had already been engaged in mining. The Igorots adorned themselves with gold necklaces, earrings, rings, and other ornaments. Employing simple tools and methods for extraction and processing, both men and women engaged in lode and placer mining (Wilson, 1965). They exchanged copper ingots and coins with jars and other wares from Chinese traders. They also bartered gold with salt, blankets and some animals with the Ilocano from the lowland provinces. During the Spanish colonial rule, it was this exchange medium used by the highlander Igorots in trading that drew successive attempts by the Spaniards to encroach into the Cordillera (Casambre et al., 1987: 154). However, it was the Americans who effectively commenced the full scale mining in the province. The Americans prospected for gold not by investigating and tracking down the mineral veins but by befriending the Igorots who eventually guided the former to the lode outcrops where traditional mine workings were found (Lopez, 1992: 11).

Small scale miners in the Cordillera are estimated to be 10,000, the majority of whom are in the province of Benguet with concentrations in Itogon where large scale mining concessions and mineral areas are found in almost all of its *barangay* (villages). Small scale mining serves as the main livelihood source for many households in Itogon and to some extent, Mankayan. For other households, small scale mining supplements regular work such as formal employment or farming (Bugnosen, 2002). For the year 2002, the Mines and Geosciences Bureau records that small scale miners in the Cordillera produced 4,350 kilograms of gold, about 66 per cent or a substantial two-thirds of the combined production of Philex (3,030 kg) and Lepanto (3,500 kg) (Gales et al., 2003).

Mining activities

There are two modes of small scale mining: tunnelling and panning. The former, which is also referred to as 'pocket mining' involves miners working on quartz veins by driving adits or caverns locally known as *usok*, or by burrowing shallow exterior workings called *abucay* (Baluda et al., 2001). Panning, which is known by several local terms, *sayew*, *barkis*, *kudkod*, basically involves retrieving gold deposits from the riverbed with the use of sluice boxes lined with jute sacks or soft cloth that will eventually trap gold particles (Chaloping, 1991). In both modes of small scale mining, the productive role of women is inevitable and considerable as they dexterously carry out the most meticulous and tiring tasks.

Hauling and crushing ores

Men, not women, undertake tunnelling as, according to custom, women are not allowed to go underground. The reason, as implied in local anecdotes, is based on the belief that the graded vein of the precious metal becomes difficult to track down. Another belief is that workers become prone to accidents when women enter underground. The practical explanation is that ore extraction demands extremely heavy manual work which is considered exceedingly arduous for women. The observance of this rule has slackened over the years, hence women and children are no longer as restricted to the tunnels as they used to be 10 to 15 years ago. In some parts of Itogon and Mankayan, there are a few small scale mining areas where women have always freely entered the tunnels. However, the tunnels where they get in are generally much shallower than the excavation workings of male miners.

Women engage in all the subsequent activities following ore extraction and transporting them from underground workings to the surface. They await the ores at the portals as men bring them out. Women put the ores into sacks and carry them to the location of crushers and ball mills which are usually located within or near the mill owner's house. From hauling and stockpiling, women move to the subsequent task of crushing. They break the ores manually with the use of sledgehammers and rubber rings to prevent the ores from splintering. The ores are reduced to at most one-fourth inch or the size of a pea. Women usually sort the ores manually while crushing is done. Thereafter, a large tubular rock called *alinteg*, which is applied as a roller, serves as a secondary crusher. Given the increasing number of jaw crushers, this implement is now seldom used.

Milling and grinding

From crushing, women proceed to milling and grinding in which the broken ore is pounded further into fine slime. An increasing number of households rely on ball or rod mills replacing the laborious task of grinding the ore by hand. Ball or rod mills are run by fuel-driven motors and sometimes by water mills. Some miners power their ball mills through electric motors (Bautista, 2004). The mills are used mainly for crumbling *paltek* (high-grade ores) or re-grinding *linang* (panned tailings). The ground ore – almost in fine dust form – is then combined with water and the resulting slime is passed through a sluice box. Depending on how much gold has been obtained, women work over the tailings twice or thrice or even more.

Gold concentration and recovery

Both men and women are involved in the recovery of the precious metal through gravity concentration which is the most common method among small scale miners in the Cordillera. More recently, a growing number of miners have adopted carbon-in-pulp (CIP) or carbon-in-leach (CIL) methods in which activated carbon is mixed in the cyanided pulp feed so as to suck in the gold in the solution. The gold-containing activated carbon is subsequently stripped of its absorbed gold by washing it with a solution processed by cyanidization. The derived solution is then subjected to electro-winning to eventually recover the gold (Akagi et al., 2000).

Panning

Women carry out three modes of panning depending on the location of their workstation (Casambre et al., 1987). First, *ampoc* involves collecting sand from the riverbed and sluicing the tailings. This is the method undertaken in drier work sites. *Ap-ap* ('laying down cloth or sacks') or *kaang* ('canal') is another method which involves the construction of water trenches on the riverbed where a continuous flow of water is available. Sacks are lined on the canal and left for one to several days, depending on the gold content, to trap tailings. In Itogon, the work sites for this type of panning are the canals through which the mine tailings of Philex pass before they are eventually piped into the company's tailings containment dam. The tailings acquired are thereafter panned. *Suyap* involves collecting the foam produced when the mining company releases tailings-containing water. The substance is subsequently sluiced and panned (Casambre et al., 1987).

Outside the province of Benguet are small scale mining areas that are relatively newer. Compared to the centuries-long track of Itogon and Mankayan, the enterprise of small scale mining in the other areas involves a lesser number of women mainly because it is still relatively sporadic. The localities where small scale mining has become increasingly important since the mid-1980s to 1990s include the more remote villages like Baay-Licuan (Abra province), Fidelisan (Sagada, Mountain Province), as well as Balbalasang, Balbalan and Pasil (Kalinga province). In early 2004, the village of Bashoy in Kabayan (Benguet province) has drawn many miners from Itogon due to some ore finds.

In Mainit (Mountain Province) where small scale mining has been practiced as early as people can remember, women's involvement in small scale mining remains very much indirect such as selling cigarettes and liquor to miners. One woman who is the wife of the financer in the operation of a few tunnels helps her husband in handling money involved in mine operations. Occasionally, there are women to whom male miners give 'dumps' which the former can re-work one or more times. These dumps have already been processed but still contain relatively substantial gold. A few women engage in panning but, generally, work in the rice fields usually takes priority. Hence, it is not unfamiliar for some women to say, 'I will do the panning later. I have a lot to do in the rice fields. The gold can wait[14].'

Small scale mining in the Cordillera has not reached the frenzy that characterized the gold rush sites in Mindanao where people, coming from many places, tend to work as fast as they could in order to acquire as much gold as they are allowed. Particularly in the late 1980s to mid-1990s, mine-related violence was becoming frequent and mercury pollution was reaching high proportions (Drasch et al., 2001; Israel, 2000; Appleton et al., 1999; UNIDO, 1998; Bayle, 1998). In a remote village of Mountain Province like Mainit and some communities within Abra and Kalinga, conventions associated with community membership as well as customs governing access to certain types of lands (such as a mine site), are keenly observed. The villagers do not welcome non-villagers because the latter are considered outsiders unless they are related to a villager by affinity. This custodial, almost jealous, practice over mineral resources has implications for the extent and frequency of women's involvement in small scale mining. Given that the villagers

---

[14] Interview with Cita (36 years old), 2 September 2003.

are relatively assured that their community is not going to draw people from outside, they do not see the urgency of working in the mines but would rather attend to their farms.

I now turn to the relationship between large scale mines and small scale mines. Given that both types of mining have the goal of acquiring the same commodity, conflicts over access to mineral lands arise and can lead to the restraint of one or the other from carrying out mining operations. Cases like this occurred in early 1989 to mid-1990 when community-based miners' associations in Itogon protested against Benguet Corporation's construction of its open pit mine (Kaniteng, 2003).

Relationship between large scale mining corporations and small scale operations

Corporate large scale mines acquired their concessions through land registration hence by legal fiat; they are the legitimate owners of the mineral areas where they are operating. This gives an 'illegal' status to the small scale miners whose operations fall within the large scale mining lease claims (Maglambayan and Cabria, 2001). Corporate large scale mines and small scale operators have carried on alongside each other, with the former owning vast tracts of lands as mineral claims, and the latter having operations bordering or located within the work sites of the large scale mines. Small scale miners in Itogon alone are estimated to be 6,421 represented by scores of mining associations. However, only two associations, which have a total membership of 123, have small scale mining permits (MGB, 2004c). There are about 56 small scale mining associations in Benguet province and all of them have yet to be formally registered. Small scale mining associations are formed primarily for the purpose of 'acquiring a personality' as well as maintaining and safeguarding tenure of small scale miners on the operation site (Casambre et al., 1987: 139).

The matter of legalizing small scale mining has been raised and debated for years. Small scale mining was given legal recognition through Presidential Decree 1899 (Establishing Small Scale Mining as New Dimension in Mineral Resources Development) issued in 1984. It provides that small scale miners who hold valid and existing mining rights and who have complied with existing mining laws, rules and regulations prior to the promulgation of the law may apply at any time for mining permits and licenses. However, the law virtually cannot find application in Benguet. As a 1987 study by the Cordillera Studies Centre noted:

> ...P.D. 1899 is a 'dead law' in the province of Benguet as it seems incapable of being implemented … the fact remains that most, if not all, of the areas in the Baguio district (which includes the municipalities of Itogon and tuba in Benguet) are already covered by valid and subsisting mining claims of operating mining companies, entities and/or individuals. Thus, there appears to be no realistic way in which small scale miners can

legitimise their activities, in the face of the unwillingness of claims owners to give their consent (Casambre et al., 1987: 140).

The 1991 Republic Act no. 7076 (People's Small Scale Mining Act of 1991) directs small scale miners to organize into formal cooperatives, pay fees and taxes and sell their gold produce directly to the government. Like Presidential Decree 1899 which preceded it, Republic Act no. 7076 requires small scale miners, if their operation sites fall within a mining claim, to enter into contractual agreement with claims owners. This means permission from the latter is needed. Small scale miners consider the presence of large scale mining concessions as a key, if not the root, problem that they face due to difficulties they experience in acquiring the claimants' consent (Kaniteng, 2004).

The Acupan Contract Mining Project (ACMP), which involves partnership between Benguet Corporation and about 1,000 small scale miners, is a novel scheme to address tensions between corporate large scale mining companies and small scale mining groups (Gales et al., 2003; Gomez, 2004). The ACMP venture binds Benguet Corporation to carry out mine planning and exploration whilst small scale mining associations undertake the required production activities. The company is currently evaluating the commercial viability of the remaining underground gold resource in its Antamok and Kelly mines for possible resumption of operation similar to the current arrangement in Acupan (Benguet, 2003). Thus far, the ACMP, now in its second year, has been a profitable project for both Benguet Corporation and the Acupan small scale miners.

Prospects for women in the mining sector

In corporate large scale mining, women's place is not limited to administrative roles such as those of secretaries and clerks. Within the company organization, women have a place in management and leadership posts. They also fill roles that are directly operations-related with the technical functions they handle.

Women are drawn to the mining profession because of their own choice as demonstrated in their taking careers in mining-related fields. Like many other professions, mining draws people largely by the promise of jobs. The increase and waning of the number of women in mining engineering and geology mirrors the rise and decline of the mining sector in the Philippines. Recruitment of larger number of workers, and therefore attracting more women to the sector, will take place only when more mines are developed and opened. Currently, the mining industry in the country is dormant and issues associated with the approval of numerous mining applications remain unresolved. Thus, it is unlikely to see a consequential increase in the number of women joining the sector in the near future. At best, the current company programs to strengthen mining companies' community relations and environmental management are the specific areas where there may be recruitment, although the numbers required for such roles are small.

In the immediate future, corporate large scale mining does not have very promising prospects in the Cordillera region due to issues associated primarily with indigenous peoples' rights, policy apprehensions concerning foreign investments, and environmental and social development concerns.

The Acupan Contract Mining Project is an operation that has provided recruitment of small scale miners. Considering that it is an underground mining operation, the workforce does not include women. The project may expand as defined by the price of gold in the international market. Meanwhile, other old mines may be re-opened leading to additional opportunities for employment in the mines. Benguet Corporation is currently assessing the commercial viability of the remaining underground resources in Antamok for possible small scale contract mining operations similar to Acupan. In addition, residents and small scale miners of Dalicno in Itogon are requesting Benguet Corporation to allow them to mine Level 875 of the company's Itogon-Suyoc mine (Sales, 2003). With the possibilities for expansion of expansion of large scale mining operations, women remain unlikely to be hired for production jobs as large numbers of male miners, who have the necessary skills and experience in jobs within underground workings, are readily at hand to be recruited. As is true in Philex and Lepanto, the place for women will be in administration, management and a few technical roles.

Among small scale miners, women have continued to predominate in the processing activities. However, with the improvements in small scale mining technology since the early 1990s, women's participation has contracted considerably. First, tramlines are now more common for transporting ores from underground workings to the surface. With the availability of tramlines, women's labour is no longer required for the hauling of ores. It is men, not women, who operate the tramlines because of the physically taxing nature of the activity. Due to the currently low grade of gold veins – the higher-grade veins have been exhausted – miners have to extract higher volumes of ore to be able to retrieve the gold and hence involving greater use of crushers as can be observed in Ucab and most areas in Itogon, the manual method of crushing being inefficient. In effect, it is now infrequent to find women performing the task of crushing which is the most laborious phase and at which women are extremely efficient. The operation of crushers is also reserved for men due to the heavy physical exertion required. Women are still the major workers in sluicing and, consequently, panning. However, the reduced amount of water in many work sites affects the frequency of their activity.

Due to lesser areas available for small scale operations resulting from decades of mineral extraction, small scale mining in Benguet province has contracted both in scale and frequency, resulting in decreasing numbers of women in the activity. The continuity of small scale mining in the province depends on the resolution of several issues. The key issues include access to mining areas that need negotiation with land-owning corporate mines, difficulties of small scale miners to meet the regulatory requirements of the Mines and Geosciences Bureau, action of relevant government agencies in regard to the proclamation of most of Itogon municipality as a watershed area thereby closing it to small scale gold mining activities.

The nucleus of mining in the Cordillera has been the province of Benguet. Whilst vast reserves in the province have already been extracted, huge gold and copper resources remain untapped in several areas of the Cordillera especially in Mountain Province and the upland portions of Abra, Kalinga and Apayao. Gold and copper reserves in the region are reckoned to be 42.87 million MT (with grade/analysis of 3.2Au/MT) and 1.40 billion MT (at grade/analysis of 0.45 per cent Cu), respectively (MGB, 2003b). These estimates do not include future discoveries under present explorations by not only the companies that have been operating in the region for decades but also the newcomer transnational companies such as Newmont Mining from the United States and Newcrest Mining from Australia. In these remote parts of the Cordillera region, the main livelihood of the people remains agriculture. However, small scale mining continues to serve as an important source of cash for many households. In these places, relatively few women are engaged in small scale mining at present but the numbers will eventually increase. Considering that small scale mining is a family-based economic activity, women will play an important role especially in the laborious processing tasks and the management of finances for mining operations.

The development and utilization of the region's vast minerals hinge on many factors primary of which is the effective implementation of community-sensitive policies and programs by both national and local governments in developing small scale mining operations as viable livelihoods for communities.

References

Akagi, Hirokatsu, Castillo, Erle S., Cortes-Maramba, Nelia Ana, Francisco-Rivera, Trinidad and Timbang, Teresa D., 2000. 'Health Assessment for Mercury Exposure among Schoolchildren Residing near a Gold Processing and Refining Plant in Apokon, Tagum, Davao Del Norte, Philippines', *The Science of the Total Environment*, 259 (1-3), pp. 31-43.

Appleton, J. D., Williams, T. M., Breward, N., Apostol, A., Miguel, J. and Miranda, C., 1999. 'Mercury Contamination Associated with Artisanal Gold Mining on the Island of Mindanao, the Philippines', *The Science of the Total Environment*, 228 (2-3), pp. 95-109.

Baluda, Redempta P., 2003. 'Philex-Padcal Operation's Environmental Management Program and ISO 14001 Certification', Paper presented at the *Philippine Society of Mining Engineers' Mid-year Mining Symposium*, Metropolitan Club, Makati, Philippines.

Baluda, Redempta P., Maglambayan, Victor B. and Murao Satoshi, 2001. *Profile of Small-scale Mining in the Baguio-Benguet Mineral District, Luzon Island, Philippines*, mimeo, Padcal: Philex Mining Corporation.

Bautista, Antonio, 2004. 'Small-Scale Gold Mining and Processing Practices in Ucab, Itogon, Benguet', Paper presented at the *International Symposium on the Diversity of Mining and Sustainable Local Development: A Meeting to Study the Business Practices of Small-Scale Gold Mining in Benguet, Philippines.*

Bayle, Romeo, 1998. 'Small-scale Gold Mining', *Community and Habitat*, (1), pp. 24-35.

Benguet Corporation, 2001. *2000 Annual Report.*

Benguet Corporation, 2003. *2002 Annual Report.*

Benguet Corporation, 2003. *Benguet Corporation Second Quarter 2003 Report* [retrieved 18 August 2004]. Available from http://www.pse.org.ph/html/ListedCompanies/pdf/2ndqtr03_BC.pdf.

Benguet Corporation, *Mining Operations* [retrieved 28 May 2004]. Available from http://www.benguetcorp.com/html/mining.html.

Bugnosen, Edmund, 2002. *Country Case Study on Artisanal and Small-Scale Mining: Philippines*, International Institute for Environment and Development – World Business Council for Sustainable Development, Mines Minerals and Sustainable Development Project.

Cabalda, M. V., Banaag, M. A., Tidalgo, P. N. T. and Garces R. B., 2002. *Sustainable Development in the Philippine Minerals Industry*, International Institute for Environment and Sustainable Development – World Business Council for Sustainable Development, Mines Minerals and Sustainable Development Project.

Casambre, Athena Lydia, Pales, Dario C., Tapang Jr., Bienvenido, P. and Paulino, Belinda M., 1987. 'Population Mobility and Development in the Mining Sector', in *Population Mobility Survey in the Baguio Area*, Area Research and Training Centre – Cordillera Studies Centre (ed.), Baguio City: ARTC-CSC, University of the Philippines College Baguio.

Chaloping, Minerva M., 1991. *Towards Sustainability of Household Well-Being: Labour Allocation between Goldpanning and Rice Farming in Dalupirip, Itogon, Benguet*, Unpublished thesis for Master of Arts in Social and Development Studies, University of the Philippines.

Disini, Artemio F., 2004. *Why Women are Not Allowed Underground at Lepanto Mines: Reply to the Query of Minerva Chaloping-March*, mimeo.

Domingo, Edwin G., 2003. 'Public Policy Initiatives in Institutionalizing CSR in the Philippine Mining Industry', Paper presented at the *World Bank International Conference on Public Policy for Corporate Social Responsibility*, Washington DC, 8-9 October.

Drasch, G., Bose-O'Reilly, S., Beinhoff, C., Roder, G. and Maydl, S., 2001. 'The Mt. Diwata Study on the Philippines 1999: Assessing Mercury Intoxication of the Population by Small Scale Gold Mining', *The Science of the Total Environment*, 267 (1-3), pp. 151-68.

Espley, Samantha, Francis, Helen and Castonguay, Marie-Josee, 2004. *Hardrock Mining and the Evolution of the Roles for Women* [retrieved 26 May 2004]. Available from http://www.laurentian.ca/wise/0286_fullpaper.pdf.

Flores, R. A. L., Ayson, J. N. R., Angeles, C. A. Jr., Estacio, R. C., Manipon, C. J. I. and Rieza, R. D. S., 2002. 'Where have the Philippine Geoscientists Gone? Emerging Career Directions of Philippine Private Sector Geology Practitioners, 1997–2002', Paper presented at the 15th *Annual Convention of the Geological Society of the Philippines*, PHIVOLCS Auditorium, University of the Philippines Campus, Diliman, Quezon City.

Gales, Diosdado C., Baywong, George B., Austria, Eduardo M. and Gacad, Felizardo A. Jr., 2003. 'Mining in Partnership: The Acupan Experience', Paper presented at the 13th *Mines and Geosciences Bureau Annual Forum*, Quezon City, Philippines.

Gomez, Angelito V., 2004. 'Benguet Corporation's Acupan Contract Mining Project (Mining in Partnership with the Community)', Paper presented at the *International Symposium on the Diversity of Mining and Sustainable Local Development: A Meeting to Study the Business Practices of Small-Scale Gold Mining in Benguet, Philippines*, University of the Philippines, Baguio City.

Gomez, Raquel P., 2003. 'Benguet Corporation 1903–2003 – A Centennial Celebration: The Power to Turn Natural Resources into Gold', *The Philippine Daily Inquirer*, 12 August, pp. A10-A11.

Israel, Danilo C., 2000. 'Mercury Pollution due to Small-scale Gold Mining: A Serious Menace!', *Philippine Institute for Development Studies Policy Notes (2000-03)*, pp. 1-4.

Kaniteng, Lomino N., 2003. 'Project Consultation: David and Goliath', in Satoshi Murao, Victor B. Maglambayan, Edmund Bugnosen and Yoshikawa Choko (eds), *Risk Communication Between Mineral Property Developers and Local Communities*, Tsukuba: National Institute of Advanced Industrial Science and Technology.

Kaniteng, Lomino N., 2004. 'Small-scale Mining in the Philippines: The Forgotten Partner', Paper presented at the *International Symposium on the Diversity of Mining and Sustainable Local Development: A Meeting to Study the Business Practices of Small-Scale Gold Mining in Benguet, Philippines*, University of the Philippines, Baguio City.

Lepanto Consolidated Mining Company, 2002. *Lepanto Profile*, mimeo, Mankayan, Benguet: Lepanto Consolidated Mining Company, October.

Lepanto Consolidated Mining Company, 2004. *Job titles of female employees, Lepanto Consolidated Mining Company – Lepanto Mine Division*, mimeo, (May).

Lepanto Consolidated Mining Company, 2004. *Lepanto Consolidated Mining Company-Lepanto Mine Division Manpower by Summary Groups Ending May 2004*, mimeo, May.

Lepanto Consolidated Mining Company, 2004. *Second Quarter Report 2004* (retrieved 9 September 2004). Available from http://www.pse.org/html/ListedCompanies/pdf/LC_SECForm17Q_2ndQ2004.pdf.

Lopez, Salvador P., 1992. *Isles of Gold: A History of Mining in the Philippines*, Oxford University Press, Singapore.

Maglambayan, Victor, and Hillel, Cabria, 2001. 'The Realities of Small-scale Gold Mining in Baguio District, Philippines', in Cordillera Studies Centre (ed.), *Towards Understanding Peoples of the Cordillera: A Review of Research on History, Governance, Resources, Institutions and Living Traditions*, Cordillera Studies Centre, Baguio, pp. 108-14.

MGB, 2000. *Gender and Development Sex Disaggregation Data Profile*, Mines and Geosciences Bureau brochure.

MGB, 2003a. *Briefing Kit on the Philippine Minerals Industry*, Quezon City.

MGB, 2003b. 'Essential Statistics CY 2002', *Paltek di Cordillera*, 3 (April-June), mimeo.

MGB, 2004a. *Industry Profile* (retrieved 24 June 2004). Available from http://www.mgb.gov.ph/indtprofile.htm.

MGB, 2004b. *Minerals Action Plan (Map): A Roadmap for the Revitalization of the Minerals Industry* (retrieved 17 August 2004). Available from http://www.mgb.gov.ph/map/map_executivesummary.pdf.

MGB, 2004c. *Macroeconomic Mining Indicators* (retrieved 24 June 2004). Available from http://www.mgb.gov/mining-indicators.htm.

MGB, 2004d. *Paltek di Cordillera*, 4 (13).

National Statistics Coordination Board, 2004. *Basic Statistics Philippines* (retrieved 3 June 2004). Available from http://www.nscb.gov.ph/secstat/d_labr.asp.

Philex Mining Corporation, 2004. *Philex Mining Corporation Second Quarter Report for 2004* (retrieved 26 August 2004). Available from http://www.pse.org.ph/html/ListedCompanies/pdf/PX_SECForm_17Q_2ndQ2004.pdf.

Philippine Society of Mining Engineers, 2003. *A Changing World for Mining Engineers: Technical Symposium Oathtaking of New Mining Engineers and Induction of Officers and Directors*, Makati, Philippines.

Sales, Tina, 2003. 'Small-scale Miners want to Mine portion of Itogon Suyoc Level 875', *The Sun Star*, Philippine Information Agency, Benguet, pp. 3, 7.

Shea, B. C., 2002. 'The Status of B.Sc. Geology Program of the College of Engineering, Adamson University', Paper presented at the 15th *Annual Convention of the Geological Society of the Philippines*, PHIVOLCS Auditorium, University of the Philippines Campus, Diliman, Quezon City, Philippines.

United Nations Industrial Development Organization, 1998. *Assistance in Reducing Mercury Emissions in Highly Contaminated Gold Mining Areas in Mindanao – Phase I*, 1998 (retrieved 19 May 2004). Available from http://www.unites.uqam.ca/gmf/intranet/gmp/files/doc/unido_philippines_project.pdf.

Wilson, Laurence L, 1932. *Igorot Mining Methods*, Catholic Schools Press, Baguio.

Chapter 10

Women in Artisanal and Small Scale Mining in Africa

Jennifer J. Hinton, Barbara E. Hinton and Marcello M. Veiga

Introduction

Women's direct participation in artisanal and small scale mining[1] (ASM) varies throughout the world. In Asia, generally less than 10 per cent of miners are women, whereas in Latin America, the proportion tends to average 20 per cent. The percentage of female artisanal miners is the highest in Africa, averaging between 40 and 50 per cent. In some regions, the ASM workforce is comprised of 60 to 100 per cent women (ILO, 1999; Amutabi and Lutta-Mukhebi, 2001; Onuh, 2002; Anon., 2002). Generally, as the degree of mechanization and scale of mining increases, women's involvement decreases (Hinton et al., 2003). Thus, women tend to be most active in small family operations, where mining supplements other income (for example, from subsistence agriculture), and where labour intensive, low-technology mining methods are employed. Women undertake a variety of activities in ASM ranging from labour-intensive mining methods to processing activities, including amalgamation with mercury in the case of gold extraction. As mineral processing is often conducted in or adjacent to the home, women and their families can be at great risk from mercury poisoning, silicosis and other hazards from ASM.

It is evident that women occupy critical roles in the ASM labour force. Despite this, there is a paucity of information on the roles of women in these communities and even fewer accounts describing the lived experiences of individual women miners. Further to this, women have received markedly less attention than men in programs seeking to improve the ASM sector. It is suspected that, as women often work part-time at informal mining operations or occupy 'ancillary roles' (for example, cooks, service providers), their participation may be viewed as being less

[1] *Artisanal mining* is used to denote all small scale as well as medium and large scale mining that may be illegal or legal, formal or informal. Artisanal mining may be better characterized by a lack of long-term mine planning and use of rudimentary techniques (Hinton et al., 2003).

relevant. Susapu and Crispin (2001) speculated that, as women are more frequently associated with transporting and processing materials as opposed to digging, they are not always identified as 'miners'.

This chapter intends to explore these issues within Africa. Further insight is provided on the evolving gender[2] roles of women in African ASM communities, the impacts of ASM on women miners and factors exacerbating and mitigating inequities in power, the distribution of wealth, and susceptibility to poverty.

Women miners in Africa

Women in ASM play pivotal roles in the mining of gold, diamonds and other gemstones, and industrial minerals, such as dimension stone, salt and sand and gravel, throughout Africa. In Ghana, women comprise approximately 15 per cent of the legal and 50 per cent of the illegal, or *galamsey* small scale metal miners, with approximately 45,000 to 85,000 women working in gold mining alone (Hilson, 2001; Hilson, 2002). In Burkina Faso and Mali, approximately 90 per cent of mineral processing activities are conducted by women, with the female participation in all ASM activities (digging, transport and processing) averaging 50 per cent (Gueye, 2001; Keita, 2001). In the Sudan, it is estimated that 35 per cent of gold miners are women in the Southern Blue Nile and Eastern Bayuda Desert regions, respectively. In the Ingessana district of Sudan, women constitute more than 50 per cent of the labour force with participation escalating to as much as 90 per cent when the men go planting (Veiga, 2003; UNIDO, 2002). In most Southern African countries, female participation in ASM averages 25 per cent (Dreschler, 2001). In the Rwamagasa District of Tanzania, however, this percentage approximates to 50 per cent (Wagner, 2003). Only 10 per cent of the miners in Zimbabwe's formal (legal) sector are women, 70 per cent of these being full or part mine owners, with the remainder engaged in ore grinding using a mortar and pestle (Dreschler, 2001). This contrasts drastically with the informal (illegal) sector, where at least 50 per cent of the more than 300,000 artisanal miners in Zimbabwe are women (Dreschler, 2001). While national averages are unknown, the majority (80 per cent) of the 2,000 panners in the Mukibira district of Kenya are women, with proceeds supporting a population of approximately 10,000 (Amutabi and Lutta-Mukhebi, 2001).

Although every ASM community differs in terms of its characteristics and conditions, women within these communities tend to be engaged in specific and often multiple roles. Typically, they are labourers (for example, diggers, panners,

---

[2] *Gender*, as applied herein, refers to the behaviours, attitudes, values, beliefs, etc., that a particular socio-cultural group considers appropriate for males and females. The authors adhere to the belief that gender roles are fluid and can shift over time, space and in different contexts (Butler, 1990).

ore carriers and processors), providers of goods and services (for example, cooks, shopkeepers) and are often solely responsible for domestic chores. Mineral processing activities range from crushing, grinding, sieving, washing and panning, to amalgamation and amalgam decomposition in the case of gold mining. Less commonly, women are concession owners, mine operators, equipment owners, dealers and buying agents.

Rock crushing and grinding

Although crushing and grinding of extracted ore can be accomplished using a number of mechanical means such as the stamp mills, when women are engaged in this activity, it is typically a highly manual activity using hammers or a mortar and pestle. The Ingessana people in Sudan and many miners in Rwamagasa, Tanzania, for instance, work in family-based operations, with men responsible for breaking rock and women responsible for grinding, as well as gold concentration and amalgamation (Ibrahim, 2003; Tesha, 2003). In addition to the labour intensive and monotonous nature of this work, inhalation of fine, crystalline dust generated from breaking and crushing rock can result in pneumoconiosis, including the incurable lung disease, silicosis (WHO, 2000). Conditions resulting from silicosis include emphysema, lung fibrosis and silica-tuberculosis. Advanced stages of silicosis have been documented among women and children as young as 14 in mining areas in Ghana (ILO, 1999).

Panners, washers and hand sorters

In gold mining regions of Africa, women are frequently engaged in panning or gold washing. Panning involves agitation of ore wherein heavier particles, including gold, settle out and lighter particles are washed away. Panning is often followed by hand picking of visible gold and amalgamation of the heavy mineral concentrate. Most of the 150,000 female artisanal miners in Zimbabwe are involved in gold digging and panning (UNIDO, 2002). Labonne (1996) documented women meticulously hand picking nuggets for countless hours in the *orpillages* of Burkina Faso. In Mukibiri, Kenya, local and indigenous women have been engaged in panning for hundreds of years, passing down their expertise to younger generations (Amutabi and Lutta-Mukhebi, 2001). Five tonnes of gold are produced annually in Kenya by panning. In Kenya, the average gold panning family income is around 5,000 to 10,000 Kenya shillings per month (US $60-120), which is considerably higher than those working in agriculture or the civil service.

Transporting ore and water

Water is a necessary component for the processing of most minerals. In many African mining communities, women are predominantly responsible for hauling water, an activity which is both time consuming and laborious. In Ingessana,

Sudan, for instance, water sources are located more than a 30 minute to one hour walk from the mine site, a situation which is exacerbated through the summer drought (Ibrahim, 2003). Due to the scarcity of water, women sometimes sell water to other mines in addition to that used by the family. Water is typically carried using plastic cans tied to strong rope and balanced across women's shoulders.

Sellers

Although revenue generated from mining is often controlled by male heads of households, particularly in the case of high-value, low-volume commodities (for example, gold), women are sometimes engaged in the selling of mined products (Hinton et al., 2003). In the Mukiberi District of Kenya, women are typically regarded as being more honest than men, and are frequently engaged in selling at the markets (Amutabi and Lutta-Mukhebi, 2001). Although women's groups have recently formed in Mukiberi to market products themselves, most men and women involved in intermediate roles such as that of buyers are predominantly of European or Asian descent.

Type of commodity

Perhaps one of the most interesting determinants of the roles and responsibilities of women involves the type of commodity mined. In the extraction of high value products, such as gold, men more commonly maintain control of the mine site and are involved in digging, whereas women are typically relegated to washing, panning and transport activities. The basis for this division of labour is presumably related to the limited physical capacity of women for pit labour. Ironically, women are engaged in digging and equally laborious activities for agricultural purposes, or in the extraction of high volume, low value commodities such as clay, limestone or dimension stones. Women are more likely to participate in greater numbers and play a more significant role in control of the land and decisions surrounding its use for low value commodities, as observed in Nigeria (salt) and South Africa (kaolinite). In Ghana, the involvement of women in industrial minerals (for example clay, stone quarries, salt) is much greater than in gold mining, with the proportion of women in salt mining as high as 75 per cent (Hilson, 2002). Women tend to dominate salt mining in many West African countries. In the village of Keana, Nigeria, exclusively women are involved in salt mining (Onuh, 2002). This process involves placing salt-rich sandy soil in pots. Water poured into the pots dissolves the salt and drains through a small hole in the bottom of each container. The salt-rich solution (brine) is collected in a hollowed-out log and heated on a fire in another pot until it crystallizes.

Differences in participation in high versus low value commodities can be viewed either as a startling indicator of gender inequity in ASM as women are proffered the control of land of lesser value or a reflection of women's entrepreneurial capacity. Nevertheless, these minerals represent an important source of livelihood for women particularly in the context of their lack of ownership rights over land (Mushunje, 2001).

Health and wellbeing of women miners

Health and wellbeing is a result of the complex interplay between factors that include, but are not limited to, income level and disparity, social support networks, education, employment, working and living conditions, physical environments, social environments, biology and genetic endowment, personal health practices, coping skills, healthy child development, health services, gender and culture (Evans, 1994; Health Canada, 2004; Marmot, 2001; Mach, 2004). The factors which influence the health and wellbeing of women miners are shown in Figure 10.1 and described further below.

The World Health Organization (WHO) defines health as 'a state of complete physical, mental and social well-being and not merely the absence of disease and infirmity'. Health is not 'an objective for living but a resource of everyday life'.

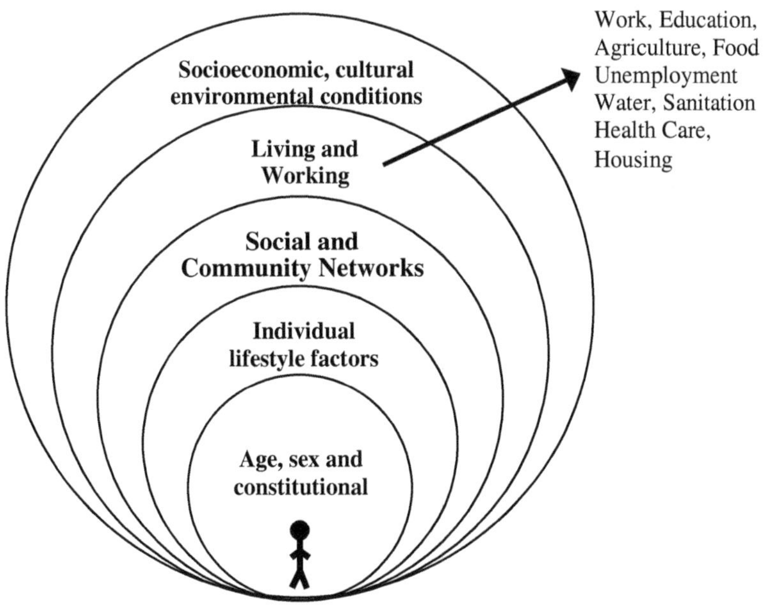

Figure 10.1 Multiple layers of the social determinants of health (after Dahlgren and Whitehead, 1991)

Poor health generates a vicious cycle – when spouses or family members are infirm and their capacity to work is diminished, a 'healthy' family member must work harder to pay for normal living expenses in addition to health costs. Ill health of a family member may initially drive women and children into mining. Arduous work, combined with inexperience in mining and lack of knowledge about chemical exposures, can further exacerbate the potential for injury or illness; thus, the cycle of ill health and poverty is perpetuated. Thus, it is critical that the multiple factors influencing women's health be explored in order to develop appropriate strategies to mitigate the negative health impacts of ASM and enhance the benefits from this activity.

Age, sex and constitutional factors

Age and gender inequities often make women vulnerable to factors that negatively impact their health and wellbeing. Girls in ASM communities are often engaged in mining at a young age. Their first exposure to mining may be as babies, where women often work with infants tied to their backs and toddlers at their side. In cases where alternative childcare or schooling is unavailable or additional supplementation of income is needed, older children accompany women to the mine site and often participate in mining. Although generally to a lesser extent than boys, girls are likely to be involved in activities such as digging, grinding, crushing and transporting materials, typically using tools such as picks, shovels and hammers and without any protective equipment (ILO, 1999). Malnutrition, thermal injuries, and skeletal damage in young children resulting from child labour in ASM has been extensively documented (Wasserman, 1999). In addition, girls are somewhat more susceptible to sexual exploitation than boys in ASM communities, and child prostitution is common, in part due to the status given to virginity, but also as young girls are deemed unlikely to carry HIV-AIDS or other sexually transmitted diseases.

Women are often the targets of acts of aggression by men, a factor which has been attributed in large part to the influence of male dominated 'mining culture' (Hinton et al., 2003). Martha Bitwale of the Tanzania Women Miners Association has described the fear of sexual abuse associated with women conducting mining and exploration in remote areas (Machipisa, 1997). A women's organization active in the region, the Mwanza Women Miners Associations (MWAWOMA) in Tanzania, reported that women working as entrepreneurs face specific challenges, including sexual harassment and abusive language (Wagner, 2003). Given the potential for acts of violence and harassment, combined with the sex trade typically present in ASM communities and the often transient nature of mining, it is not surprising that many ASM regions are plagued by high incidences of HIV/AIDS and other sexually transmitted diseases. At least 70 per cent of women interviewed in a major mining area of Kenya reported at least one incidence of venereal or sexually transmitted disease (Amutabi and Lutta-Mukhebi, 2001).

Individual lifestyle factors

The complex task of balancing mining with a multiplicity of other responsibilities can create severe stress for women in Africa. Stress creates both negative psychosocial and physiological health impacts. In Rwamagasa village, Tanzania, many of the ASM enterprises are family affairs, usually consisting of a husband and wife, children, and relatives (Wagner, 2003). In this case, Wagner (2003) observed that women often have a markedly increased workload as mining is conducted in addition to their regular household duties. Domestic responsibilities for women miners can include childcare, the production, processing, preparation and storage of food, as well as the provision of fuel, sanitation, cleaning the house and laundry. Women are often also predominantly responsible for hauling water. In Rwamagasa village, Tanzania, women haul water 50 per cent of the time, considerably more than men or children (Wagner, 2003). As well, women typically undertake a considerable portion (in some cases the majority) of agricultural duties.

Other lifestyle factors which may influence the health of women miners include alcoholism, which may be more prevalent in ASM areas, and nutrition, which may improve due to access to funds to buy food, or worsen, due to dietary changes or reduced access to traditional foods. Increased female participation in ASM may result in decreased attention to agricultural crops, reducing food security and resulting in poorer nutrition. Alternately, participation in other traditional occupations (for example crop production) may continue, resulting in a substantial increase in the work burden of women (Çagatay, 2001). In fact, one of the most significant gender inequities relates to time. Women are often exclusively responsible for domestic responsibilities even when they participate in the labour market, whereas men tend to be responsible for predominantly economic activities (Veiga, 1997). This reality, coupled with differing spending habits, generally means that women's participation in activities that generate revenue can be equated with increased labour hours.

Social and community networks

The health and well being of women in ASM communities is strongly impacted by social networks and structures. The development of supportive relationships can significantly reduce social isolation, which is an established determinant of health (Bower, 1997). The effects of migration and social disconnection may weigh heavily on the psychological well-being of women. In Mashonaland, Zimbabwe, a study of effects of rural economic development on women's blood pressure revealed notably elevated levels in mining areas in comparison to areas where large-scale agriculture and traditional economic activities took place (Hunter et al., 2000).

The contribution of women to social capital, which features elements of social organization, networks, norms, and trust, is a vital issue in ASM communities. In

Kangaba, Mali, Beatrice Labonne (1998) described the commitment of Malinke women to community health. Normally, Malinke women are required to turn over all gold recovered to their spouses. However, during the inter-season when agricultural responsibilities had concluded and full scale *orpaillage* activities had not yet resumed, men rest and women commence mining activities early. In this 'moonlighting' period, women are allowed to retain the profits from gold mining in order to purchase food and spices, children's clothing and medicine. Labonne (1998) documented women assisting elderly women with the cultivation of gardens, and with revenue planning behind hiring a midwife for the community.

When united, women are more likely to broach topics of rights and safety, may be in a better position to adopt alternative mining methods and livelihoods, and access credit. Connecting women with Non-Governmental Organizations (NGOs) and strengthening linkages between individuals or groups of women can markedly improve women's access to resources and lead to their empowerment. Within a formal network or organization, women are better equipped to respond to negative impacts of ASM and take advantage of opportunities in this sector (Wiego, 2002). As women are increasingly being financed through micro-credit programs, the formation of organizations or cooperatives may even better situate women to receive financial support from charitable organizations or formal lending institutions.

Living and working conditions

The occupational health and safety issues that plague ASM can primarily be attributed to: the informal and often illegal nature of this activity, economic demands that result in inadequate equipment and neglect of safety measures, and a frequent lack of expertise and insufficient training. Although the chemical dangers, in particular, those associated with mercury and cyanide misuse, first come to mind, most occupational hazards are a consequence of poor physical conditions, such as ground failure, shaft collapses and machinery accidents. Hydraulic monitoring[3] of secondary deposits can also be extremely unsafe, as there is potential to undercut hill slopes and induce landslides (Hinton et al., 2003). Poor lighting and ventilation, electrocution and explosives misuse are other frequent causes of accidents. Although accidents are underreported due to the illegality of ASM, ILO (1999) states that reported non-fatal accidents in ASM are still six to seven times greater than in formal, large-scale operations.

Women who work in artisanal mines face additional illness, injury and stress from dust and noise pollution, as well as extreme exertion from highly labour-intensive jobs. For instance, several hours of digging, carrying large weights great distances, and bending over in awkward positions (for example, during panning or

[3] Hydraulic monitoring involves the high pressure application of water to 'fluidize' metal-bearing highly weathered soils (Hinton et al., 2003).

scavenging for gems) can result in painful, chronic injuries (for example, lower back pain) and fatigue. In addition, the inhalation of fine, crystalline silica dust, which is generated from breaking and crushing rock, can result in pneumoconiosis, an incurable lung disease (WHO, 2000).

As women are most commonly involved in the processing aspects of ASM, they are also highly susceptible to chemical dangers, particularly in association with mercury misuse in gold mining. Mercury is widely recognized as one of the most toxic metals known to man. Mercury vapour released during amalgam decomposition poses a serious hazard to women and others in close proximity to gold shops and amalgam decomposition sites. In many regions, gold decomposition takes place in the home (using the kitchen stove) or in small sheds adjacent to processing sites. Women can have elevated levels of mercury accumulation in hair and exhibit symptoms including kidney pain, respiratory problems, and dizziness (Sass, 2002). Chronic exposure to mercury vapour can also result in gingivitis and muscular tremors. Mild cases of mercury poisoning have many psycho-pathological symptoms, such as depression and exaggerated emotional responses, which can be mistaken for alcoholism, fever, malaria or other tropical diseases. Exposure to acute levels can produce dysfunction of kidneys and urinary tract, vomiting, and, potentially, death.

Women involved in reworking tailings may simultaneously be exposed to multiple pollutants. Cyanide, which is used as an alternative to mercury in recovering gold, is an effective asphyxiant and hydrogen cyanide gas (HCN), and can be fatal to humans at concentrations around 250 parts per million in air. Chronic exposure to low concentrations of cyanide has been linked to neuro-pathological lesions and optical degeneration (Potter et al., 2001). In addition to cyanide and mercury, women reworking tailings may be exposed to highly toxic metals, including lead, cadmium and arsenic.

As most ASM communities are located in sparsely populated, rural, and often, remote, regions, living conditions are often dictated by this reality. In comparison to their urban counterparts, women in rural communities typically have less access to services, such as health care and education; a reduced dependence on the cash economy for food and non-food purchases, and fewer health problems due to contamination (for example, air, water pollution) from overcrowding (Ruel et al., 1999). In hastily-established ASM communities, residents may experience the worst of both urban and rural environments, being able to access exclusively poor services, becoming increasingly dependent upon the cash economy, and exposing themselves to pollutants (Shoko, 2002).

Socioeconomic, cultural and environmental conditions

Socio-cultural context, income, education, and the natural environment are significant influences on the health and well being of women in ASM communities. Socioeconomic position or status is one of the most significant determinants of health and, across nations and cultures, it is widely recognized that health status generally decreases with socioeconomic status, in other words the rich and powerful live healthier, longer lives than the poor and disempowered (NALD, 2001).

Although there are multiple components of socioeconomic status, income levels are commonly used to bridge the gap between status and health. Income inequality is a major determinant of poor health, mainly as it is a proxy for conditions associated with inadequate material goods (for example, food) and socioeconomic status (Hinton et al., 2003). Despite the deplorable working conditions, the economic benefits accrued from ASM can, in theory, positively impact women's health through access to health services and the capacity to purchase food. Women's incomes are often considerably higher than those obtained through similar work in rural or urban areas. Despite this, although women undertake the same labour as men, considerable inequities in pay exist. In order to address this economic imbalance, women in Guinea – where men are paid four times more for the same quantity of gold – often undertake a 'troc', or trade of sex for additional money or gold (USAID, 2000). In some cultures, inequities in pay scales are exacerbated by the fact that women do not always have control of their earned income.

Many studies indicate that the revenue generated and managed by women in ASM contributes more directly to the well-being of households then that of men. Specifically, the income generated by women is more likely to be directed towards improving the quality-of-life in the family – through education, food, agriculture, etc. – whereas men tend to spend revenue on gambling, prostitution and alcohol (Hentschel et al., 2002). Sociological assessments in communities in Rwamagasa village in Tanzania and the Ingessana region of Sudan indicate that men predominantly control the family income from mining and make most household decisions (Wagner, 2003; Mtetwa and Shava, 2003; Al Medani, 2003). Among the people interviewed at Rwamagasa village, 60 per cent of males head the household and, in the cases where women are heads, they are either single or widowed (Wagner, 2003). The majority of study participants (86 per cent) in Ingessana District, Sudan, indicate that decisions are mainly made by men, while a minority (15 per cent) contends that decisions are made by consensus (Al Medani, 2003). In Kangaba, Mali, although women and men work side-by-side with men gold digging, Malinke tradition dictates that women turn over all gold to their husbands (Labonne, 1998). The same tradition typically applies to women in Kenya, although there are a number of cases where women manage family finances and give their husbands an allowance (Amutabi and Lutta-Mukhebi, 2001).

Cultural beliefs and traditions strongly influence interactions between individuals and groups, the nature of community organization and societal rules or norms. Although these factors are critical in terms of women's socioeconomic status and overall health and wellbeing, they have been poorly studied in the context of ASM. In an assessment of women's participation in tin mining in the Jos region of Nigeria, Ogbe (2001) ascertained that the role of a woman is largely determined by her husband – in this case, mining is predominantly conducted by women. Although many differences exist between the 374 different cultural groups in Nigeria, women are generally subordinate to their husbands and in-laws, with little autonomy (Ogbe, 2001).

A significant barrier to women's equitable participation in mining involves cultural taboos and superstitions. In Zambia, for instance, Namakau Kaingu, mine owner and Chairperson of the SADC Women in Mining Trust, described challenges in gemstone mining due to the belief that women should not approach a gemstone mine as the spirits of the stones would drive the gems deeper into the earth (Synergy Africa, 2001) or would disappear altogether (Kaingu, 2003). Some believe this is particularly significant during menstruation. Disappearance of the stones can be averted with the slaughter of a goat or cow, and the calling of the spirits of ancestors (Kaingu, 2003). In N'tulo, Mozambique, women are believed to attract bad spirits and, therefore, are banned from working in the mines (Dreschler, 2001). They are, however, permitted to sell food and beer. In nearby Manica, women are prohibited from digging trenches but can transport and wash the gold-bearing ore. Due to the belief that they bring bad luck, women have been banned or deterred from working underground in countries throughout the world (Robinson, 1998).

Education, a major driver of socioeconomic status, can be critical in transforming gender stereotypes and empowering women to derive greater benefits from ASM. It has been demonstrated that high levels of maternal education can cause good population health even at low income levels (Evans, 1994). In many countries, girls are systematically discriminated against in terms of education – this is mainly attributed to family expectations concerning domestic responsibilities, in addition to social and cultural barriers. A survey of the Ingessana people of Sudan indicated that most residents feel that the education of women is not important, as illustrated by an illiteracy rate in women of 99 per cent (Al Medani, 2003). Amutabi and Lutta-Mukhebi (2001) determined that families in Kenya are more likely to send boys to school than girls, as has been observed in many other African and South East Asian countries (UNESCO, 2000). In some cases, girls are removed from school at the onset of puberty, due to mistrust of male teachers and students and fears of unwanted pregnancies (UNESCO, 2000). When girls do enter the school system, they tend to be responsible for considerably more extracurricular chores than boys, thereby resulting in longer hours worked daily and less attention paid to homework. Gender inequities may be further exacerbated by educational practices and beliefs that reinforce gender stereotypes. For instance, educators may associate men with public life and corresponding lifestyles (work, politics, power), and link women with domestic responsibilities.

Finally, environmental conditions are critical to the health of women and families in ASM communities. In the case of gold mining, organic forms of mercury – specifically, methylmercury – is of greatest concern in terms of environmental exposure. Due to its tendency to increase in concentration upward through aquatic food, individuals reliant on fish in mercury impacted areas may be at risk. Women of childbearing age and their children are particularly susceptible, as methylmercury readily crosses placental barriers and is considered to be a developmental toxicant (Grandjean, 1999). In addition to mercury, exposure to other potentially toxic metals (for example cadmium, lead) can also occur through inhalation of dust or consumption of contaminated drinking water and food. In addition to chemical contamination, ASM can detrimentally impact ecosystems through deforestation and the modifications of hydrologic systems, for example, through silt accumulation in rivers or construction of water reservoirs (Akagi and Naganuma, 2000). Deforestation can significantly impact women and families, due to the importance of forests for fuel and, sometimes, food and medicine.

Conclusions

Gender equity is evidently a critical component in achieving women's health and well being within ASM communities. At the World Bank Roundtable on Regularizing ASM (1995), Mrs. Ofei-Aboagye of the Ghana Institute of Management and Public Administration stated that primary constraints to the effective participation of women include 'legal and social taboos; widespread illiteracy; and organizational, technical, and financial constraints'. These issues are inextricably linked to gender inequity. Inequities have been evidenced by limited political power, access to resources (capital, information, education, etc.), mobility, and basic human rights. Thus, it is not surprising that 'while the positive impacts (of ASM) are hardly felt by women, they are hard-hit by the negative impacts' (Traore, 1997). Dreschler (2001) further articulates that 'women (in Tanzania) make up 50 per cent of the work force but do not receive 50 per cent of the rewards'.

In order to mitigate the negative repercussions of ASM on women, it is critical that their participation in this sector be understood. Key findings from this review include the following:

1. Women working in mines are most commonly involved in transporting ore and water, washing, or seemingly 'delicate' work such as panning, amalgamation and amalgam decomposition. Women also function as small operators, and mine owners but this tends to be more of an exception than the rule.
2. In extraction of high value products, such as gold, men more commonly maintain control of the mine site. Women are more likely to participate in greater numbers and play a more significant role in control of the land, and decisions surrounding its use for low value commodities, such as industrial minerals.
3. Women's direct participation in ASM generally increases with the decreasing scale of the operation.

4. Outside of mining, women often assume key roles in the provision of goods and services (for example, cooks, sex trade workers, shop owners, housekeepers). Moreover, within the community, women are of fundamental importance in terms of food security, are critical to community stability, cohesiveness, and morale, and act as primary agents in facilitating change.

Both direct and indirect involvement of women in ASM in Africa is believed to be on the rise. This can be attributed to a number of factors, including: escalation of rural poverty from droughts and/or structural adjustment programs resulting in a greater need to supplement incomes; outward migration of skilled male miners from ASM areas due to increased large-scale mining development in other regions or in pursuit of other opportunities in urban areas; evolving cultural norms with respect to gender roles; lack of employment in other sectors; and high birth rates and growth of extended families[4].

Given this reality, it is critical that ASM be transformed into an activity that empowers women and supports the development of healthy families in Africa. In order to respond to inequities in political power, access to resources (capital, information, education and training, etc.), mobility and basic human rights, combined with the presence of legal, socio-cultural, and financial constraints, the following are recommended:

1. Commitment to gender mainstreaming in government agencies, and appropriate recognition of women in policy frameworks, particularly in relation to land ownership rights.
2. Adoption of strategies inclusive of, and accessible to, both women and men, and which support women's participation in political decision-making.
3. Elimination of discrimination from educational systems, and provision of support for families sending children to school.
4. Formal incorporation of gender issues, and the adoption of holistic approaches to ASM communities though technical assistance and community development programs.
5. Promotion of micro-credit and other programs that provide financing for women.
6. Execution of educational campaigns that target women in order to mitigate specific health risks, and which provide the means for them to access other resources.
7. Implementation of programs to train women in various aspects of mining, as well as in marketing, management and bookkeeping.

---

[4] With escalating mortality rates in many countries due to AIDS and other causes, families are increasingly inheriting their relatives' children, resulting in larger families and a greater economic strain (Wente, 2002).

8. In-depth research on women's involvement in ASM communities, and the differential impact of current practices and technical change on the lives of both women and men.

ASM has the potential to be an important catalyst for entrepreneurial activities, and the formation of strong, resilient communities throughout Africa, but progress will be severely limited until all stakeholders have the necessary tools to affect positive change and challenge existing inequities. Women can play a vital role in this transformation, in part, through the development of sustainable livelihoods. Providing women with access to information, credit, and training combined with an amenable policy framework are key steps to facilitating this change. The implications these programs and initiatives have for the future of women's health and wellbeing in ASM communities in Africa is critical, and the need to urgently support development in these areas is evident.

References

Akagi, H., and Naganuma, A., 2000. 'Methylmercury Accumulation in Amazonian Inhabitants Associated with Mercury Pollution', Proceedings of US-Japan *Workshop on Human Health Effects of Low Dose Methylmercury Exposure*, National Institute of Minamata Disease, Minamata, Japan, November 28-30, pp. 3-145.

Al Medani, K. A., 2003. *The Blue Nile Socio-Economic Report, Sudan*, Report to Global Mercury Project, p. 14.

Amutabi, M. and Lutta-Mukhebi, M., 2001. 'Gender and Mining in Kenya: The Case of the Mukibira Mines in the Vihiga District', *Jenda: A Journal of Culture and African Women Studies*, vol. 1 (2), p. 23.

Anon., 'Gemstone Miners US Fair Bound', *Times of Zambia*, posted at AllAfrica.com, 2002, http://allafrica.com/stories/200212060252.html.

Butler, J., 1990. *Gender Trouble: Feminism and the Subversion of Identity*, Routledge, New York and London.

Çagatay, N., 2001. *Trade, Gender and Poverty*, United Nations Development Programme, p. 45.

Dreschler, B., 2001. S*mall-scale Mining and Sustainable Development within the SADC Region*, Report to MMSD, p. 165, http://www.iied.org/mmsd.

Grandjean, P., 1999. 'Methylmercury as a Developmental Neurotoxicant', Proceedings of *Mercury as a Global Pollutant – 5th Internaional Conference*, Rio de Janiero, Brazil.

Gueye, D., 2001. *Small-Scale Mining in Burkina Faso*, Report to MMSD, p. 35, http://www.iied.org/mmsd.

Hentschel, T., Hrushka, F., and Priester, M., 2002. *Global Report on Artisanal and Small-scale Mining*, Report to MMSD, p. 67, http://www.iied.org/mmsd.

Hilson, G., 2001. *A Contextual Review of the Ghanaian Small-scale Mining Industry*, Report to MMSD, p. 29, http://www.iied.org/mmsd.

Hilson, G., 2001. 'Small-scale Mining and its Socioeconomic Impact in Developing Countries', *Natural Resources Forum*, 26 (1), pp. 3-13.

Hinton, J. J., Veiga, M. M., and Beinhoff, C., 2003. 'Women in ASM: Gender Roles and the Road Ahead', in G. Hilson (ed.), *The Socio-Economic Impacts of Artisanal and Small-Scale Mining in Developing Countries*, Swet, The Netherlands, Ch. 11.

Hinton, J. J., Veiga, M. M. and Veiga, A. T. C., 2003. 'Clean Artisanal Gold Mining: A Utopian Approach?', *Journal of Cleaner Production*, vol. 11 (2), pp. 99-115.

Hunter, J. M., Sparks, B. T., Mufunda, J., Musabayane, C. T., Sparks, H. V., and Mahomed, K., 2000. 'Economic Development and Women's Blood Pressure: Field Evidence from Rural Mashonaland, Zimbabwe', *Social Science and Medicine*, 50, pp. 773-795.

Ibrahim, M. S., 2003. *Information about Ingessana Hills Artisanal Gold Mining Sites, Sudan*, Report to Global Mercury Project, p. 16, http://www.unites.uqam.ca/gmf/intranet/gmp/documents/documents.htm.

International Labour Organization (ILO), 1999. *Social and Labour Issues in Small-scale Mines*, Report for discussion at the Tripartite Meeting on Social and Labour Issues in Small-scale Mines, ILO, Geneva.

Labonne, B., 1996. 'ASM: An Economic Stepping Stone for Women', *Natural Resources Forum*, vol. 20 (2), pp. 117-122.

Labonne, B., 1998. 'Of Gold and Women', *DESA News*, vol. 2 (1), February.

Labonne, B., and Gilman, J., 1999. 'Towards Building Sustainable Livelihoods in ASM Communities', Proceedings from the *Tripartite Meeting on Social and Labour Issues in Small-scale Mines*, Geneva, Switzerland, 17-21 May.

Machipisa, L., 1997. *Rocky Path for Women Miners*, Interpress Service, Project Underground, http://www.moles.org/projectunderground/motherlode/gold/women5.html.

Maponga, O. and Ngorima, C. F., 2003. 'Overcoming Environmental Problems in the Gold Panning Sector through Legislation and Education: The Zimbabwean Experience', *J. Cleaner Production*, v. 11, pp. 147-157

Mtetwa, C. and Shava, S., 2003. *A Sociological Survey of Small-Scale Artisanal Gold Mining In The Kadoma-Chakari Area*, Report to Global Mercury Project, p. 40, http://www.unites.uqam.ca/gmf/intranet/gmp/documents/documents.htm.

Mushunje, M. T., 2001. *Women's Land Rights in Zimbabwe*, Report for the BASIS Management Entity, p. 19, http://www.wisc.edu/ltc/live/bassaf0106a.pdf.

Ofei-Aboagye, E., 1995. 'Second Session: Organizational, Social, and Women's Participation Issues', Proc. World Bank Roundtable on Regularizing Informal Mining, Washington DC, May 17-19.

Ogbe, F. M. D., 2001. *Success Stories of Women in Natural Resources Management in Africa: South West Nigeria*, UNU/INRA, ACCRA, p. 51, http://www.unu.edu/inra/success_stories/nigeria.pdf.

Onuh, B., 2002. 'Salt Women of Keana', *Newswatch*, 2002, http://allafrica.com/stories/200211190755.html.

Potter, J., Smith, R. L., and Api, A. M., 2001. 'An Assessment of the Release of Inorganic Cyanide from the Fragrance Materials Benzyl Cyanide, Geranyl Nitrile and Citronellyl Nitrile Applied Dermally to the Rat', *Food and Chemical Toxicology*, 39, pp. 147-151.

Ruel, M. T., Haddad, L., Garret, J. L., 1999. 'Some Urban Facts of Life: Implications for Research and Policy', *World Development*, vol. 27 (11), pp. 1917-1938.

Sass, J., 2002. *Women, Men, and Environmental Change: The Gender Dimensions of Environmental Policies and Programs, Population Reference Bureau*, http://www.measurecommunication.org.

Shoko, D. S. M., 2002. 'Small-scale Mining and Alluvial Gold Panning within the Zambezi Basin: An Ecological Time Bomb and Tinderbox for Future Conflicts among Riparian States', in G. Chikowore, E. Manzungu, D. Mushayavanhu and D. S. M. Shoko (eds), *Managing Common Property in an Age of Globalization*, Weaver Press, pp. 67-85.

Synergy Africa, 2001. *Small-scale Gemstone Miners Cry for Help*, http://allafrica.com/stories/200212060252.html.

Tesha, A. L., 2003. *Information about the GMP Project Site in Tanzania*, Report to Global Mercury Project, p. 13, http://www.unites.uqam.ca/gmf/intranet/gmp/documents/documents.htm.

Traore, A. P., 1997. 'Socio-economic Aspects and Gender Issues Related to Artisanal Gold Mining', Proceedings of the UNIDO Expert Group Meeting on *Introducing New Technologies for Abatement of Global Mercury Pollution Derived from Artisanal Gold Mining*, UNIDO Headquarters, Vienna, July 1-3.

UNESCO, 2000. 'Women and Girls: Education not Discrimination', *World Education Forum*, Dakar, Senegal, April 26-28. http://www2.unesco.org/wef/en-press/press-kit_wome.shtm.

UNIDO, 2003. *Final Proposal: Removal of Barriers to the Introduction of Cleaner Artisanal Gold Mining and Extraction Technologies*, Project Number GLO/01/G34, p. 86.

USAID, 2000. *Mining for Gold in Siguiri: A Close Look at a High-risk Population, Report on Guinea*, US Agency for International Development, http://www.usaid.gov/gn/health/news/001226_mines/index.htm.

Veiga, M. M., 1997. *Introducing New Technologies for Abatement of Global Mercury Pollution in Latin America*, Rio de Janeiro UNIDO/UBC/CETEM/CNPq, Rio de Janiero, Brazil, p. 94.

Veiga, M. M., 2003. *Personal Communication: ASM in the Ingessana District of the Sudan*.

Wagner, S., 2003. *Socio-Economic Survey Of Rwamagasa Mining Site In Geita District*, Report to Global Mercury Project, p. 40, http://www.unites.uqam.ca/gmf/intranet/gmp/documents/documents.htm.

Wasserman, E., 1999. 'Environment, Health and Gender in Latin America: Trends and Research Issues', *Latin American Research Issues, Environmental Research Section A*, 80, pp. 253-273.

Wente, B., 2002. 'It's the Women Who'll Save Africa', *Globe and Mail*, December 14, p. A21.

Wiego, S., 2003. *Women in Informal Employment: Globalizing and Organizing*, www.wiego.org.

World Health Organization (WHO), 2000. *Silicosis*, Fact Sheet no. 238, http://www.who.int/inf-fs/en/fact238.html.

World Health Organization (WHO), 2001. 'Advances in Behavioural Medicine', *The World Health Report – Mental Health: New Understanding, New Hope*, http://www.who.int/whr2001/2001/main/en/chapter1/001b2.htm.

Chapter 11

Women in the Mining Industry of Contemporary China

Linqing Yao

Introduction

Chinese women were for a long time at the bottom of the social hierarchy and completely excluded from social and political life. Even though the situation has changed a lot and women now appear in any social activities, men still play the dominating role in China. Up to now, women are considered as a weak group with older and younger people together. It can be demonstrated from the role that women are taking in social activities.

Mining and quarrying were thought of as masculine areas in traditional China. Whether in big or small mines, miners were almost invariably male. Women who worked in mines formed an extraordinary minority. After the formation of the People's Republic of China, women workers were encouraged to engage in every sector to contribute to overall society. In the mining industry, women were not only participating in surface work but also underground until the law of Provision for Protection of Working Women was passed in 1992, which prohibits women from working underground in mines. Whilst the fulfillment of the law is for protecting women's rights, it also deprived women of some working opportunities in mines. The specific focus on protecting women's biology, meanwhile, reflects traditional images and medical beliefs about women as well as fluctuating Chinese state policy toward women (Woo, 1994).

Women workers are seen as less reliable, less efficient, and more expensive than male workers (Yanjie Bian, 2000), so gender discrimination is obvious in workers rights in contemporary China and women's social position has even declined in some areas. Women are considered to be unsuitable for work in the heavy labour sectors so that the proportion of female employees in the Mining and Quarrying sector is the lowest among all 16 sectors. Economic reforms have resulted in large numbers of workers being laid-off, and women being first to be laid off has become a common phenomenon. This situation is worse in Mining and Quarrying, where the impacts on women workers are more severe. Even so, as women are powerless in this sector, the situation did not attract attention from the government or from resident Chinese academics.

Most of the data in the chapter are from China's yearbooks. It should be noted that the published data on Chinese female miners are limited and hard to collect, so I have had to use data concerning the entire Mining and Quarrying sector as the frame for the whole chapter, with emphasis on data for the mining industry. The reasons for the lack of data lie in the fact that China's statistics are often at a macro level, sometimes even contradictory between different published sources, because the types of units and the categories of employment are complicated and confusing as they have been changed many times. Consequently detailed and systematic information for all kinds of employees is hard to find. For example, the China Statistical Yearbook and the China Labour Statistical Yearbook focus on urban employees, and ignore rural employees such as people working in Township and Village Enterprises, for which very limited statistical work was done, despite the fact that employment of Township and Village Enterprises keeps increasing.

Some data are collected from sector yearbooks. According to the definitions of the National Statistics Bureau of China, the Mining and Quarrying sector is composed of coal mining and processing, petroleum and natural gas extraction, ferrous metals mining and processing, non-ferrous metals mining and processing, non-metal minerals mining and processing, other minerals mining and processing, and logging and transport of wood and bamboo. Those sub-sectors are administered by relevant independent government departments, and their part of the sector's yearbook is organized and edited by relevant departments. But in these yearbooks, much attention is paid to economic benefits, outputs, efficiency, technical items, etc, rather than on employment issues, so the data about female employees are few. Again some data do not exactly correspond between yearbooks.

In the past, the Chinese government controlled and administered business affairs closely. For example, there was a labour department under the Ministry of Coal Industry, which governed all the business units of the coal production system, dictated to the business units, and gathered information concerning employment within them. China has undergone massive government institutional changes in the past few years. Central government decentralized management to local government, and reduced the administrative interference to business units. Before 1997, coal mining and processing was controlled by the Ministry of Coal Industry; ferrous metals and non-ferrous metals mining and processing by the Ministry of Metallurgical Industry; and petroleum and natural gas extraction by the Ministry of Petroleum Industry. The ministries had strong powers to control business units with most of the big decisions being made by them at that time. There was a big institutional change from 1997 to 2000 when, under the Ministry of Energy, the Bureau of Coal Industry and the Bureau of Petroleum Industry were set up instead of the respective formal ministries, whilst the Ministry of Metallurgical Industry changed into the Bureau of Metallurgical Industry under the State Economic and Trade Commission. The power of those bureaus to control business units was diminished, but they still had influence over their administration. Another government institution reform occurred in 2001 removing some powerful administrations and setting up new institutes such as the Office of Coal and Construction Material and the Office of Metallurgical and Non-ferrous Industry. They were designed to supervise and adjust the mining industry by macroeconomic policy, not by dictating directly. The result of this huge flux is that no government department intends to collect the details of miners and few officials know anything about them or are willing to make available what they do know (Gunson, 2000); even a lot of older information was lost because no department takes responsibility for that and so no official cares about it.

As already noted, China's enterprises are categorized into several types by ownership, state, urban collective, private, township and village owned (TVE), etc. For the purpose of promoting economic growth and absorbing the surplus rural labour force and discouraging excessive urban migration, the TVE boomed rapidly and came to dominate China's rural economy after the economic reforms. For example, one quarter of gross domestic product and one third of farmers' income came from TVE in 1999. TVE contributed significantly in the Mining and Quarrying industry. In 2000 there were 55,141 TVE in the Mining and Quarrying industry employing 2,936,683 people, of which 16,302 TVE were in coal mining with 1,311,032 employees (CTVE, 2001). This number does not include small private mines and illegal mines. In fact, the data on employees of TVE are probably suspect as the employees working in TEV are not considered as formal employees unlike people working in State-owned or Urban Collective-owned units.

Further research needs to be done to deeply understand the issues related to women miners, for this topic has been neglected in contemporary China. This can be done only through extensive fieldwork collecting the primary data first-hand, as it is difficult get detailed information from published data and the real data are usually unavailable.

Women at work in contemporary China

The significant roles of Chinese women in society and economy have emerged in the middle of the twentieth century. In Chinese history, under the Confucian doctrine *Nanzunnubei*, that is, 'women are inferior to men', Chinese women had been put at the bottom of the social hierarchy. There was the oligarchy of gender, which made fathers, brothers, husbands and sons the oligarchs over the mothers, sisters, wives and daughters, and ordained all men sovereign, all women subject. For national salvation and self-emancipation, Chinese women waged a dauntless movement that lasted for over a century. Major social changes affecting women's lives took place between 1949 and the end of the 1970's under Maoist philosophy. There was a surge of mass movements throughout the country to quickly change the backward economic and cultural outlook left by old China and to eradicate the antiquated social system and outmoded customs that fettered, discriminated against and humiliated women (IO, 1994). At that time, equality between women and men was an official state policy as the representation of successful communism. 'Women can hold up half the sky' was a dominant slogan that consequently mobilized women. The principle of 'equal pay for equal work' was emphasized by government. The Chinese Communist Party had made great political efforts to raise the status of women. Therefore, to a certain extent in contemporary Chinese society, women enjoy political, economical, cultural and legal rights. It was evident to observers in China in the 1970s that most women in China perceived their situation and the conditions in which they performed their productive and reproductive activities to have been much altered and improved by the communist government (Croll, 1983). For example, in China during the 1970s, the number of women working in top central political bodies was high according to international standards (IO, 1994). After the economic reforms of the post-Mao era, the Chinese government has paid attention to developing its legal system to embody women's rights rather than rely on political power. While clearly delineating the various concrete rights and interests of women, China's legislation amplifies the responsibilities of state organs in protecting women and clarifies the special status and role of the women's federations and other women's organizations in lawsuits thus establishing a comprehensive legal mechanism for protecting women (IO, 1994). 'Chinese women have begun to work through the legal process to achieve their equality and will likely continue to do so' (Woo, 1994).

Women's status changed fundamentally after the Chinese revolution and they swarmed into every sector, devoting energy to building a new socialist China. This can be demonstrated by the change in the number of women in the labour force from 0.6 million female employees working in State-owned units (SOU) in 1949 to 1.85 million in 1952 and by 1960 it reached about 10 million. After that, it decreased during 1960 to 1963, and rose again from 1964. When the Chinese economic reforms began, the numbers of women workers in SOU in China was about 23 million in 1979. After the economic reforms, China began to experience an economic boom, and its employment developed rapidly[1]. Because of the decentralization and the encouragement of private economic endeavours, the proportion of SOU and urban collective-owned units (UCOU) has dwindled and TVE and private business units have flourished. Employment all over the country is constantly expanding, but it has decreased in urban units. Total employment was about 681 million in 1995 and 737 million in 2002, corresponding to a stable proportion of around 56-57 per cent. The employment in TVE has also prospered from about 92 million in 1997 to 133 million in 2002. Meanwhile the trend of employment in urban units fell, the turning point being 1995. The fall was dramatic

---

[1] In the period of planned economy, private business was restricted and state-owned and collective-owned units could represent all the Chinese workplace. After the economic reforms, many different types of units appeared, of which main types are state-owned units, urban collective-owned units, TVE and private businesses. TVE and private businesses expanded very quickly and state-owned and urban collective-owned units declined.

during the time between 1997 and 1999. Female employment in urban units experienced the same trend as the total employment, going down from 1995 and increasingly so in 1997–1999. The numbers of employees in urban units dropped from 153 million to 110 million, with females decreasing from 59 to 42 million between 1995 and 2002 (Table 11.1). The opposing trends of employment change in rural areas and urban areas are the result of Chinese economic reforms. More and more workers are laid off from traditional formal units[2], but thriving township and village enterprises and private businesses absorb a large number of the surplus labour force leading to a significant change in the nature of employment for both males and females.

Table 11.1 Employment in urban units (million)

	1995	1996	1997	1999	2000	2001	2002
Total	153	152	150	121	116	112	110
Women	59	59	58	46	44	42	42

Source: China's Labour Statistical Yearbook, 2003

All the urban workplaces (units) of China are divided into three types: Enterprise, Institute and Organization, and also into three types according to ownership: State, Collective and Other. The female proportion in such units amounts to 38 per cent, with the highest being in Institutes and the lowest in the Organizations. In terms of ownership, more women are in the 'Other', whereas those in the State units comprise the least portion. Overall, people usually prefer to work in Organizations and SOU, which currently have the lowest female proportion and which is continually reducing. To compare the data of 2001 to 2002, it can be found that the female proportion in all units has declined except for State-owned Organizations, which increased slightly, from 25.1 per cent to 25.2 per cent. For the category Other-ownership Institute, females dropped from 60.0 per cent to 46.3 per cent (Figure 11.1). This indicates that gender discrimination against women's rights to work not only exist but are also getting worse.

Some work categories or positions were thought of as unfit for women under Chinese traditional beliefs, a view strengthened by the Women's Rights Protection Law. Thus gender segregation by occupation is not just an occasional phenomenon. From the distribution of female employment within sectors[3], the

[2] State-owned units, urban collective-owned units and some other urban ownership units are traditionally regarded as formal units in China.

[3] There are 16 sectors according to China's Statistical Bureau: Farming, Forestry, Animal Husbandry and Fishery; Mining and Quarrying; Manufacturing; Electricity, Gas and Water Production and Supply; Construction; Geological Prospecting and Water Conservancy; Transport, Storage and Communications; Wholesale and Retail Trade, Restaurants; Financial Intermediation and Insurance; Real Estate Activities; Social Services;

largest number of women (about 13 million) are to be found in Manufacturing, accounting for 31 per cent of all urban female employees; the second is 'Education, Culture and Arts, Radio, Film and Television', about 7 million. The Mining and Quarrying sector ranked tenth together with the sector of Construction, and its female employees number 1.37 million or 3.3 per cent of total urban female employees in 2002.

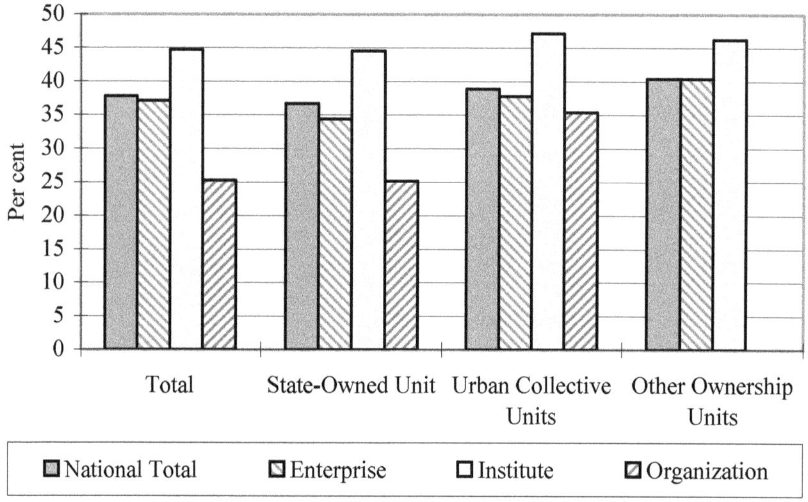

Figure 11.1 Proportion of female employment in urban units grouped by Enterprises, Institutes and Organizations in 2002 (per cent)

Source: China's Labour Statistical Yearbook, 2003

In recent years, the distribution of female employees among the sectors has changed significantly. Whilst Manufacturing still has the largest number of female workers, it has gone down from 42.1 per cent in 1995 to 30.8 per cent in 2002. The Mining and Quarrying sector has shown a downwards trend at a similar pace in its share of female employees (4.1 per cent in 1995 to 3.3 per cent in 2002). Women are tending to go into the Financial Intermediation and Insurance, Real Estate Activities, Social Services, Health Care, Sporting and Social Welfare, Education, Culture and Arts, Radio, Film and Television sectors. Although Real Estate Activities has the second lowest proportion of females, it is rising quickly, being twice what it was in 1995 (Table 11.2). In fact, female employees are growing in

Health Care, Sporting and Social Welfare; Education, Culture and Arts; Radio, Film and Television; Scientific Research and Poly-technical Services; Government Agencies, Party Agencies and Social Organizations; Others.

the tertiary sector and declining in the secondary, both in number and proportions a result of China's economic structure transition.

Table 11.2 Proportions of female employment in every sector over total female employment in urban units (per cent)

	1995	1996	1997	1999	2000	2001	2002
Mining & Quarrying	4.1	4.0	3.8	3.8	3.5	3.3	3.3
Farming, Forestry, Animal Husbandry and Fishery	4.3	4.0	4.1	4.1	4.4	4.3	4.1
Manufacturing	42.2	40.7	39.3	33.4	32.3	31.2	30.8
Electricity, Gas & Water Production and Supply	1.4	1.5	1.6	2.0	2.1	2.2	2.2
Construction	3.5	3.5	3.5	3.3	3.3	3.2	3.3
Geological Prospecting & Water Conservancy	0.6	0.6	0.6	0.6	0.7	0.7	0.6
Transport, Storage & Communication	3.8	3.9	4.0	4.3	4.4	4.4	4.4
Wholesale & Retail Trade, Restaurant	14.6	14.5	14.4	11.4	10.5	9.3	8.4
Financial Intermediation & Insurance	1.9	2.0	2.2	3.0	3.2	3.5	3.8
Real Estate Activities	0.5	0.5	0.5	0.7	0.8	0.9	1.0
Social Services	3.7	3.8	4.0	4.6	4.8	5.0	5.2
Health Care, Sporting & Social Welfare	4.2	4.4	4.6	6.0	6.3	6.7	6.9
Education, Culture & Arts, Radio, Film and Television	10.0	10.6	11.2	14.8	15.6	16.7	17.1
Scientific Research & Polytechnic Services	1.0	1.1	1.1	1.2	1.3	1.3	1.3
Government Agencies, Party Agencies & Social Organization	4.1	4.4	4.4	5.8	6.1	6.5	6.5
Others	0.4	0.7	0.9	0.8	0.9	1.0	1.1

* Percentage = the number of female employees of particular sector / the total number of female employees in urban units

Source: China's Labour Statistical Yearbook, 2003

The major occupations of women in urban units in 2000 were working in production professional and sales and services sectors (Table 11.3). Compared to men, women are more represented in sales and services. Currently many Chinese women have a good education, but because of gender segregation, they find it more difficult to get managerial positions, so they tend to work as professional staff. The apparent proportion of professional women is higher than men's, 23 per cent versus 17 per cent, however there are almost twice as many male employees as there are females in absolute numbers.

Table 11.3 Proportion of urban female workers by occupation in 2000 (per cent)

Gender/ Occupation	Managerial/ Administrative	Prof.	Clerical	Sales & Service	Production
Male	8.0	16.7	11.6	21.4	34.3
Female	6.1	22.8	8.9	30.8	22.0

Source: The report of Second Survey of Chinese Women's Social Status

Before the Chinese revolution, women workers earned much less than men. For example, in the cotton textile industry in 1929, women's wages were roughly 10 to 45 per cent lower than men's (Bobby Siu, 1981). The policy of 'equal pay for equal work' for men and women was enshrined by the People's Republic of China. Women were encouraged under the slogan 'Anything a man can do, a woman can do also', however, the wage differentials between men and women never disappeared so there is some argument that the Chinese state never equalized labour market opportunities between men and women (Bian et al., 2000). Gender discrimination in employment is ubiquitous and worse in present-day China than before. Although 'equal pay for equal work' is still implemented by the Chinese government, male preference obviously occurs in the process of recruitment and promotion. Once women have a job, they can get the same basic salary as their male counterparts, but the problem of the 'male preference' custom means that women's working rights are greatly limited. Looking at the working status of women overall, they are generally in subordinate and unimportant positions and get fewer promotion opportunities and lower average earnings than men. The average annual earnings of female employees in the urban area was 7,409.4 RMB (~US$900) in 1999, only 70 per cent of that for men. When compared with the data for 1990, this represents a 7.4 per cent increase in the income gap. Almost half the women had an income lower than 5,000 RMB. The proportion of women in the lower income bracket was 19.3 per cent higher than men[4]. The gender inequalities in income are caused by women generally working in the lower ranks, with fewer chances to be promoted. Occupation segregation make women gather in low

[4] Chinese Women's Federation collaborated with China's National Statistic Bureau, December 2000, The Second Survey of Chinese Women's Social Status.

income occupations and sectors. Yanjie Bian's study shows that when men change occupation, work unit, or both, they have the clear advantage of ending up in administrative, market activity and market potential jobs rather than other jobs, and, in contrast, while women in any job change category have an advantage of moving into clerical jobs, they have no opportunity to move into administrative jobs, and their chances of moving into jobs with market involvement are less than their chances of moving into jobs with no market involvement or decision making (Bian et al., 2000).

In a market economy wage differentials depend more upon education and work experience. Because women have lower average education than men, they receive less money. The Labour Insurance Regulations, first instituted in the 1950s and reinstituted in 1979, differentiate between men and women by requiring women workers to retire at age 50, as compared to age 60 for men. This prevents women from getting more chances of increasing their salary.

After the reforms, the economy developed rapidly and with the changes in the labour system, women's working status dropped to some extent, and gender discrimination increased. This manifests itself not only in segregation by occupation and gender income gap, but also in the female employment rate. From 1990 to 2000, the employment rate of women and men both dropped, but for women the reduction was bigger than for men. At that period, the employment rate of women cut down from 76.3 to 63.7 per cent, and of men from 90 to 81.5 per cent[5]. So far, the economic reforms have expanded inequalities between men and women. Facing free competition and economic benefit, women have been confronted with new challenges and discriminations, especially in terms of working rights.

Employment in mining and quarrying and its change

China is relatively well-endowed with mineral resources and has a large mining industry. It is a major player in the international mineral market. There are 156 varieties of mineral deposits in China, with tin, tungsten, titanium, antimony, and tantalum resources being the largest in the world, and coal, bauxite, molybdenum, niobium, silver, and lithium resources are the second most abundant. China is unique in the scale and diversity of its mining industry. It is one of the world's largest producers and consumers of coal, and also a major mineral exporter, ranking among the top countries in sales of tin, lead, zinc, magnesium and a host of minor metals and industrial minerals. Total trade in mineral products exceeded 800 billion RMB a year in recent years (Asia Pacific Bulletin, 2003). The gross industrial output value of Chinese mining and quarrying was 540 billion RMB in 2000, accounting for 6.3 per cent of total Chinese gross industrial output value.

[5] Chinese Women's Federation collaborated with China's National Statistic Bureau, December 2000, The Second Survey of Chinese Women's Social Status.

Coal mining is the major part of China's mining industry, being now the largest in the world. Chinese early industry originated from the mining and transport industry. Consumption and production of coal as primary energy has been around 70 per cent of total energy for decades in China and it is expected to be 60 per cent in 2010 and no less than 50 per cent in 2050[6]. China has 11.6 per cent of world coal reserves, the third biggest in the world. Consequently coal is regarded as a strategic material in China.

The coal output was just 25.96 Mt (million tons) in 1948, and 32.40 Mt in 1949 from about 200 coal mines. The mining industry developed rapidly thereafter and coal output has ranked the first in the world since the 1990s. Increasing demand for coal after 2000 led to around 1,600 Mt being mined in 2003, nearly 50 times that in 1949. It included 800 Mt produced by State-owned mines controlled by central government, 289 Mt produced by State-owned mines controlled by local government and 520 Mt produced by TVE and private mines.

There are mainly three types of coal mines: state-owned with central government control, state-owned with local government control and TVE[7]. Currently there are more than 3,000 state-owned mines, including 106 large state-owned mines controlled by the central government, which provide more than 50 per cent total coal output. There were only about 10,000 TVE in 1980, but they expanded quickly and reached 80,000 or so in 1997. It was a sort of malignant swell that caused many problems for the coal industry. From December 1998, government forced the closure of some small mines that were illegal or substandard, the number of TVE mines declining to 25,000 in 2003[8].

In 1978, just before the economic reforms, about 95 million were employed in the SOUs, with 6.5 million in the Mining and Quarrying sector. The trend of employment in the Mining and Quarrying sector is similar to the employment in SOUs as a whole as set out earlier in this chapter, increasing until 1995 when a decline set in. By 2001, there were about 5.4 million workers in SOUs in this sector (Figure 11.2). The proportion of employment in this sector over total employment constantly decreased, from 7 per cent in 1978 to 5 per cent to 2001. In the coal industry, the trend of employment mirrored that of the Mining and Quarrying sector, the number of employees of SOUs rose in the 1980s and early 1990s, and then fell thereafter. The peak of 5.45 million was in 1993, declining to 3.7 million in 2001.

---

[6] Zhongguo Meitan Ziyan Gaikuang, June 2004, http://www.cctd.com.cn/html/detail/02/09/12/1040278.html.

[7] Because private mines have similar characteristics to TVE – small scale and regarded as informal, and it holds little proportion in mining industry – in most Chinese statistics, private mines are accounted in township and village mines.

[8] China's Coal Industry Statistical Yearbooks, Beijing: Meitan Gongye Chubanshe.

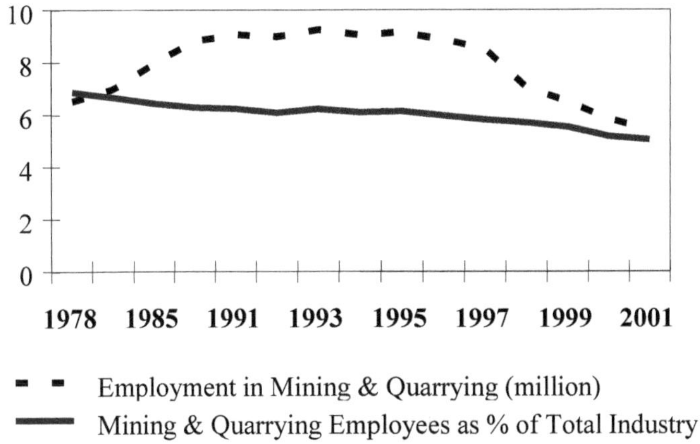

Figure 11.2 Employment trends in SOUs in Mining and Quarrying and its percentage over total employment in SOUs

Source: China's Statistical Yearbook

The downtrend of employment in stated-owned mines is mostly due to four factors:

1. China has drastically changed its economic structure favouring tertiary industry whilst the proportion of primary industry and second industry has been cut back.
2. A large number of employees were dismissed or laid off as state-owned mines ran into the red, financially.
3. The Mining and Quarrying sector is labour intensive, so it received priority in automatization.
4. Expansion of TVE and irregular competition from them meant that state-owned mines lost some parts of their proportion in terms of employment.

Women's work in the mines

Chinese women were seen to be inferior to men in the traditional society. Men worked outside to support their family and women worked inside and did housework. This formula has been in place for thousands of years. Women appeared in the labour market at the early stage of industrialization in China. There were about thirty thousand women workers around 1912 to 1916 (Liu Jucai, 1989). Most of them worked in the textile industry. A 1947 study showed women workers in the cotton textile industry in Shanghai as making up 65 per cent to 70 per cent of all women workers in manufacturing industries (Zi, 1947). Men comprised the overwhelming majority of the mining workforce at that time. In the 1930s, there were about 20,000 to 30,000 miners working in big mines, and tens of thousands of part-time miners working in small mines (Laite, 1991). Whether in big mines or small, miners were almost invariably male. Female miners who worked underground formed an extraordinary minority, but they could, indeed, be found occasionally. For example, the Weekly Newspaper of the Mining Industry recorded women miners in Xi County, Hunan province, who worked with men together underground, and all did the hard work as men, pulling out coal from the coal pit[9]. This was an uncommon phenomenon in old China. Some women did assistant types of works, such as conveying, on the ground in some small mines and even their number was small. Why did Chinese women play an insignificant role in the early stages of the country's mining industry? The reasons are as follows.

The availability of abundant cheap male labour made it easy to employ a large number of mature male workers thus reducing the chances of a woman to be employed in the mines. Additionally, the factors of culture and custom in traditional China ensured that mining was regarded as a masculine work. Miners usually worked without clothes in the pits in very harsh working conditions. Consequently, it was inconvenient to have both sexes working together. Another factor was the conventional superstitions. In many places, people deemed it ominous to have women present in the mining tunnels. The limits of women's physical abilities because of footbinding, a custom in old China, worked against their mobility. When girls were seven or eight years old, some of them were even younger, their feet were bound into *Sancunjinlian*, which were very small and shaped triangle feet. This custom, though abolished in 1911, was still practiced in some regions until 1930 (Verschuur-Basse, 1996). Women suffered from it for all their life to please the false aesthetic values of men. Footbinding made heavy work impossible for women. This is reflected in the fact that as footbinding prevailed in Huabei region but was unpopular in Huanan region, women miners were mostly found in Huanan region.

A law was promulgated in 1925, limiting women and children working in mines[10]. However, this was not main reason of the lack of participation of women in mining.

In the Mao era (1949–76), the policies emphasized the belief that 'women are the same as men'. The 'equal right' was over-stressed and applied to every aspect of social life, ignoring the biological differences between men and women. Women and men during this time were viewed as unisex, and women's issues were seen as societal, not biological (Woo, 1994: 184). Hence with that kind of conception of 'equal right', a large number of women came into the work sphere of heavy labour industries.

This phenomenon was more prevalent in the late 1950s and early 1960s. Some young women, in a fanatic and revolutionary mood, volunteered to do tough work, such as altitude work, heavy work in the pit, poisonous and noxious types of work,

---

[9] Kuye Zhou Bao, p. 1213, 14 May 1936.

[10] Zhonghua Minguo Xianxing Fagui Daquan, 1933, p. 748, Shanghai: Shangwu Yinshuguan.

etc., which was previously considered unsuitable for women. The media glorified women's public roles as proletarian fighters and as 'Iron Girls' (Honig, 2000). Women appeared underground to work with male miners as part of the 'equal rights' policy, especially in the big mines. They set up work groups named 'Iron Girls Well-sinking Team' or 'Female Work Team', or other similar names, intended to demonstrate the special character of the new generation women. Even so, most woman miners usually acted as assistants in coal mines. The Iron Girls represented only a very tiny percentage of the female population and left unaltered (and uncontested) the status of the vast majority of women. During that period, women as real miners working underground were still a minority, and indeed they had more of a symbolic value of the emancipation of women from the ties of tradition.

The Chinese government did not order women to work underground. In fact, woman miners had many troubles when they had to compete with the men in the tough work. Their health was damaged by the demands on their strength and the harsh worksites. Some of them were affected by gynaecological diseases, which dampened down women's enthusiasm and made them aware of the realities of work in the mines. A couple of accidents and disasters occurred in the mines leading to female casualties. At the end of 1960s, an accident occurred in the Taoyuan mine of Shaanxi Tongchuan coal company, killing a woman miner who drove a tramcar underground. This accident raised the issue of women in the mines with the result that the government ordered a stop to employing further women miners. But there were still a small number woman continuing to work in pits. In 1982, there were 190,442 women miners, representing 4 per cent of all miners[11]. Another national earthquake disaster happened in the 1980s, which claimed the lives of all 28 women working in a mine in Jixi city. Then the 'Provision for the Protection of Working Women' was issued for enforcement in 1992, prohibiting the employment of women working underground in mines. The 'Labour Law' promulgated in 1994 strengthens that ordinance, banning the engagement of women in mining jobs under the mine well. Since then, women miners have been deployed at the surface.

Thus most of the women miners working underground had already disappeared, but employing women illegally as miners still exists in certain areas. These only become public knowledge when big mine disasters occur, such as the one in Baixing mine, on February 23, 2004. Baixing was a small privately owned mine located in Jixi city in Helongjiang province. Here a mine gas explosion claimed the lives of 37 miners of which two were women. Subsequently, the administration stopped the production of all the small private mines temporarily for security inspection in Jixi city. As a result, 39 woman miners were found working underground in a state-owned mine. These women worked in mines to earn a livelihood and support their families. How many women miners are still left in

[11] Women Study Institute of China Women's Federation and The Study Institute of Shaanxi Women's Federation Zhongguo Funu Tongji Ziliao 1949–1989.

China? No one knows exactly, including the government officials. The participation of women in mining continues to be intentionally shrouded, keeping them in the shadows.

Why do some mine owners continue to take the risk of employing women workers in underground mines? Do women want to be miners? There were a total of 5 women miners in Baixing mine before the accident. They all came from poor families, so they voluntarily did men's work in dangerous and rough conditions, in order to get more money to support their poor families. One example is Wang Jingmei, one of the victims in the gas blast. Her husband was also a miner in Baixing mine before but he had been injured in a mine accident and became disabled in 1996; he got a small amount of compensation, only 180 RMB (~US$ 25) per month. Before becoming a miner, Wang had worked in a small factory. The factory went bankrupt in 1998, and Wang lost her job. The financial condition of this family became perilous with the only income being the husband's compensation and not enough for the sustenance of the entire family. Their daughter was a university student and her tuition fee was more than 10,000 RMB every year. Wang had made a great effort to get the job as a miner working underground. She wanted this job very much, even if it was illegal and dangerous, because she could earn a higher salary than working on the ground. The job was not easy to get, since there were many rivals. There were too many poor families and unemployment in the mine area. She had to beg people to get that job. In fact, the manager of Baixing mine had intended to sack the five women before the mine accident, but those five women remained through many kinds of relationship, as they could not get a better job. They needed the job more than the job needed them as there was widespread unemployment and many laid-off workers in this part of China, which is an old industrial area with many struggling state-owned enterprises in a poor state. As employers, they would certainly have liked to use women's labour in the mines as miners because of the lower cost. Even working underground, Wang Jingmei's salary was just 300 RMB per month, which was lower than male workers, but which was high for women workers in Baixing mine. Wang did even the tough and dangerous jobs as her family was still saddled with a debt of 40,000 RMB. This is why Wang didn't care more about her health and life to risk doing illegal work. The government prohibits women working underground in mines, for the purpose of protecting women's perceived virtues. However, since everything has more than one simple dimension, this law also makes a woman lose the opportunity to gainfully find employment in mining work and turns her into a lower waged, illegal worker.

Due to the perceived unsuitable nature of work in the Mining and Quarrying sector, numbers of female employees have always been on the low side compared to that in other industry sectors, the proportion being 25.3 per cent in urban units in 2002, the third lowest of the official 16 sectors, much less than 37.8 per cent of total women urban employees. Before the economic reforms, businesses had to get permission from the government labour department to recruit labour. To enhance the status of women and protect women's rights, the government set a quota for their employment. This ensured that women could enter every sector and helped balance the gender proportion. The proportion of female employment in the Mining and Quarrying sector was always kept over 20 per cent.

Within the mining industry, even though the women workers' proportion is significant, it does not imply that women miners are ubiquitous. Far from it, for, as a result of enforcing the law of women's rights protection, women miners have disappeared from China's mines except for illegal ones. In the mining industry, women mostly work in unit-affiliated branches such as byproduct factories, clubs, libraries, hospitals, sanitariums, epidemic prevention stations, nurseries, schools, etc. The SOUs had put great emphasis on workers' welfare before economic reform, so a lot of affiliated branches were run. A big SOU was like a small self-contained society. There still are many affiliated branches which are run for workers' welfare in the mines, and many state and collective-owned mines run, after the economic reforms, some new business branches besides coal production which are called the tertiary business sector of the company or unit involved. There are two main purposes for operating such businesses: one is to expand the business and get more profit; another is to absorb the surplus labour force of SOU employees, most of which are the laid-off workers, or the family members of the employees, in order to appease their own employees. In fact, they absorb many female workers.

Because it is thought that only a few jobs are suitable for women in collieries, there are not many working there. Some do office work, and others do some assistant and service work in collieries, such as in the pithead bathhouse, miner's lamp depot, refectories, etc. Although the proportion of female workers in Mining and Quarrying was the third lowest among all the sectors, in 2002 the proportion of middle-level female technical personnel was about 35 per cent of all mining workers. Clearly, women who are working in the mining industry are not in physical work, yet their contribution to the mining industry of China cannot be denied.

Among the Mining and Quarrying sector, the proportion of employment in UCOUs is 37 per cent, much higher than 25.2 per cent in SOUs and 23.2 per cent in Other Owned Units (Figure 11.3). This is because some UCOUs are not in business in the area of mining and quarrying, even though they are in the sector of Mining and Quarrying. They engage in tertiary businesses. In 2000, the employees of SOUs in the coal industry were 3.42 million, of which women were 0.75 million, with a proportion of female employment of 22.1 per cent. The breakdown by Enterprise, Institution and Organization was 21.8, 30 and 25 per cent respectively.

The trend in female employees in the Mining and Quarrying sector paralleled the trend in all sectors in urban areas. The number of female employees in the Mining and Quarrying sector steadily increased until 1995, decreasing thereafter, indeed falling by more than 40 per cent in 2002 (Table 11.4).

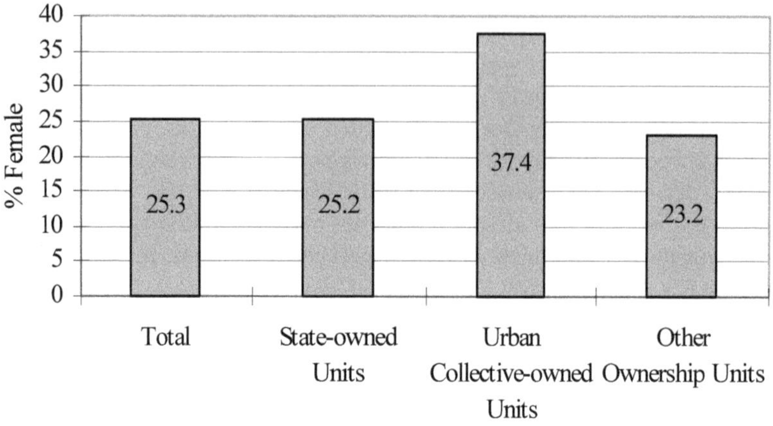

Figure11.3 The proportion of female employment in the Mining and Quarrying sector (per cent)

Source: China's Labour Statistical Yearbook

Table 11.4 Female employment in Mining and Quarrying (million)

Date	1994	1995	1996	1997	1999	2000	2001	2002
Female Employment	2.29	2.39	2.33	2.23	1.74	1.53	1.39	1.37

Source: China's Labour Statistical Yearbook

One cause of the decline is resource exhaustion. A report issued by China's Intendancy Committee of State Property shows that 3080 state-owned large enterprises closed because they ran out of resources, or because the companies were experiencing other difficulties, largely financial. There are another 2,500 state-owned, resource-exhausted, large mining enterprises and other state-owned large enterprises already facing bankruptcy at present (China Coal, 2004). That is part of the reason why Chinese state-owned enterprises have dropped in numbers, both in terms of scale and employment numbers. China has 178 mining cities, more than one fourth of the total number of cities in the country, among which 10 per cent are facing resource exhaustion, and are in dire need of transformation (People's Daily, 2004). Among these cities, 29 have been engaged in the mining industry for less than 20 years, 131 have a history of 20 to 50 years, and 18 are with a mining history of over 50 years. The cities with the longest tradition of mining and old stated-owned mining companies had large numbers of employees, many of whom have lost their jobs. The employee numbers in the Mining and

Quarrying sector decreased from 9.14 million in 1995 to 5.22 million in 2001, a 43 per cent reduction with a corresponding impact on the level of female employment.

From 1985 two new items could be found in China's Coal Industry Statistical Yearbook, under the categories of 'Employment of the Young Unemployed', and 'Employment of Family Members of Miners', and they disappeared in 1992. Here the 'young unemployed' normally implies the young unemployed who had never had a job and are related by blood – usually offspring – to formal workers of the mining enterprises. When they became mature, the enterprises became responsible for helping them to look for jobs and get established in society. In regard to 'Employment of Family Members of Miners'[12], the family members of miners were mostly miners' wives and some were mature-aged children of the miners, who had just moved from rural areas to cities. In the planned economy period, people could just get work from the local labour department. The government had full power to deploy labour resources and had a responsibility to offer jobs to urban residents. Even in the 1980s, this continued in some areas, especially in old traditional industries. The appearance of these two items in the statistical yearbook demonstrates that the employment was still based on the planned economy system in the coal industry at that time. The pace to move to a market economy was slow in mining areas, because of the heavy burden of dealing with the unemployment there left over from the times of the traditional economy.

The term 'Employment of Young Unemployed' was common throughout China at that time. It was merely tougher to implement in old mining areas because the unemployment rate was much higher with its many associated social problems. The term, 'Employment of Family Members of Miners', was a special issue in the Mining and Quarrying sector. It is closely related to the so-called Hukou system, which divides people into two groups, namely urban and rural residents, issuing residency permits to that effect. In effect it offers a method for controlling rural to urban migration.

Because of the large difference in the economic status of urban and rural areas, Hukou was very important to Chinese people. Those with urban residence permits can get permanent jobs and health care and other kinds of social welfare, making urban life better and more attractive to rural residents. The Hukou system is still in existence but reformed to allow greater migration from the countryside, along with a reduction in privileges for urban residents. Although the Hukou is not as important as before, the difference in the treatment of urban and rural residents in contemporary China makes Hukou a matter of concern. Hukou is usually inherited

[12] There were two groups of people working as miners in collieries. One was formal workers, who were permanent, another was peasant workers, who were temporary and would change regularly. The treatments were different between two groups. The formal workers were urban residents and they could enjoy all the welfare delivered by enterprises; by contrast, peasant workers could just get limited welfare. Here 'miners' refers to the formal workers.

from the mother, meaning that if the mother is an urban resident her children will be urban residents, whatever the children's father's Hukou.

In mining, Hukou manifests itself strongly in the apparent choices of partners. Because the miners' work is dangerous, tough and of low income, urban girls are usually unwilling to choose them as husbands, so most of them had to marry rural women. Since the miner's wife was a rural resident, their children would be rural residents so that they too could not get formal jobs arranged by local government, a house allocated by the company, and many other kinds of benefits that the urban families could have. As there was a large proportion of such mining families, it became a common problem within the coal mining industry. In 1984 a conference was held jointly by several government departments, including the Public Security Department, the Ministry of Commerce and Trade, the Ministry of Labour and Personnel, the Ministry of Coal Industry, in order to address this problem. At that time in coal SOUs, there were 2.8 million rural miners (927,000 families) of which 1.8 million were involved in actual underground mining. A policy was made at that conference that miner's family members could be changed into city residents when they met certain standards. There were 370,000 such families which could meet the standards at that time, amounting to a total of 1.2 million people who would get permission to become city residents.

Then a new difficulty arose for coal companies. When those people settled down as at least nominal city residents, the coal companies were dictated to help them to get jobs. To encourage and supervise this process, the program 'Employment of Family Members of Miners' was established in the former Ministry of Coal Industry. This work was regularly carried out from 1985 and finished in 1992. There were 76,683 family members of miners who got jobs in 1985, with the peak being 168,252 miners' family members being found jobs in 1987. In 1992, the last year of this program, the number was 120,086. Because most of the family members of miners were female, the proportion of female employment increased in the coal industry. For example, the proportion of female employment of SOUs in the coal industry kept going up during 1982 to 1987, from 14.8 per cent to 16.6 per cent[13]. An efficient way of absorbing the unemployed was to set up collective-owned businesses, a frequently used practice at the time. A large number of women were arranged into collective-owned businesses, which were run by state-owned coal companies. At that time, the ratio of female workers in the Mining and Quarrying sector was higher for collective-owned units than in SOUs and other ownership units.

[13] Data from China's Coal Industry Statistical Yearbook.

Table 11.5 Average earning of State-owned units by sector in 2002 (RMB)

Sector	Rank	Average Earning
National Total	n.a	12701
Mining and Quarrying	14	10580
Farming, Forestry, Animal Husbandry and Fishery	16	6234
Manufacturing	13	10825
Production and Supply of Electricity, Gas and Water	4	15636
Construction	12	11139
Geological Prospecting and Water Conservancy	10	12219
Transport Storage, Post and Telecommunication	3	15758
Wholesale and Retail Trade & Catering Services	15	9371
Finance and Insurance	2	18313
Real Estate Trade	6	14144
Social Services	11	12067
Health Care, Sporting and Social Welfare	5	15121
Education, Culture and Arts, Radio, Film and Television	9	13237
Scientific Research and Polytechnical Services	1	18792
Government Agencies, Parth Agencies and Social Organization	7	13858
Others	8	13258

Source: China's Labour Statistical Yearbook

The data concerning women's income in the Mining and Quarrying sector is unavailable, because, theoretically, women and men get equal pay and hence there is no need for official departments to collect such information. It can be inferred from the whole sector's average earning. In 2002, the average earning of state-owned units in the Mining and Quarrying sector was 10,580 RMB, ranked 14th in all of 16 sectors, 2,121 RMB less than the average earning of national total stated-owned units (Table 11.5). Among the Mining and Quarrying sector, the highest earning was 19,605 RMB, got by employees of Petroleum and Natural Gas Extraction; the employees of Coal Mining and Processing earned the third one, 10,108 RMB in 2002 (Table 11.6).

At the time of the planned economy, the principle of 'distribution according to work' was implemented, and the Labour and Personnel Department of government made the wage scales for all sectors. Because miner's work is tougher, harder and more dangerous than others, their payment (salary and subsidies) was the highest in China and the average income of the Mining and Quarrying sector was one of the highest among all sectors.

Table 11.6 Average earning of mining and quarrying employees in SOUs (RMB)

Sector / Year	2002	2001	2000	1999
Coal Mining & Processing	10108	8868	7592	6686
Petroleum & Natural Gas Extraction	19605	18636	16091	14450
Ferrous Metals Mining & Processing	11830	9855	8991	8545
Non-ferrous Metals Mining & Processing	9164	8260	7714	7073
Non-metal Minerals Mining & Processing	8111	7190	6367	5931
Other Minerals Mining & Processing	8477	6968	7123	6067
Logging & Transport of Wood & Bamboo	4896	4662	4537	4287

Source: China's Labour Statistical Yearbook

After entering the market economy, the situation has reversed. The old traditional industry lost its advantage and SOUs stagnated that made the rank of average earning of Mining and Quarrying sector sink to the bottom. The worst thing is not only the payment being low but also not always paid on time. In bad financial periods, it was common that even the stated-owned mines could not pay their workers regularly and normally. However, the stated-owned enterprise workers were promised they would get all of their income when businesses ran into good status. In recent years, the economic status of the coal industry has risen and paying of arrears largely eliminated in state-owned mines.

Some data about the income gap between women and men in individual mining units can be derived from the work of Guo Rongxing (2004), who studied two state-owned mines (here called J and L) belonging to the same big coal mining group in Hebei province in October 2000. The study showed that men's average income was 1244 RMB in J mine and 991 RMB in L mine, versus women's 1029 RMB in J mine and 577 RMB in L mine. This means women's average income is 83 per cent of men's in J mine and 52 per cent in L mine which can be compared with a value of 70 per cent found for women countrywide by the Second Survey of Chinese Women's Social Status carried out by the Chinese Women's Federation in collaboration with China's National Statistic Bureau. It should be noted that informal workers were excluded from the survey. Most women in mining are working in assistant and service roles on the ground, for which the salary is much lower than key positions and underground positions. It is likely therefore that in the majority of mining units the gap is likely to be bigger than the national average.

In some small mines, the situation is worse – awful and dangerous working conditions, poor protection, simple and crude machines, and low payment. In small mines working underground, the miner's salary is about 500-1000 RMB a month, and women workers who work in assistant type roles on the ground is about 200-500 RMB, which varies widely from region to region. In fact, there is only a small number of women working in private small mines, as women hardly find work opportunities in them.

In state-owned mines, workers can get reasonable health care according to government policy. For example, when a miner gets black lung, he can get medical treatment free of charge, and will be given an easy job on the ground. If the miner is seriously ill and loses his working ability, he can retire in advance of the certain retirement age. If a black lung sufferer becomes disabled and has to depend on others for personal care, he can even get some money for nursing costs. In some big mines, miners suffering from an occupational disease would be sent to a sanatorium to recuperate. Women get the same health care as men and, for special physiological reasons, they can even get extra treatment. In the big state-owned mines, female employees are organized to get regular health inspections, usually once every two or three years. Women can get special treatment during five periods of pregnancy, delivery and nursing.

In small private mines, or TVE, the situation is very different. Because most workers in these units are temporary, they can't acquire the same treatment as those of state-owned enterprises. Some of them would hardly get any health care. All-China Women's Federation conducted a survey that shows that 32.1 per cent of male employees and 37.8 per cent of female employees were not offered health care by the units who employed them (ACWF, 2001).

State-owned enterprises offer a more regular and legal environment due to an integrated organizational structure, as well as a much better social security system. In big state-owned mines, they usually have a women's committee under the leadership of the union to protect women's right. The women's committee has many functions, such as training women, appraising women through comparison, helping the poor women workers' families, organizing some cultural activities for women. For instance, the women's committee of the West Mountain Coal Corporation pushed the enterprise to issue a rule that if women get the top three places in the skills competition of the enterprise, they would get a bonus, 10,000 RMB for the champion, 5,000 RMB for second place and 3,000 RMB for third[14]. A second example is the women's committee of Datong Coal Corporation, which organized its female employees and male employees' wives to build a security volunteer team that offers services in the areas of mine safety and security. The

[14] Women's Committe of West Mountain Coal Corporation, Shishi Shuzhi Gongcheng Tisheng Nvxing Shuzhi, Journal Dangdai Kuanggong, Volume 3, 2004, Beijing.

team has 903 female volunteers[15]. Women's committees' influence on women varies from unit to unit. Some of them have put in a lot of effort to guarantee women's rights, and some of them offered smaller services, but even so they can still do a little to help women at least, much better than without them.

Table 11.7 Educational level distribution of employees in Mining and Quarrying in 2001 (per cent)

	Total (%)	Illiterate/ Semi-literate (%)	Primary (%)	Junior Secondary (%)	Senior Secondary (%)	College (%)	U/G (%)	P/G (%)
Total	*1.8*	*1.0*	*1.7*	*2.4*	*1.5*	*0.6*	*0.5*	*0.2*
Male	2.6	2.4	2.7	3.7	2.0	0.8	0.6	0.3
Female	0.7	0.3	0.5	0.7	0.9	0.4	0.4	

Note: For each of the columns, 100 per cent represents the whole of the national total employees in that particular category

Source: China's Statistical Yearbook

Table 11.7 shows the distribution of educational levels of employees in the Mining and Quarrying sector in 2001, at different levels of education, ranging from illiterate/semi-literate through secondary school to tertiary educated. Each figure on the table represents the proportion of employment of a particular educational level of the Mining and Quarrying sector in national total employment of relevant educational level. Among the national total employees, the employment of the Mining and Quarrying sector occupied 1.8 per cent, which included male 2.6 per cent, and female 0.7 per cent. Taking those numbers as base, comparing male with female educational level distributions, it can be found that there was a much bigger proportion of male employees in the educational level of illiterate/semi-literate and primary school than females (2.4/2.6 and 2.7/2.6 versus 0.3/0.7 and 0.5/0.7 per cent). The highest proportions of women employees are those with secondary school education. That implies that the entrance threshold of the Mining and Quarrying sector for women is higher than for men, and women have fewer chances to get jobs in mines than men. The relative proportion of tertiary educated women is also lower than men's, which means the sector has no strong attraction to higher educated women.

[15] Women's Committe of Datong Coal Corporation, BaiLi Meihai Niangzijun, Journal Dangdai Kuanggong, Volume 3, 2004, Beijing.

Conclusion

Although women are prohibited from working underground in the pits, they are present in a significant proportion in the mining industry and contribute greatly by their diligence and intelligence. They hold positions throughout the organization, for example, in Jincheng Coal Group Corporation, some are taking on important management work. Cao Xueqing, in her thirties, is the vice director of the Mechanic Shop of Wangtaipu mine. Another example is Wang Jinyin who has worked in the company since 1992 and became the vice director of Jin Cheng Haisi Medical Company. The West Mountain Coal Corporation organized a skill competition meeting in 2002, in which women got seven gold medals out of a total of 52.

Women's educational levels tend to be higher than men's in the Mining and Quarrying sector which explains why the proportion of female technical personnel is higher at 35 per cent. The percentage of female technical personnel in all female employees in the coal industry is 35.6 per cent compared to the overall average of 22.1 per cent.

Women play a significant role in the Mining and Quarrying sector, not only through their own employment, but because of their husband's. Though women cannot work as miners underground themselves, as miners' wives they can influence their husband's behaviour and inspire their husbands' working spirit. The wives' influence is recognized in China. For encouraging them to support their husbands' work, many state-owned mines organize rewards events such as selecting 'good wives' or 'good families', etc.

Chinese women in the mining industry are facing a big challenge. As has been demonstrated earlier, the sex discrimination is prevalent in the industrial rights for women in China. In the reform era, Chinese employers see women workers as less reliable (married women with children have the potential to interrupt work schedules in order to fulfill child care responsibilities), less efficient (married women are less amenable than men to overtime assignments), and more expensive (work units must provide maternity leave with full pay and child care services to married women with children) than men (Bian et al., 2000). Chinese women are becoming disadvantaged. The mining industry is often thought of as a men's arena which makes the issue worse. Women are laid off disproportionately highly in relation to men, proving this view. For example, in Fengfeng Coal Group Corporation, there were 76,949 employees, including 19,509 women of which 5,113 were laid off, accounting for 26.2 per cent of the female workforce. In Chenghe Mining Company, there were 2,823 women workers, and 810 of them were laid off, the proportion being 28.7 per cent. The ratio of female laid off workers is far higher than the average ratio of laid off workers in state-owned mines, which is 15 per cent.

The Chinese government is endeavouring to put equal rights for women and men in action. However, the regulation of protecting and ensuring women's rights is not yet entrenched, and the traditional old concept of women being inferior is deeply rooted in people's minds. The belief of 'regardless of what a woman did or what roles she might undertake, she nevertheless retained a changeless womanhood as the core of her identity' (Hinsch, 2004) is still popular, so that the introduction of the market economy has put strong pressure on women. Chinese women are now suffering new forms of discrimination.

The problem of working rights of women is more grievous in the mining industry. It is an old industry in China; some old state-owned mines carry heavy historical baggage, such as resource exhaustion and large numbers of surplus workers who are not easily redeployed. There was a certain period when the whole sector of coal industry was stagnant, state-owned mines faced huge economic pressures, conditions deteriorated and women were often laid off first (this happened in other sectors as well). There was even a consideration about whether it should let women go home and leave work positions open for men in society.

Poverty and unemployment are the primary reasons for the existence of illegal female miners. They exist not only in TVE and small private mines, but also in state-owned ones. Most of the mining regions in China are poor areas, where the social security system is not good, and unemployment rates are high. Many women would rather do any job that they can get money from, and would not consider whether it is legal or suitable. It is just a struggle for survival. Government let women leave their work positions underground in order to protect them, but poverty and unemployment harasses them more than working underground. The rule of protecting women's rights therefore makes an irony of the situation. The only way to protect women is by improving their economic status, so as to ensure their basic survival rights; then illegal female miners would disappear.

The government should do more for women's roles. Chinese women have been kept in an inferior position for thousands of years. The deeply rooted concept that 'women are inferior to men' is not easy to dispel. Although regulations and rules designed for women's protection have been gradually promulgated, they are not enough. The legal environment in China is still unsatisfactory; so many women still can not receive practical protection from a bad situation. Government should push for more powerful measures to reduce sex discrimination perhaps by affirmative action such as stipulating sex ratios among employees for different types of units.

In the mining industry, the old state-owned mines had, in the past, made great contributions to the national economy, so their burden should be taken on partly by state, otherwise they could not stand up. As old and exhausted enterprises, they by themselves are not able to accelerate or extend their businesses to absorb surplus employees, even though they are devoted to doing so. A good example is Fengfeng Coal Group Corporation which made a great effort to assist its laid off women workers in finding new jobs. Some of the laid off workers were trained in doing house services and got such jobs in other cities, with the help of the enterprise. It was taken as a good example in re-employing laid-off women workers, and was recommended to other enterprises. However, even though the enterprise worked hard to do that, only 40 of the 5,113 laid off woman workers got this kind of new job (Liu Fang, 2004), far less than enough.

The Government is in debt to old state-owned mines, and is responsible for doing more work to help the stagnant old mines. While the government pays more attention to new economic reforms, the existence of old industries should not be neglected.

References

ACWF, 2001. *Survey of the Implementation of the Law of the People's Republic of China on the Protection of the Rights and Interests of Women*, All-China Women's Federation,. www.china.com.cn/chinese.zhuanti/241810.htm.

Beijing Women's Federation Researcher Group, 2003. 'Beijing Xiagang Nvgong Zaijiuye Xianzhuang ji Liangxing Bijiao', *Population Research*, March.

China Coal, 2004. *Quanguo 2500 Zuo Dazhong Guoqi Kuangshan Ying Pochan Guanbi*, 23 April, www.chinacoal.gov.cn/coal/jryw/031121x2.htm.

China Concept Consulting, 2004. *Mining Modernization*, http://www.china concept.com/en/resources/government/kyxdh.asp

China's Coal Industry Statistical Yearbooks, various years. Beijing: Meitan Gongye Chubanshe.

China's Labour Statistical Yearbooks, various years. Beijing: Zhongguo Tongji Chubanshe.

China's Statistical Yearbooks, various years. Beijing: Zhongguo Tongji Chubanshe.

Chinese Women's Federation and China's National Statistic Bureau, 2000. 'The Second Survey of Chinese Women's Social Status', *Funv Yanjiu Luncong*, May.

Croll, E. J., 1983. *Chinese Women since Mao*, London: Zed Books Ltd.

Croll, E. J., 2001. 'New Spaces, New Voices: Women Organizing in Twentieth-century China', in Ping-chun Hsiung and Ceciliq Milwertz with Bed Chan (eds), *Chinese Women Organizing*, Oxford: Oxford International Ltd, pp. 25-40.

CTVE, 2001. *China's Township and Village Enterprises Statistical Yearbook, 2001*, Beijing: Zhongguo Tongji Chubanshe.

Du, Fangqin, 1988. *Nuxing Guannian De Yanbian*, Henan: Henan Remin Chubanshe, pp. 28-212.

Edwards, Louise, 2004. 'Constraining Women's Political Work with Women's Work', in Anne E. Mclaren (ed) *Chinese Women – Living and Working*, New York: Routledge/Curzon, pp. 109-110.

Goodman, David S. G., 2004. 'Why Women Count', in Anne E. Mclaren (ed) *Chinese Women – Living and Working*, New York: Routledge/Curzon, pp. 19-41.

Guihua, Yang, 1995. 'Strengthening Social Control and Protecting Women's Employment', in Tan Lin and Li Jianxin (eds) *Women and Sustainable Development*, Tianjin: Tianjin Kexue Jishu Chubanshe, pp. 60-74.

Gunson, A. J. and Yue Jian, 2001. *Artisanal Mining in the People's Republic of China*, Commissioned by the Mining, Minerals and Sustainable Development Project of the International Institute for Environment and Development, http://www.iied.org/mmsd/mmsd_pdfs/asm_china.pdf.

Guo, Rongxing, 2004. Shengchan *Jishu Tiaojian Chayi yu Qiye Zhigong Shouru Fenpei.*

Hinsch, Bert, 2004. *Women in Early Imperial China*, New York: Rowman & Littlefield Publishers Inc.

Honig, Emily, 2000. 'Iron Girls Revisited', in Barbara Entwisle and Gail E. Henderson (eds), *Re-Drawing Boundaries: Work, Households and Gender in China*, Berkeley: University of California Press, p. 98,

Hu, Jihai, 1931. *Journal: Special Issue on Mining Industry*, Guangzhou, p. 311.

IO, 1994. *The White Paper of Chinese Women's Status*, Beijing: Information Office of the State Council Of the People's Republic of China, June, p. 1.

Jiang Yongping, 2003. 'Guanzhu Laodongli Shichang zhong de Xingbei Pingdeng', *Funv Yanjiu Luncong*, March.

Jiang Yongping, 2003. 'Liangzhong Tizhi xia de Zhongguo Chengshi Funv Jiuye', *Funv Yanjiu Luncong*, January.

Johnson, Kay Ann, 1983. *Women, the Family and Peasant Revolution in China*, Chicago: The University of Chicago Press, Ltd, pp. 155-177.

Jun Shan, 2004. *China's Policy on Mineral Resources*, http://chineseculture.about.com/library/china/whitepaper/bl2003mineral04.htm retried on 1st July, 2004.

Kuye Zhou Bao, 1936. 14 May, p. 1213.

Laite, Tim, 1991. *Zhongguo Jingji He Shehui Zhong De Meikuangye*, Beijing: Dongfang Chubanshe, p. 216.

Liu Dezhong and Niu Bianxiu, 2000. *Zhongguo de Zhiye Xingbei Geli Yu Nvxing Jiuye*, Beijing: Funv Yanjiu, no. 4, pp. 46-48.

Liu Fang, Dianran Xiagang Nugong Xiwang Zhi Huo, 2004. *Journal Dangdai Kuanggong*, Beijing, 3.

Liu Fuguo, Jixi Yunan Nukuanggong Diaocha, 2004, http://www.china-woman.com/gb/2004/03/27/zgfnb/zhxw/7.htm.

Liu Jucai, 1989. *Zhongguo Jindai Funv Yundong*, Beijing: Zhongguo Funv Chubanshe, p. 271.

Liu, Shisheng and Liu Yang, 1999. *Zhongguo Funu Tongshi*, Qingdao: Qingdao Chubanshe.

Ma Gengcun, 1995. *Zhongguo Jindai Funu Shi*, Qingdao: Qingdao Chu Ban She.

Mao Jian, 1991. 'Zhongguo Funv Jiuye Xianzhuang Yu Zhanwang', in *Zhong Mei Funv Wenti Yataohui Lunwenji*, Beijing: Zongguo Funv Chubanshe.

People's Daily, 2004. 16th February, Beijing.

Siu, Bobby, 1981. *Women of China, Imperialism and Women's Resistance 1900–1949*, London: Zed Press.

Tan, Lin, 1995. 'A Penetration of the Situation of Women's Status in China: The Equality and Inequality between Men and Women', in Tan Lin and Li Jianxin (eds) *Women and Sustainable Development*, Tianjin: Tianjin Kexue Jishu Chubanshe, pp. 24-36.

Tan Sheng, 2000. *Gongye Fazhan yu Funv*, Beijing: Funv Yanjiu, no. 2, p. 40.

Tao Chunfang and Jiang Yongpin, 1993. *Zhongguo Shehui Funv Shehuidiwei Gaiguan*, Beijing: Zhongguo Funv Chubanshe, pp. 80-101.

Verschuur-Basse, Denyse, 1996. *Chinese Women Speak*, Westport: Praeger Publishers.

Vibekw, Hemmel and Sindbjerg, Pia, 1984, *Women in Rural China – Policy Towards Women Before and After the Cultural Revolution*, London: Curzon Press Ltd.

Watson, Rubie S., 1994. 'Girls' Houses and Working Women', in Maria Jaschok and Suzanne Miers (eds), *Women and Chinese Patriarchy*, London: Zed Books Ltd, pp. 27.

Weekly Analysis of Breaking Issues in Asia Pacific, 2003. 'China Looms Large in the Global Mining Industry', *Asia Pacific Bulletin*, 4th April www.asiapacificbusiness.ca (retried on 1st July 2004).

Women's Study Institute of China Women's Federation and The Study Institute of Shaanxi Women's Federation, 1991, *Statistical Data of Chinese Women 1949–1989*, Beijing.

Woo, Margaret Y. K., 1994. 'Chinese Women Workers: The Delicate Balance between Protection and Equality', in Christina K. Gilmartin, Gail Hershatter (eds), *Engendering China: Women, Culture, and the State*, Boston: Harvard University Press, p. 284.

Xinping, Gan, 1995. 'Women's Employment: In the Background of China's Economy and Culture at Present', in Tan Lin and Li Jianxin (eds), *Women and Sustainable Development*, Tianjin: Tianjin Kexue Jishu Chubanshe, pp. 45-59.

Yanjie, Bian, Logan, John R. and Shu, Xiaoling, 2000. 'Wage and Job Inequalities in the Working Lives of Men and Women in Tianjin', in Barbara Entwisle and Gail E. Henderson (eds), *Redrawing Boundaries: Work, Households and Gender in China*, Berkeley: University of California Press, pp. 112-113.

Zhang Yufa, 2003. *The Change of Chinese Women's Suffrage in the First Half of the 20th Century*, Taibei: Zongyang Yanjiuyuan Jindaishi Yanjiusuo Chuban, pp. 39-71.

Zhong Mei, 2004. *Funu Wenti Yantaohui Lunwenji*, Beijing: Zhongguo Funu Chubanshe.

Zhongguo Meitan Ziyuan Gaikuang, June, http://www.cctd.com.cn/html/detail/02/09/12/1040278.html.

Zhonghua Minguo Xianxing Fagui Daquan, 1933. Shanghai: Shangwu Yinshuguan, p. 748.

Zhu Chuzhu and Jiang Zhenghua, 1991. *Zhongguo Nuxing Renkou*, Henan: Henan Renmin Chusanshe, pp. 117-143.

Chapter 12

Women in Small Scale Gold Mining in Papua New Guinea

Geoff Crispin

Introduction

Papua New Guinea (PNG) is a culturally diverse country. A population of approximately 5.5 million is spread across a wide variety of cultural parameters with over 700 language groups contained within a population of approximately 5.5 million. This incredible cultural diversity is further complicated by geographical parameters, ranging from the elevated Highland areas in the west, to the riverine Sepik Provinces, Fly River basin in Western province to the Island groups from New Ireland, Manus, New Britain, Bougainville, and various coastal groups and the numerous island communities in Milne Bay Province. There are few roads and most transport is by sea, aircraft of various types or by walking.

This cultural diversity means that the status and role of women varies. Some parts of the country are patrilineal and others are matrilineal. The economic and social power of women therefore is not the same throughout the country but this is still overlaid with the problems of domestic violence, rape, HIV/AIDS and lack of economic equality between women and men, and lack of adequate opportunities to obtain that equality.

The focus on mining in PNG tends to be on the large mining activities such as Porgera, Lihir, OK Tedi, Tolokuma and the prospects of Hidden Valley, Kainantu and Fredia River (Figure 12.1). The SSM communities and its contribution to both rural and national PNG have been largely overlooked in the past.

Since the early part of the last century the importance of small scale mining (SSM) to rural communities' economic activities and survival has increased. PNG's independence in the 1970s increased the scale of SSM more quickly. SSM has now grown to the stage where it takes place in all provinces of PNG including the National Capital District (Susapu and Crispin, 2001). The financial returns from SSM underpin the economic activity or are an essential addendum in many parts of rural and remote PNG.

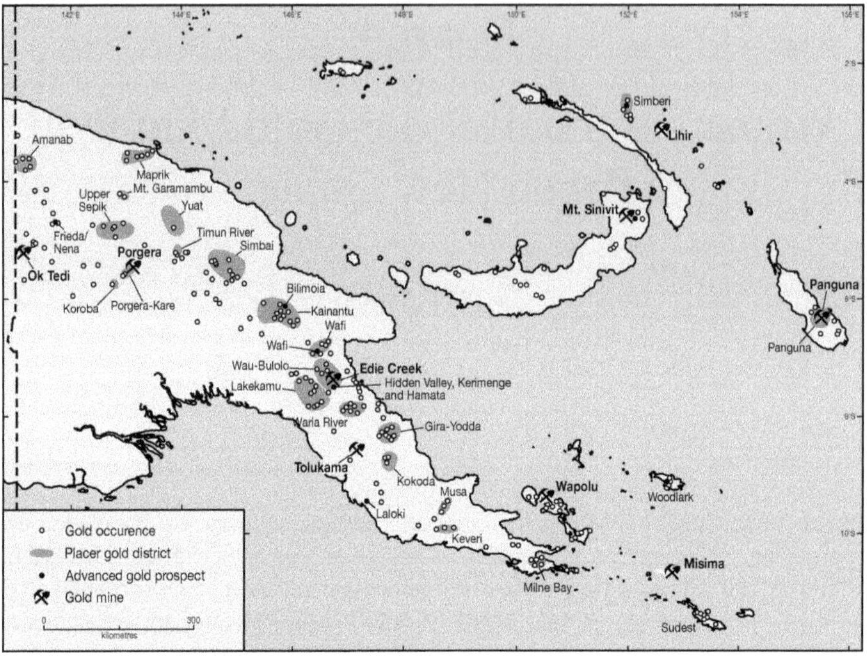

Figure 12.1 Goldfields of PNG (after PNG Chamber of Mines and Petroleum)

In the long term, the sustainability of these SSM activities and on-going positive contributions to development are subject to a number of managing parameters. These include limits to resources, environmental impacts, occupational health and safety issues (especially use of mercury and future possible cyanide use), legal and legislative frameworks, relationships with existing or proposed large mines and the level of mechanization of the mining processes.

The Asian Development Bank (ADB) rated PNG on the same level as Burundi and Mali using the human poverty index with 31 per cent of the population living below the poverty line of $1 USD/capita/day. The richest 10 per cent account for 36 per cent of consumption and the poorest 50 per cent account for just 20 per cent. The report further suggests that there is a dualistic pattern of growth with part of state revenues used to support the growth of a large public service. The subsistence sector has shown little growth (ADB Report, 2000).

The challenge is to make sure that all sections of the population share in an equitable and just economic growth that is sustainable. The role played by SSM becomes important in meeting this challenge as it provides a vital source of income for many rural-based communities particularly in view of the central government's increasing inability to service the rural and more remote areas of the country. SSM thus provides an excellent community-based mineral output area. However, the access for women to the economic benefits, skills involved, and awareness of possible health and safety issues involved in SSM is important for community-based mining to be sustainable and make an ongoing contribution to economic survival and development.

A brief overview of the small scale mining sector

The Department of Mining (DoM) collected data on an informal basis over the three years 1998–2001 during the Australian Government's Technical Advisory Support program (AusAID) and the picture that emerged at the time was highly complicated and variable. The AusAID Project 1998–2001 provided the basis for an informal census of the economic base of SSM, population of SSM communities, how much gold was being produced, distribution of the gold mining communities, problems faced by the SSM communities healthwise, and social and environmental issues therein.

The social change generated by mining has resulted in a change from a subsistence-based economic system to a cash-based system in some areas and a mixture of the two systems in others. These changes to the economic and social structure mean that cash becomes essential particularly to access medical supplies, travel and social responsibilities.

In PNG most of the rural incomes are from agriculture and many of the people are paid in kind or non-wage rewards for their activities. Approximately 50 per cent of the population receives some kind of monetary benefits and the rest are mainly subsistence farm/fish dependent (ADB, 2000). This fact by itself shows how much the SSM activities have changed the social organization in the gold mining areas with cash usually readily available all year round from gold sales. Mining has become such an important way of life in some areas that many people do not garden extensively any more. Their incomes are derived solely from mining. In PNG this is unusual as home gardening has been the primary source of food for the families for almost all communities. Most of this gardening is carried out by women.

When people need money for school fees, medical expenses, social obligations or travel they mine some gold in areas where SSM is well established. This means that people are full-time miners around the beginning of the year when school fees are due and again around Easter and Christmas times as many PNG people follow the Christian religion. The need for cash is not met by the limited number of opportunities thrown up by the stagnating economy in the rural areas continually ailed by the challenges of negotiating extensive and often unbridgeable distances over hills, forests and the sea. Yet this need for cash arising from changing aspirations and consumptions of the people is what drives the SSM in most places.

However, there is a sizeable proportion of the population that mines on a part-time basis all year round. Yet another section of the population has agricultural interests such as growing coffee or copra. No mining is carried out when crops are to be harvested until that work is finished. Thus agriculture remains the primary, and, above all, often the preferred, source of income for most rural communities in the gold-bearing areas.

The difficult terrain and limited resources in PNG means that a complete picture of the population and who and how many are mining and what percentage of their income is due to mining is difficult to ascertain, despite the extensive informal survey by DoM. Some areas have been visited and estimates have been made by DoM staff. In the Wau area in Morobe Province the latest census (2001, unofficial figures) indicated a population of about 25,000. The unofficial figure used by the Department of Mining based on observations of the local community over a number of years is that in this area approximately 75 per cent of the population is involved in mining at some time or other. This includes women and school age children.

In the Porgera/Mt Kare area in Enga Province downstream from the Mt Kare prospect, virtually all of the 2,000–3,000 people were mining except for very small children. In the Sepik area it is difficult to estimate the numbers, but around Maprik, the largest single centre away from Wewak (the capital), probably about 10-15 per cent were full-time miners. In some of the villages around Kainantu in Eastern Highlands Province such as Bilimoia about 75 per cent of the population is involved in such mining on a full-time basis.

The information in Table 12.1 was first presented at the Geology Exploration Mining Conference in Port Moresby by Tongo and Crispin in June 2001 and updated as information became available. Word of mouth indicated that there are more populations mining than were visited by the DoM during the three years from 1998–2001. It has been estimated that in PNG's case a further 420,000 people are dependent in some way upon small scale mining. This includes other businesses including transport, supply of food, equipment and other services. This figure represents approximately 9-10 per cent of the population (European Union Interim Report of SYSMIN eligibility of PNG, 2001).

The official figure for alluvial gold production in PNG in the years 1998–2001 was approximately K60 million Kina (Kina to US$ varies from 0.40 to 0.25K). The observations around the country by the outreach team suggested that the total alluvial gold produced was probably in the range of 150-180 million Kina – which is at least two or three times the official figure (Susapu and Crispin, 2001).

Much of this gold may have been exported illegally. These figures were based on a gold price of approximately US $270 per ounce. In the intervening period from 2001 to 2004 the gold price has increased to over USD $400 per ounce. This means that the value of the gold mined assuming that the production estimates remained constant may have increased to K250-300 million Kina per year. Part of this estimated value is directly returned to the rural people and would provide a sizeable proportion of the income of rural families.

Table 12.1 Estimated mining population in PNG

Province	Estimated mining populations
Bougainville	900-1000
Central	2-300
East New Britain	50-100
East Sepik	10-12,000
Eastern Highlands	1000
Enga	4-5,000
Gulf	3-4,000
Madang	2-3000
Manus	20-50
Milne Bay	1000
Morobe	15-20,000
National Capital District	Uncertain
New Ireland	4-500
Oro	4-500
Southern Highlands	2-300
Simbu	Uncertain
West New Britain	A few only
West Sepik	3-5,000
Western	50-100
Western Highlands	50-100
Total	**41,270-53,950**

Role of women in SSM

The estimated mining population is 60,000. Out of this 50 per cent were estimated to be women and children. This estimation is further proportioned with 20 per cent women and some 30 per cent school age children under 16 years. These figures are from the informal survey carried out by the DoM, but the actual proportions and numbers may vary from place to place. In some areas very few women are seen working at mining but in others, especially where families are involved, there may be several generations involved from children, to grandmothers and close relatives.

Only a handful of women were identified as miners in their own right although this does occur in a few areas. More commonly, women are engaged in transporting the dug materials to preferred sites for panning and sluicing and final separation of gold. Women do some of the digging as well. In the Sepik at one site the women moved around the mining area picking up very small flakes of gold from both worked and unworked areas. These scavenged gold flakes were stored in small bottles. It is impossible to say whether the women are able to keep this gold for themselves or have to hand it over to the men, but that might be the case given the prevalence of male domination in the country.

Women tend to stay away from the outreach programs for such miners. A good example of the varying participation of women in the outreach program was that in one area in the Sepik the women remained on the opposite side of the road to the presentation where many men and children were participating in the discussion and viewing the videos presented. This is not an uncommon behaviour for women to shy away from participating in public processes. In Milne Bay, however, when a small group was presented with the same basic information they were very vocal in asking questions especially on the issues of the use of mercury and its problems. In this case they indicated that they would be asking questions of the health workers about their knowledge of the problems.

The first educational outreach trip where the booklets were available was into the Waria River valley close to Wau in geographical terms but required a short plane flight over the mountains into the next valley. We visited several centres and in one of these villages, Asama, the headman bought one copy of each of the first three booklets and immediately the women gathered around in a group to share the information, and one of them proceeded to read out some of the contents to her friends. My understanding was that they were going to use it in their village school as part of their local educational program.

Many a time during the outreach trips, children were the only ones who could read (usually Tok Pisin and sometimes English), so they had to translate the contents of the booklets into Tok Ples (the local language) for the adults.

In the Wau/Bulolo area of Morobe Province several women were seen working on gold processing such as panning and sluicing. These women were usually accompanied by small children who were either helping out or playing in and around the location of mining activity. The stronger and bigger ones among them were also engaged in digging, transporting, panning, sluicing and final separation of gold. They also used it as part of their 'play' activities and in one scene several children could be seen sitting in the sluice box and appearing to be suggesting that it was a transportation vehicle.

Rarely were women seen involved in the 'cooking' of the amalgam. This appears to be almost exclusively a male area of work, probably a result of the exclusively male SSM team in the majority of the program. In many cultures it would be rare for females to openly discuss their activities with unknown males so that may have restricted observations. The other alternative is that the males kept control of the end product to either hold the money coming in or they were the ones transporting the gold to the gold buyer for sale through countryside where *raskol*s (bandits) were an ever present problem. To the credit of the DoM, when this possible shortcoming of male dominance was discussed at length with the senior management, a professionally qualified female was employed on the SSM Branch team when the next opportunity was available.

When filming the first of the three videos, we intended to film women in all sorts of processing of gold including working and panning. When this was being carried out the senior male from the local village kept on criticizing the woman shown on the video repeatedly, until we were forced to ask him to leave to allow the filming to continue. It was unclear if it was simply a gender thing or jealousy because the woman was going to be part of the video instead of him.

During the outreach activities word of mouth indicated that there was one village on an island in Milne Bay Province where only the women were mining. This could not be verified by a visit, nor was the reason for the reported exclusively female mining activities explained.

I will now briefly discuss the role of funding bodies and associated projects in this chapter. As indicated by Table 12.2, both women and men attended the DoM Workshops on SSM.

Table 12.2 People attending DoM workshops on SSM

	Numbers	Per cent
Men	2,910	40
Women	2,040	28
Children	2,244	32
Total	7,194	100

These figures are taken from DoM reports of outreach visits. A large number of workshops were held throughout PNG provinces for miners and non miners. This table shows the numbers recorded by DoM officers on field reports where they were accessible. Some reports where lost, stolen or unable to be located. These figures do not include contacts at the various provincial shows and school visits. For example:

1. For World Environment Day in 1990 in Morobe Province, DoM officers spoke to over 1,000 children in a number of schools, many of whom said they had mined gold.
2. During several days at the Morobe Show in Lae each year many hundreds of people looked at booklets, watched videos and discussed SSM with officers of the DoM.

The figures in Table 12.3 are from DoM reports on outreach in Morobe Province. Together, Tables 12.2 and 12.3 reflect the estimation of the DoM officers that overall the number of miners was approximately 50 per cent men, 20 per cent women, and 30 per cent school age children.

Table 12.3 Workshop attendees at various villages who described themselves as miners in Morobe Province only

	Numbers	Per cent
Men	360	40.5
Women	164	19
Children	360	40.5
Total	884	100

What can be done in assisting women in SSM in PNG

Women make up a significant proportion of the estimated visible mining population and they may make up an unseen proportion as well. To assist them to make a continuing and possible expanding contribution to the SSM with safety and a concern for the environment then they have to be included in educational programs and gain access to support systems as well as financial services. Because of the cultural variations it is important that women be included in the professional teams carrying the various educational activities so that the message gets through to everyone.

The educational and support activities can also be carried out through the various women's groups such as suggested in the Draft report for the SYSMIN project funded by the European Union (SSM section by Burke). The report noted that it was:

> ...widely agreed that during the coming years, after the Porgera Joint Venture closes, Porgera will become a very significant small-scale mining area. When this happens the Porgera Women's Association will have a pivotal role in sustaining social order. In the light of the negative impacts of the 1989 Mt Kare gold rush the PWA will need tangible support to prepare for this. The Association gives skills training, advice, encouragement and support to the women of this mining district.

Health issues are of particular importance, and at the time of the AusAID project medical staff had not been trained to recognize the symptoms of mercury poisoning and had no drugs to treat it. The point was made to the relevant authorities and to the local communities that there should be some attempts to address this shortfall. The health issues are vitally important for women as they are the ones who would most likely spend the most time in huts that may be polluted by mercury accumulation caused by burning amalgam indoors. Such possible full time exposure to mercury vapour could endanger their health and the health of unborn children as well as existing children.

Some progress was made in the health awareness of miners generally in and around the Wau area but only to the point of changing some habits: burning amalgam outside huts, keeping upwind of fires when doing the same procedure and keeping children away from the smoke generated (Crispin, 2003).

The Asian Development Bank project on microfinance in PNG has the opportunity to assist women miners through their pilot scheme in Wau. It is located in the midst of the major SSM field in PNG. There were no major banks represented in this town to support financial services. The micro-bank pilot scheme was aimed at assisting the established small business people to grow and support their local economies.

Miners are all small business people and some of these including women would be capable with assistance to grow their income to support their families and possibly expand by being able to access slightly better technology, for example, better sluice boxes and water pumps to improve gold recovery. There are a number of projects aimed at assisting SSM in PNG including the EU SYSMIN project mentioned above and the Japanese Social Development Fund.

The future of PNG in rural areas will depend on SSM for some time to come. Women play an important role in this activity. Recognition of their contribution and support for maintaining that role and expanding women's economic power base will enable the support of rural communities that have a small share in the little government financial support that trickles down for basic infrastructure.

References

Asian Development Bank, 2000. *Report RR:PNG 32472*, Microfinance and Employment Project, Manila: ADB.

European Union, 2001. Draft Report on the Accessibility of PNG to SYSMIN Funding.

Gavu, V., Tongo, G. and Lole, H., 1999. *Mercury-safe Uses in Small Scale Alluvial Gold Mining*, Department of Mineral Resources, Port Moresby.

PNG Department of Mining (Mineral Resources), 1998–2001. Various informal reports, unpublished.

Subasinghe, G. H. N. S. and Okada, S., 1998. *The Role of Mercury in Alluvial Gold Mining: Handling, Toxicity, Pollution and Possible Alternatives*, Report, Department of Mining Engineering, University of Technology, Lae, PNG.

Susapu, B. and Crispin, G., 2001. *Country Study Report on Small Scale Mining in Papua New Guinea*, Mining Minerals and Sustainable Development Reports, London.

Tongo, G. and Crispin, G., 2001. *Small Scale Mining Update*, Proceedings of the GEM 2001 Conference, AusIMM (Port Moresby Branch), pp. 5-7.

Chapter 13

The Invisible Work of Women in the Small Mines of Bolivia

Els Van Hoecke

Introduction: The mining industry in Bolivia

Bolivia is known worldwide for its tin, silver, gold, zinc, lead, wolfram and antimony, as well as a large range of other mineral deposits. The country lies between the two Andean mountain chains, the Cordillera Oriental and the Cordillera Occidental, up in the high plains and valleys. The presence and importance of state-owned enterprises in the mine industry has greatly diminished in recent years, as a result of new economic and political schemes resulting in privatizations and mine shutdowns.

There are a few large multinational companies in the country mainly oriented to intensive exploitation, use of high technology, more alteration of the surrounding environment, and small use of labour. The small scale mines use little technology, as do the mining cooperatives where miners perform hard physical work, confronting problems arising out of the harsh conditions. The origin of mine cooperatives goes back to the 1930s when mining was in crisis; their growing number had a close relationship with levels of unemployment, as was the case in 1985 when state mines shut down operations and the large number of workers who lost their jobs had to join the cooperatives. Currently, the mine cooperatives have around 53,000 members which can be compared to 4,000 workers in medium scale mining, whilst small scale mining and joint venture mining have only 3,500 and 750 workers respectively.

Due to the growing number of conflicts between mine enterprises and the communities, new models of relationships between the two sectors have come up, in which the entrepreneurs would assume an ethical conduct reflected in concepts of environmental and social responsibility to guide the interaction, taking into account the region's development.

Since 2003 ownership of the mine deposits has come under discussion, triggered on one hand by the occupation of several mines by the workers, albeit on a small scale, and an increase in mineral prices on the other hand, which opens new horizons for the weakened mine economy of the past. After twenty years of crisis, a light illuminates the end of the tunnel, and a feverish interest in mining is beginning to emerge in all mining centres. Many small mines are reopening, the cooperatives have begun to close ranks in view of the growing demand for jobs from persons who want to work in the mines' dark galleries; others who moved away from mining are willing to return. A horde of migrant workers is filling the small mining towns' narrow streets hoping to get work and participate in the bonanza which, as usual, comes accompanied by the opening of entertainment places such as bars and karaoke clubs.

The trouble is, neither the small scale miners nor the mine cooperatives are in a position to take advantage of the opportunity, lack of mechanization being a major problem. The small scale mines need to modify their work methods to obtain better results regarding productivity, thus improving its workers' family life, as well as the life of the region's inhabitants as a whole.

Who are the women mineworkers in Bolivia?

Who are these women who have been doing an invisible work for more than a century? Who are these women who crush rocky stones looking for tin, the mineral that made Simon Patiño one of the wealthiest men in the world? Who are these women sitting on the cold floor in picturesque groups, dressed in coloured clothes, who bend down and move monotonously and regularly, one hand armed with a hammer falling on the metal stone piece secured by the other hand (Mitre, 1981). For centuries mining activity has been considered throughout the world as an activity reserved for men only. Using an expression common in Latin America, this is an activity for *machos*[1]. Based on this assertion, it is worth asking, what is the contribution of women and children to an activity that is said to be reserved for men? Mining is an industry where prejudice about physical and intellectual capacity, in addition to well-known cultural prejudice such as 'women at the mine is synonymous with bad luck', or the supposed inability to respond to extremely stressful situations (fear), promotes inequality. It is believed that the *tio de la mina*[2] usually hides the mineral in the presence of a woman. However, the crisis of the last 20 years has caused women to work inside the mine just like men. The *palliri*[3] women's presence is found throughout Bolivia's mining life. The conditions under which these women were and are forced to work constitutes a real 'national shame'. The work performed by the *palliris* and *guardas*[4] (caretakers) ranks lowest in the work scale. They are brave women working in the formal or informal sector,

---

[1] Spanish adjective denoting masculine virility.

[2] Mine workers believe in the 'mineral owner' represented by the devil or *tio de la mina* (mine's uncle).

[3] *Palliri* is a woman who works at the mine site picking out ore remnants by hand. The word comes from *quechua* meaning 'to pick out'.

[4] Women hired by the cooperative to watch over and take care of the site.

who lack capital and have no industrial security or technology. They have been tirelessly working like this for more than a century.

Ignored by all, these poor women cling to life with resignation, and tirelessly working like ants, they remove soil without pause, break hard stones, transport heavy metal sacks, never complaining, in order to support their children, several in most cases, and to take advantage of the sole right they are entitled to, survival (Wicky, 2002).

Technology is too grandiose a word to describe the *palliri* women's work, most of them members of a cooperative, who pick ore out manually from mucks or pre-concentrated ore remnants. By 2004, there were around 1,500 of these women who were affiliated to 65 mine cooperatives. Within the mining society the *palliri* is considered as a poor woman who, by being a widow, or abandoned by her husband or male companion, a single mother, or for many other reasons, has been forced to do this work. And she herself confirms it when she says she is doing this work to support her children, 'a single mother has to work even as a *palliri* to support her family…'.

The *palliri* women learn their job through observation. They know the minerals by grade, colour, smell, and even taste. They classify the ore according to grade to be later delivered for grinding.

To learn the 'job' is still a task these woman do by themselves, without any guidance. They are fully aware of the physical characteristics of the ore they rescue, pick out, or concentrate; they even know in their own language some of the mining technical terminology; for example, to 'burn the excess of bronze' which actually is the chemical reaction of sulphur acid, copper, and other impurities present; 'to amalgamate and recover fines,' which actually refers to mercury amalgamation precisely to recover microscopic gold pieces.

These women also know the impurities present in the extracted ore; words such as 'veins', 'black', 'bronze', 'copper', etc., refer to the visible physical characteristics of antimony, arsenic, and other compounds which usually contaminate these minerals. Likewise, they know the grades of the mined ore and whether it reaches the grade required for the concentration stage. It takes many years of hands-on experience to acquire this knowledge (Cepromin, 1996).

The *palliri* women sell their produce to commercial companies or to *rescatiris*[5], women who act between the *palliris* and the commercial companies. The cooperatives sell their ore to commercial companies, private enterprises which export ore directly, or after concentration to upgrade it to reach the level required by international markets. The *rescatiris* (retail buyers) buy the ore at lower prices than the commercial companies, paying cash to avoid any paperwork. The ore delivered by these women is subject to excessive discounts on account of humidity and impurities, and they are easily cheated through the amounts of discount.

[5] *Rescatiris* (or rescuers) are middle women between the *palliris* and the commercializing companies.

To reduce the impact of discounts, the women make their own previous calculation of the ore they will deliver, on the basis of the mineral's weight. For this purpose, they have requested the non-governmental organization Cepromin to receive basic mathematics courses and to be taught how to read and write.

The *relaveras* are women who work with shovels to gather tin remnants from the sand of contaminated rivers downstream of the mills. The *veneristas*[6] are women who work in the fields mineralized by alluvial sedimentation brought down from the deposits above but only the leftovers are for these women. The *guardas* watch over the cooperatives' mine entrances. These women live at certain levels on top of the Cerro Rico de Potosi mountain over 4,000 meters above sea level. They are loyal women who should prevent the *juqueo*, ore and materials stealing. They live right there around the mountain in humble adobe constructions, without electricity or water which they have to seek elsewhere.

With the rise in mineral prices at the beginning of 2004, the Cerro Rico, already over exploited for more than 500 years, is once again over populated. Those are the so-called *juqueadoras*[7], women who work illegally in an extremely harsh way. They go into the mine around 11 o'clock at night to clandestinely transport the ore taken out by mine workers, thus helping to steal the mineral from the company's galleries. They also enter into prostitution because of their desperate economic situation. Young eighteen-year-old women enter mine adits to have a night of pleasure: 'The sound of music is heard from the parties they have there, where alcohol consumption is high'[8]. These young girls offer their services in exchange for mineral, and they are paid by the men with three to five loads of ore, which means they receive between 20 to 60 Bolivian pesos per night. They leave at 2 or 3 o'clock in the morning with full backpacks. The number of women making prostitution their way of living is rising, particularly as a result of the increase in mineral prices, which has been followed, not unsurprisingly, by an increase in the incidence of venereal diseases.

Women work inside the mine on equal terms with men; they extract ore just like the men and are exposed to dust and toxic gases as well as chemicals used in the mineral refining process. Poor ventilation and humidity, temperature changes inside and outside the mine, affect all the workers equally irrespective of their gender.

Women's jobs are varied; some control the mills' concentrating tables; others work as cleaning women or doing domestic chores in the offices and homes of the companies' executives. They are also nurses, schoolteachers, and secretaries. Women also take on the roles of housewives, mothers and wives. Since the 1980s, the housewives have joined the productive labour force leaving the domestic chores distributed among their children, an average of 5 to 6 per family. The

---

[6] Women working at the 'veneros' or mineralized camps.

[7] 'Juquear' means to steal certain quantities of mineral at odd hours.

[8] Unless otherwise stated, quotations are from surveys and interviews conducted by the author.

production work to back up the husband includes such chores as *pallar* (pick out) ore, grind, wash, load, vent, pack, weighing, and even inside the mine work is done by these women when they are not busy with other tasks. An activity requiring more skill on the part of these women is to operate the buddles, small stationary hydraulic concentrating machines.

In spite of their participation in the productive process, neither the work of these women nor that of their children is recorded as work and is not paid for as such. The man gets paid more because of higher production. Therefore, these women are statistically invisible. This is very unfair, unless the cooperative's member salary is considered as a family income, not just the man's salary. For survival, in addition to their work as *palliris*, the women also sell beverages and candy. Women overtax themselves with responsibilities, looking for additional jobs to increase the husband's salary, or to have enough to feed the family, since usually she assumes the role of head of the household: '…there were some little old women among them, holding the hammer with trembling hands, and knocking their fingers time and again' (Cajías, 1997). To go from housewife to widow is almost unavoidable in the mine environment. The husband's abandonment or death finds these women almost always unprepared, without any documents to process the widow benefits, without a job, or previous working experience and little information about the husband's activity. 'I've worked since I was fifteen, then I stopped working for a while. After my husband's death I had to work as a *palliri* again, I am working for 16 years now'.

There are no regulations governing women's work. Many widows complain of the obstacles they encounter in obtaining a pension, because their late husbands failed to make the required social contributions. Others are unable to follow the procedure to obtain a pension because they are illiterate, ignore the administrative and legal procedures, or because they do not have an identity card or the documents necessary to get one. *Palliri* women who are members of a cooperative do not have this difficulty since they are required to have an identity card before joining the cooperative.

The *barranquilla* activity, or gold washing, concentrates the work of many women, the *barranquilleras*[9], in the gold mining area. This is an independent and marginal work without any planning, and one which tends to degrade the deposits. The *barranquilla* working woman scrapes soil up with a crowbar and a rustic hard vegetable fibre instrument called *sajraña*, then she puts the soil into a hollow dish where through circular movements in running water soil and sand are discarded leaving physical gold at the bottom. This way she supports her family while the husband is busy in 'mine activities', but her work is at the bottom of the ladder in gold mining activity and the lowest paid. The average daily earnings in this activity is equal to 2 *palos* (sticks). Each 'stick' is worth 100 milligrams of gold, and it costs little more than a dollar (Cepromin, 1996).

---

[9] Women who work in gold mining.

Women help to alleviate poverty without any recognition. 75 per cent of the *barranquillera* women are young, less than 40 years old; half of them are single, widow or separated, that is, women who are heads of household. This work is essential as the only way to generate income to support the family. Illiteracy is high since there are no schools at camps sites. The *barranquilleras*, the *palliris* not affiliated to the cooperatives, and the women helping with the *juqueo* are women working in the informal and illegal sector. Here again statistics do not reflect these women's contribution to the informal mining activities. They are in total abandonment.

Child labour

There were always boys and girls working at the mine mucks[10]. The following excerpt comes from the book 'En las Tierras de Potosi' (At the Potosi surroundings):

> There were young girls, 10 or 12 years old, who were the most enthusiastic and active workers. Many have bandages around their fingers, or ugly scars caused by the hammer or rocks.
>
> Mendoza, 1976

Child labour at the mines is considered the worst of child exploitation. There, boys, girls or teenagers risk their physical integrity and suffer psychological and moral harm. This is recognized under the Bolivian Boy, Girl and Adolescent Code (Law N° 2026) as well as by the International Convention 182 of the International Labour Organization recently signed by the Bolivian state. These children are involved in mining activities working either for a third party or within the family environment looking to increase the family income to improve their living conditions.

These boys and girls, who should be enjoying the rights inherent to their youth, are rather sacrificing hours of sleep to wake up at early hours in the morning in very cold weather to go work at the mine centres. There are no Sundays or holidays for these children, they have no social security, no right to vacations, Christmas bonus or dismissal pay; indeed they are seldom directly paid at all, except perhaps when they help the *palliris*. They are the innocent victims of society.

Children, by the time they are seven years old, start performing several activities from selling ore to even working in the mine itself. Inside the mine they handle the metal cars transporting the mineral. At many cooperatives, mine workers are still carrying the ore on their backs, and young boys are generally selected for this work which requires a lot of physical effort taking the ore to the mill along dangerous narrow and slippery trails. Boys also work as drill-worker

[10] Earth portion around the mine containing mineral remnants.

helpers, handing over tools, cleaning the place around the mine lode, and picking out the ore extracted. Another job performed by these boys during the drilling process is to prepare the dynamite charge which is no doubt the most dangerous of all, but which they are very skilful at in spite of their young age.

Outside the mine these young boys and girls manipulate, over an iron plate, a very heavy tool called *quimbalete* (a hand crusher) weighing more than ten kilograms. They work several hours with this instrument until obtaining a very finely ground mineral. It is a slow and tedious job because the *quimbalete* is heavy and the children are small. It is also a dangerous task since the children stop the instrument to remove the mineral with their own bare hands and then continue grinding, a process in which the risk of getting crushed by the machine is high. Children use chemical reagents such as mixing xanthate with acid to recover better ore. Inhalation of this reagent may cause headaches and a sense of weakness and lack of strength. The children said that drinking a glass of milk (which most of the time they do not have at home) afterwards is beneficial for health. Any contact of this mixture with the skin may cause malformations and burns.

Mineral concentration is done by these children using several tools such as buddles, chisels, drills, hammers and others, which these boys and girls manipulate with a wooden stick in up and down movements. After mineral concentration there remain some remnants called *cajas* or *mud* by mine workers. These are processed again by continuously washing with water to remove any waste and thus recover residual ore. This job is generally done by girls with skilful hands. The last stage of this process is putting the mineral in sacks, a job done with the children's help using shovels. The sacks generally have 25 pounds capacity, and once filled they are carried to the cooperative's dumpcart. The girls participate in this activity, mending the torn up sacks to be reused (Cepromin, 2002a).

Most of the children work after school which leaves little time for school homework thus resulting in low school performance. Most of the children, boys as well as girls, labour for 2 to 3 hours, while others put in more than seven hours. Although payment is very low, it still contributes to the family income. Most of the children are paid in cash for their work, but others receive payment in kind only, that is, housing, clothes, and food. In gold mining, women 'train' their daughters in the *barranquilla*[11] work and to learn the procedure to take care of and handle tools, and this is transmitted from one generation to the other. Men have the chance to perform other jobs such as the use of drills, shovelling, wheelbarrows, or to be loaders, trammer, ferryman, and others. Women only have the exclusive option of the *barranquilla* work.

---

[11] Work done to rescue ore by washing placers at river banks.

Housing: Mine camps

Camps were created around the mine sites; artificial, temporary and inhuman. 40 per cent of the women own their small houses in the mine districts whilst the rest of the families dwell in houses owned by the company or the mine cooperatives. These houses are of medium quality, two or three rooms used as one or two bedrooms and a kitchen. Many of the districts have electricity but some do not have this as a permanent service. Access to sewage, drainage, and waste disposal is not always satisfactory.

Most of the women dream of having their own house outside the mine area. They lack a sense of belonging, their feeling is to be there 'for the time being' only and this, of course, contributes to the lack of motivation in helping to improve conditions in the community or district. In fact, few families now live in the mining area as this is seen as a work place only. The primary cause for migration is to find work: whole families migrate to other mine centres looking to improve their living conditions.

The *barranquillera* women live in improvised houses made of wood, *charo* (a bamboo-type long lasting cane), and palm leaf roofs. When production is good, galvanized corrugated sheets are used for the roof, the house has more rooms and it is built on a better site. Housing, however, is always temporary, as the camp may move because of, for example, changes in the river bed.

Equal rights, is that possible?

Under the Supreme Decree 21060[12] in 1985, the so called 're-localization' of workers took place, whereby more than 20,000 workers lost their jobs resulting in unemployment and consequent mass migrations and disintegration of families. The mine worker has had to look for another job in other cities. In some cases he would leave the family behind, and he would, maybe, visit them once a year. In other cases, the man being alone would find another woman in the tropics or other country towns; he would then assume a double family life along with the first wife and children, having to be responsible for two households. Generally, though, he would remain with the second wife, never returning to his district of origin.

The woman with her children stays close to the mine. She has no choice but to assume her responsibility to work as a *palliri* (hand picker) or as a small merchant while her children, still very young or just reaching adolescence, begin to work in the mine. What is even worse, these women no longer have a point of reference towards a 'normal family', since the man, the father, is no longer present. The lack of the father's physical presence as the 'protector' is traumatic for some youngsters with consequences for their behaviour. The women without husbands end up with

---

[12] A law containing several economic structural changes aimed at controlling the process of hyper-inflation the country was going through.

married men, who keep on turning between their first wife and this second woman. Situations like this cause innumerable conflicts and frequent family quarrels where everybody intervenes and the children are often the victims.

Children who were born around 1985 and who are now 15 to 20 years old have few if any family values, so their parents temporary or permanent separation is a normal fact they have witnessed since their early age: '...why should it be different, when we have always seen things like that...' The parents' attitude towards their children has also changed; now when a young boy or girl decides to leave home, the pain and sadness for the separation is minimal: 'that is life, ... he (she) has decided to leave'.

At present, lack of education and health services at mine communities furthers family disintegration. The women visit the mine community for 2 or 3 days, then return to the city to be with their children who have had no choice but to continue their studies in the city; there are only first-grade or basic schools at the mine communities. In cases like this, the parents prefer to have the children away from home either to protect them from working in the mine or to pass on their responsibilities to the school because of alcoholism. The survey regarding this aspect has yet to be completed, so the data obtained from youngsters attending SENTEC (a Technical Learning Institute in the city of Oruro), should not be generalized. Thus, there are some questions for teachers at mine centres, for example, How to bring back family values? What strategy to apply to recover the sense of belonging within a family? A large number of women are heads of household:

> I was 24 years old when I had my first child who later died of malnutrition. It was a painful experience but I learned not to count on men. I got my 3 daughters through high school, just by myself as I had no husband. I am a responsible mother, I still have 3 younger daughters, 13, 11, and 8, to get through school...

More responsibility, but less instruction

Everything has changed since the mines' privatization. During the time most of the mines were owned by the state, a teacher was proud of having the chance to teach at one of the mine districts' schools because the level of education there was classified as 'very good'.

Education has been rapidly left behind. Mines were no longer important for the future. Hospitals were left with aspirins only, no doctors, no nurses. It can be said that nobody thought mining would recover its place some day; the state was the principal actor which does not count with mining anymore, forgetting that most of the country's foreign exchange came from mining. Many women have attended only basic school, which means they have had the chance to study up to 14 years of age. Others have gone through the third or fourth grade, or only the first or second. A change could be seen among youngsters who finish high school. Illiteracy was predominant. A systematic work and analysis (Cepromin, 2000), showed that after 5 years of continuous work with women at mine districts, the illiteracy level among them went down 13.6 per cent. However, it should be noted that in many families the husband and/or father did not let the mother or daughters attend the courses to learn how to read and write. As a consequence, disinformation about their rights continued to be widespread among women; they continued to be misinformed about the country's situation or their husbands' jobs, because those were 'men's affairs'.

Presently, illiteracy is large in those districts among families of mine labourers; they are peasants trying to work as miners, not affiliated to the cooperatives but hired by their members. These labourers also work as temporary workers in the mine enterprises. They have no social rights such as a widowhood pension. Such a situation becomes dramatic when a woman becomes a widow; she is left without any legal and economic support.

No respect for women's working areas

An area of work is assigned to women members of the cooperative, but without any other rights; they receive no technical assistance, nor financing or training. Preferred areas of work are *mucks* or *cajas*, places where mud rolled down from mills and ore fallen between loading and unloading are picked out by women workers. A good example of their lack of rights is what happened in 1996 at Cerro Rico de Potosi when they were left without their working area because those areas were within the mucks assigned to mine enterprise COMSUR, owned by former President, Gonzalo Sanchez de Lozada, expelled from Bolivia in October, 2003. The enterprise needed to open a road, and without notice the spots where the *palliris* worked were blasted. Faced with such a loss the women tried to claim justice but without any result. Similar situations also take place may times when members of the cooperatives themselves require spots to develop their work and just take them without any consideration for the *palliri* women.

Equal earnings?

At one of the cooperatives, women working in groups were earning the same as men… for some months. However, there were complaints that such equality should not occur. So, women's earnings were reduced to 75 per cent of what men earned. Equality lasted only three months:

> I met my husband, he said we were going to marry, but I was fifteen and didn't want to marry. I went to La Paz, then to Atocha, and from there to Tasna. I worked with my uncles at the mine, but then I thought it was strange I received only a pair of shoes after working for a long time… I worked so hard for just a pair of shoes? Looking for another

job I went to Tasna Rosario and worked at the grocery store, delivering bread. Afterwards I went to Uyuni.

Cepromin, 2002b

Sanitation, health, and safety

Risk factors for women working in mining are mainly humidity, low temperatures, and atmospheric pressure, plus general under feeding which lowers defences, and the deeply rooted consumption of alcohol. Lack of instruction and poor knowledge about diseases caused by improperly cleaned food is a major issue.

Women suffer diarrhoeic and acute respiratory diseases, plus illnesses causing head and back pains, cancer of the uterus, rheumatism, and polycythemia are very common. Back pain is the result of the hard work they perform while manipulating and lifting heavy loads, or injuries caused by falling rocks. The use of a hammer or stone to grind the mineral causes hand cuts, injuries, and bruises, pain and deformations of the bones, nervous disorders such as lumbago and even spinal cord hernias. There is a risk of pneumoconiosis (silicosis) due to the constant exposure to dust during loading and unloading of mineral without any respiratory protection.

Contamination of water due to mining is inevitable because of the poor exploitation methods used. The *palliri* women work with their bare hands in polluted rivers, sometimes wearing old gloves or none at all. It is common to see them working close to, or in, waters contaminated by mine discharges, organic waste, community rubbish and drains. The Bolivian national network of mine working women published various press articles to sensitize public opinion over the consumption of contaminated water. In several mine communities the shortage of water for domestic use is a big problem, washing is done with *copagira* or water from the mine, which leaves metal particles on the clothes causing diseases by contact with the skin.

Minerals such as lead, antimony, arsenic, copper and chromium are present in the water, all of which are toxic. Uptake of lead by children can impair their intellectual development and the toxicity of arsenic is well-known. The use of mercury in gold amalgamation in concentration activities and by the *barranquillados*, mainly carried out by the riverside is the cause of the high level of pollution in rivers flowing into the Amazon basin.

Air pollution results from dust produced in treatment and transportation of minerals, together with poor ventilation inside the mine, which affects the lungs. Because of the exhaustion of mineral-containing raw material at their work areas, presently the *palliri* women at Potosi's Cerro Rico are working in the activity called *picha*, scavenging the remnants left at drains after the mineral is loaded causing a lot of dust; this work is generally carried out with the help of children who are also exposed to the same dangers as the *palliris*.

Women are excluded from the so-called 'industrial safety': very few of them are familiar with safety implements such as boots and helmets usually worn by the men. Many of them have seen or 'lived' an accident but, in their own words, nothing has been done to avoid those accidents.

Sanitation is another problem. Pigs freely meander around the community, and no cleaning of the market is practiced. The *barranquillera* and *palliri* women have their meals right at the work site, with the consequent lack of comfort and cleanliness; the distance between the work site and these women's homes is to be blamed for this. The same takes place in the gold mining sector where meals are served at the work site where mosquitoes, flies, and other insects, endemic to the hot and humid subtropical areas, are abundant.

Most common among diseases affecting young working boys and girls are acute respiratory diseases as well as diarrhoea, illnesses typical of cold weather where housing and living conditions are precarious. Other complaints concern muscular pains, due to use of the *quimbalete* (hand crusher) and the heavy ore loads they transport on their backs from the mine entrance to the mill and from there to dumpcarts; bone pains because of the constant contact with cold water when washing ore; respiratory problems caused by inhaling chemical reagents during the process using xanthate.

Through conversations with mine women, it was found that consumption of alcohol is high in every family activity, be they social, civic-cultural, or sporting. Actually, any event is an excuse to drink alcoholic beverages for three continuous days, the first day is to celebrate, the second to get over from the intoxication, and the third day, to finish up in 'good' shape. The price of alcohol is cheap, 6 to 8 Bolivianos per litre, but of course the alcohol content is high, 90 per cent, that is methylated spirit, in other words alcohol destined for other uses. The miners are taking drinks that are slowly killing them.

The number of women drinking alcohol is rising; worse still, young children of 12 to 13 years old are beginning to drink, so when they reach adolescence many of them are already alcoholics. Alcohol consumption has become a cultural trait: '...everybody drinks, that is normal, it has always been'. The list of problems caused by alcohol consumption is long, with domestic violence high on the list (Testimonies: Cepromin and Sentec).

Cultural and social values: The importance of mine identity

'I like to be with my people, they are the fellow working women at the mine districts. I am a mine woman, I feel it inside me...' Women who live in the mining areas may not be considered as rural, indigenous, or peasant women, yet they are not like the women who live in nearby urban areas. Mine women form a sector with its own special characteristics and even a culture of its own; who also have different needs and demands which should be identified and visualized in order to be met specifically.

People at the mines have developed their own uses and traditions inherent to their surroundings, such as legends like the *tio de la mina*[13], the *ch'alla*[14], as well as other cults, and rituals such as to help them to find abundant ore, for instance. These people are different from the rest of Bolivia; they are individuals of strong character, fighters, and workers with a defined identity. In health and education fields some particularities have also been detected such as the 'mine disease' and others.

Cultural characteristics

The mine is the kingdom of the *Tio* (the mine's uncle, or the 'devil') who guards the wealth deep down in the earth. He has the power to decide upon the destiny of those who invade his territory. The souls of brother workers who died by accident are supposed to be wandering around the gloomy mine galleries; a flow of air, a stone falling from the dome, the noise of dripping water, all are signs of their presence. For this reason, just as they do with the *Pachamama* (mother-earth) the miners pay tribute to them, represented by clay statues placed in niches along the audits, which should always have coca leaves, brandy, and cigarettes in exchange for their protection, to sustain the ore veins and to maintain a large production. Any illness believed to have been caused by the *Tio* or the *Pachamama* is considered as a penalty for an act of disrespect to deities and it is accompanied by the failure to find ore shoots.

Sexuality is omnipresent in all mining activities. Anything goes if it will help to find a lode. It is said women are forbidden to enter inside the mine for many reasons; for example, the *Pachamama* becomes jealous and besides, the miners ask themselves, how could a woman do the same job as us? They cause the disappearance of ore shoots or veins, so men have to protect them from women; as soon as a woman looks at an ore vein, the same disappears: 'I was doing very well, but then I let my wife enter my place of work and... ciao ore shoot! After that I'd never again let a woman enter the spot where I'm working'.

The entrance of women into the mines in Potosi disturbs the sexual relationship between men and the *Pachamama* Mountain: 'Men belong to the *Pachamama*, if we, women, enter the mine, she may feel jealous; the *Pachamama* makes ore appear only for men'. There was a time when women entered the mine, as has been the case in later years.

> Some men believe that, when they are not making enough (money) they said a woman should enter the mine and have sex. This is what the *Tio* wants. There was this woman who has sex with anyone who had money. She liked sex and worked at night. The guys who had some bucks – those who were doing well – would take her to their spot to make

---

[13] Supposed to be the mine's owner, literally 'the mine's uncle'.
[14] A ritual to bring good luck to the place.

love. But she was very clever, and she began asking for some quantity of ore in retribution, she would give herself right there. Then she was earning more than men, each week she took out 2 to 3 sacks of ore without working. And the men who were with her began to make more (money) also; they said they offered it (to have sexual intercourse) to the *Tio*. I think they are thinking of the *Tio* when doing that; it seems it provokes something, and that's why they always gain... These two women had decided to enter into an abandoned mine, and to work there 'just like men do' because open pit mining did not pay, earnings were miserable. Then, men began asking themselves, their first question was, are they producing ore? How are they working at the mine? They are not really women; they're weird, these two women who work inside the mine until late at night. It is said they're half man, half woman (Absi, 2003).

And the social attributes

An inquiry (Cepromin, 2000) showed that women who had participated in literacy courses gained the respect and recognition of their families; likewise, their interest in participating in meetings and assemblies increased with a predisposition to assume responsibilities. To learn how to read and write means for women not to be humiliated or deceived any more, to be able to help the children with their homework, to be able to sign, to stop being fearful, to demand women's rights, to value themselves and notoriously develop self-esteem. More and more these women are rejecting injustices in their everyday lives, domestic violence, or to be excluded from taking decisions whether within the family, or at public instances such as local organizations.

The women's organization: A history of struggle and conviction

Mine housewives' national committee

Women have organized housewives' committees since 1961, calling themselves 'the husbands' allies' and supporting the men, whether as dependants or independently of their union centre.

Four mine working women started the country's return to democracy in 1978 with a hunger strike that toppled a seven-year dictatorship. Their strike grew with new picket lines in solidarity and later on gave way to democracy. Throughout Bolivia's history, the participation of women in social struggle has always had clear political contents (Cajías and Jimenez, 1997).

Women have always been very brave; they participated in the front line carrying their children in demonstrations organized to demand employment for their husbands after the mines' shutdowns under the so-called 're-localization' when mine workers were massively dismissed from the state-owned enterprise COMIBOL as well as from private mine enterprises as a result of the enactment in 1985 of Supreme Decree N° 21060 which put into effect the 'Structural

Adjustment Policy'. The women were resolutely present regardless of heavy rain or hot sunshine.

Till then, the housewives' committees' objectives primarily referred to those general issues, generally to accompany their husbands' social struggle by being in the front line in demonstrations. But they soon began to feel responsible for the fight to improve their living conditions, demanding schools, health care and food supplies. The defence of their husbands' employment became a third objective.

Paradoxical as it may seem, it was during the 're-localization' process that these housewives succeeded in organizing the Mine Housewives' National Committee CONACMIN under which all mine housewives' committees within the nationalized and private mine centres are gathered.

The mine working women, after major difficulties – during the Twentieth National Mine Congress held at Mina Matilde in April, 1984 – were able to convince the leaders of the Mine Workers Federation of Bolivia to let them organize the First National Conference of Mine Housewives which, when gathered in commissions, discussed the political, economic, social, and cultural problems facing the country at the time. However, it was not until the first General Assembly on January 20, 1986, that the first mine housewives' union board was formed when the national federation came into being. The women discussed the serious problems of the moment at the first national assembly conducted by male leaders of the FSTMB[15].

After attending several congresses, invited as observers only, the women were, in 1991, able to get approval of the housewives' committees as members with full rights of FSTMB, the main mine workers' federation. In spite of the several obstacles encountered, the women were able to write up their bylaws in 1996, and organize their national assemblies in 1988, 1991, 1993, 1996, 1998, and 2004, thereby conquering and defending their space within the union world. The 6-year lapse between the sixth and seventh general assemblies was due to the weakening of union organizations in Bolivia, such as FSTMB and even CONACMIN. This was achieved mainly due to the conviction that women have (Cepromin, collective testimonies, 1990–2005).

To better explain what a national organization of mine working women means, the following are some statements made during recent assemblies: '…Never thought to be a leader was such a difficult task…but not impossible, because when you made up your mind (about being a leader) you get there…'; '…they (the male workers) don't want to give us what we really deserve…' and '…the government does not think of us, we are hungry, we want to eat, we want to live…'

Women were discussing issues such as the dismissals with benefits, the closure of the 'Totoral' mine, the misleading voluntary retirement scheme, the suits against union leaders at Chuquihuta without respect for the union's legal rights. They also denounced the government's attitudes, which openly defended private enterprises interests.

[15] Spanish initials of the National Federation of Mine Workers of Bolivia.

Among some important names in the history of mine working women there is that of Domitila Chungara (Viezzer Moema, 1977), who since the creation of the first housewives' committee at the 'Siglo XX' mine district has been a permanent defender of human rights. In 1962 when the first committee came into being at the time of several social conflicts between mine workers and the government, she said the housewives' committees were organizations fighting together with the workers for the same causes.

Another was Norma Salguero, the first chairman of CONACMIN who made history through her strong interventions at mine congresses and assemblies demanding justice and equal participation for mine women, thus gaining the respect of several male leaders. Right now she's assisting the new National Committee of Housewives whose term started in March 2004.

The change of women's role under neoliberal policy in 1985

In the difficult times that followed the Structural Adjustment Policy, women concentrated on collecting funds for the common pot or organizing activities for the workers' children. The areas where thousands of families had settled were in miserably precarious conditions. Worse still was the situation at the local schools which all of a sudden saw the number of students swell enormously with a large young population coming from the mines. There were cases where many families of *relocalizado* workers settled at 'El Alto'[16] had to send their children to the city of La Paz looking for something to eat; many girls turning to prostitution as a last solution. The woman's role changed, she had to leave home and work outside to replace the man's salary. He, a miner, was out on the street desperately looking for a job, turning to drinking, and, in several instances, deserting the family.

To better illustrate this situation, the following comments are from women whose husbands lost their jobs:

> Our *compañeras* (fellow women) gave birth and still go to the marches. And they have other children, three, four, five years old who want things they see on the street, ice cream, soft drinks, but we cannot buy them anything because we do not have money.

and:

> We were threatened by the police, they wanted to reprimand us. They beat us with sticks from every side. My *compañeras* did not pay attention; then, from the corners they brought dogs trying to scare us, the children started to scream and cry watching the dogs, even my fellow sisters were scared, but still they did not back off. We decided to sit right there and turn our backs; we started to cry in desperation and began to sing the National Anthem, trying to protect ourselves that way. But it was as if the dog guessed it

[16] A city located near the capital city of La Paz.

> should not bite us, instead one had bitten a policeman's hand. Then a lieutenant came and said, 'you may form a commission of 5 persons, they will talk to the Vice Minister' so headed by doña Norma de Salguero we went to talk to the Minister.

By July, 1988, the *relocalizadas* mine women had already held their first National Meeting in the city of Santa Cruz.

The secretary for palliri women's affairs

The last part of a bulletin published in 1971, written by the *palliri* women themselves but signed by the 'Union of the Siglo XX Mine Workers' and the 'Union of Radio and Television Workers of Bolivia, Siglo XX subsidiary', stated: 'The signatures below are of honourable individuals who felt sympathy toward our problems. We do not sign ourselves, for fear of reprisals'.

The *palliri* women worked for the 'Catavi' mine enterprise, but they were not recognized as being due social benefits:

> Under any temperature, we must go up the mucks and for many hours pick out rocks containing ore remnants. Then we put them on our backs in worn out sacks given by the company, and carry them for half a kilometre up to our spot. There, when the machine-driven *chancadora* (the grinding machine) starts, we load them again to have our rocks ground. Mechanization is an unknown word at the mucks; our tools are limited to our hands, some shovels or pieces of them which we use to clean our work spot removing dumps, rough rocks without ore falling down the mucks as we keep digging to recover mineralized stones. Sometimes we have to shovel during full days before getting out the dumps and moving some one hundred meters, and removing the muck little by little.
>
> Brochure, 1971

The *palliri* women, already organized, were able to write the following:

> It is difficult to imagine any of this; still more difficult to believe in this century of technology, two hundred Bolivian women still work as hacks at a lost spot of the prestigious mine enterprise Catavi.

And with some irony they go on saying: 'Maybe some day some one may dare to paint all of this in a book which may be titled "Systematic annihilation of women in Bolivia" or, "Tears and blood at the Siglo XX mucks"...' Since 1992 FENCOMIN, the National Federation of Mine Cooperatives of Bolivia, has had a position at its board of directors called 'Women's link', but it is headed by a man! Nevertheless, women are gaining space gradually, even though today they have only one representative in the board of directors, the secretary for *palliri* women's affairs.

Regional *palliri* women workers' associations began to organize less than a decade ago, when the number of working women within the cooperatives increased markedly. The VIIIth National Assembly of *Palliri* Women was held in November,

2003. 85 *palliri* workers met to discuss the country's situation, gas for Bolivians, the future constituting assembly, social security, work and production areas, and women's participation in the cooperatives' system. The resolutions included (Document, 2003):

1. To call for urgent attention to diseases affecting the cooperatives' female members, such as arthritis and rheumatism, as well as to train women in first aid and traditional medicine.
2. To propose a *palliri* woman for the Vice-Ministry of Women Affairs in charge of decision-taking and policymaking in favour of women of this sector, until now abandoned and forgotten.
3. To modify retirement age to 45 years for women and 55 for men working at the mines, taking into consideration the professional risk involved. It should be pointed out also that most women work as *palliri* from when they are 15 years old.

Cooperatives' housewives

Currently, the housewives or wives of cooperative members are still in the process of organizing themselves, so only a few regional organizations of women exist. The Cooperatives' Miners' Federation, FENCOMIN, does not pay much attention to these women because they are not members of the cooperatives, just housewives. However, the housewives who work as *palliris* picking out or concentrating ore, besides other tasks supporting their husbands' paid work should be taken into consideration. On top of all this, the women have to take care of the house, feed the children, etc. which means the division between the two groups of women (the *palliris* and the housewives) is artificial. It must also be noted that macho attitudes still prevail within these organizations and there is still a long way to go for women to participate, take decisions, and lead their organizations.

The pathway to solution

The lack of work tools adversely affects the *palliri* women's income. The objective of a Cepromin project is to buy wheelbarrows and tools collectively for groups of three or four women who would pay in monthly instalments according to their economic capability; after a while the tools will become their property. This way these women will also be able to increase their earnings since they won't have to wait in a queue to use the cooperative's wheelbarrows any more. It is also a show of solidarity, and they certainly feel proud of having their own revolving fund to buy other tools.

In order to prevent, progressively reduce and ultimately eliminate child labour, processes of information collection and reflections on possible interventions have been initiated at different levels and with different actors to establish alliances in looking for viable solutions to this issue. To highlight this issue for inclusion in the national agenda, it is urgent to gain public awareness at national, regional, and local levels. To come up with new productive initiatives to improve the families' income is one way of changing attitudes regarding child labour in the mines little by little. Improving the miners' labour practices through some degree of mechanization thus obtaining better incomes for families may have positive results in diminishing and even eliminating child labour.

As to industrial safety, campaigns to raise the issue of improving the miners' living conditions should be a first step. To organize on-site training of workers in accident prevention is of vital importance as well safety information for the mine population including women, children, women workers, *palliris* and housewives. Industrial Safety Committees should also be included in the mine cooperatives' boards with the participation of women. All of these are tasks that can change the hazardous environment at the mines.

It is important to intensify efforts to improve the quality of drinking water at the mines; to launch educational programs on public sanitation related to the discharge of waste water and drinking water supply. It is advisable to treat water containing heavy minerals and micro-organisms hazardous to health. Bleach (a drop per litre) is used to eliminate microbes or through the SODIS method which is a homemade method of purifying water through disinfecting it with solar energy (there is plenty of sun in Bolivia), thereby eliminating pathogenic micro-organisms and improving the water's poor quality for human consumption. Water decantation is done by filtering water which is a more complicated method but still within the reach of mine women.

Harmful emissions of mercury may be greatly reduced by applying simple recuperation procedures which at the same time reduce operational costs. To find a solution for waste disposal should be everybody's concern at the mines. A change in attitude is necessary in carrying out cleaning actions. Many waste elements such as paper, non-synthetic cloths, plant and wood residues may be recycled.

The creation of a common fund for camp and community improvements when good mineral prices prevail requires a changed awareness but it is not an impossible task. A statement by a dedicated woman shows a road to solidarity and a concern for the welfare of the family, the community, the region, and the country:

> As a health defender, I am monitoring and controlling who in my area takes vaccination shots, and who have sick children and want to have them vaccinated. At board meetings I suggest what can be done about this, because without the board's and the fellow sisters' help it can't be done, they only listen to the board's chairman, isn't it true? So far, we have cleaned the streets, we want to plant trees and plants. I hope we will be agreed on this at the next meeting.

If actions like this could be multiplied, the mine districts will become pleasanter places to live in. To make visible mine working women's invisible work would represent an invaluable contribution to humanity.

References

Brochure, 1971. *Una Abominable Vergüenza Nacional: Las Palliris Desmonto Siglo XX,* Oruro, Bolivia: Quelco.

Cajías de la Vega, Magdalena and Jiménez Chávez, Iván, 1997. *Mujeres en las Minas de Bolivia,* La Paz, Bolivia, Ministerio de Desarrollo Humano.

Cepromin, 1990. *La Conciencia, el Comité y el Sindicato*, Cepromin La Paz, Bolivia.

Cepromin, 1991. *Testimonio Colectivo: I Ampliado de Mujeres Mineras: Organización del Comité Nacional de Amas de Casa*, Cepromin La Paz, Bolivia.

Cepromin, 1991. *Testimonio Colectivo: II Ampliado de Mujeres Minera, La Paz octubre 1988*, Cepromin, La Paz, Bolivia.

Cepromin, 1993. *Testimonio Colectivo: III Ampliado de Mujeres Mineras, La Paz septiembre 1992*, Cepromin, La Paz, Bolivia.

Cepromin, 1994. *Testimonio Colectivo: IV Ampliado Nacional de Mujeres Mineras La Paz diciembre 1993*, Cepromin, La Paz, Bolivia.

Cepromin, 1996. *El Trabajo Invisible de la Mujer Minera Boliviana*, Cepromin, La Paz, Bolivia.

Cepromin, 1997. *Testimonio Colectivo: V Ampliado Nacional de Mujeres Mineras, La Paz septiembre 1996*, Cepromin, La Paz, Bolivia.

Cepromin, 1999. *Comité Nacional de Amas de Casa Mineras: Nuestro VI Ampliado, La Paz 11 al 13 de noviembre de 1998*, Cepromin, La Paz, Bolivia.

Cepromin, 2000. *Mujeres Mineras y su práctica Educativa en Alfabetización: Sistematización del trabajo 1995–2000*, Cepromin, La Paz, Bolivia.

Cepromin, 2002a. *Diagnostico Situacional de Niños y Niñas Trabajadores Mineros de Siglo XX Llallagua*, Cepromin, La Paz, Bolivia.

Cepromin, 2002b. *Por qué y para qué tanta Lucha?*, Cepromin, La Paz, Bolivia.

Cepromin and IDRC-CRDI, 2003. *Informe de Investigación, casos: Proyectos San Cristóbal y San Bartolomé*, Cepromin, La Paz, Bolivia.

Documento, 2004. *Resoluciones Emanadas del VIII Encuentro Nacional de Mujeres Palliris de Bolivia, Pulacayo 10 al 12 de noviembre de 2003*, Red Nacional Mujeres y Minería – RNMM.

Mitre, Antonio, 1981. *Los Patriarcas de la Plata: Estructura Socioeconómica de la Minería Boliviana en el Siglo XIX*, IEP Ediciones, Lima, Peru, p. 51.

MMSD y IIPM, 2002. *Minería, Minerales y Desarrollo Sustentable en América del Sur*, Uruguay, IDRC – IIPM.

RIMM Autoría colectivo, Memoria Conferencia Internacional 'Mujeres y Minería' Conference Report II: International Conference, 'Women and Mining' – RIMM, La Paz, Bolivia, Red Internacional Mujeres y Minería – RIMM, 2002.

SOCAVON, 1983–1996. Various articles revista nr 22, 28, 36, 39,41, 44, 52, 131, Cepromin, La Paz, Bolivia.

Wicky, Jean-Claude, 2002. *Mineros: Bolivia*, Lunwerg Editores, Spain.

PART IV
Global Processes, Local Resistances

Chapter 14

Gendered Labour in Peripheral Tropical Frontiers: Women, Mining and Capital Accumulation in Post-Development Amazonia

Jeannette Graulau

Introduction

Globalization of capital has triggered a resurgence of mineral and resource extraction in peripheral tropical regions. Increasing mineral demand in advanced Western economies and continuation of industrialization processes in the developing world have pushed mining capital deeper and wider in the tropics in a more aggressive search for profits. Consequently a fierce global expansion of mining investments, markets and operations have characterized recent decades.

As the logic and functioning of mining capital become global, the logic of small scale and peasant mining also becomes global. Increasing poverty in the developing world provokes a heavy influx of men to urban economies, forcing women and children to informal sectors of production such as peasant mining. This gendered labour force sustains production of industrial minerals such as gold and silver in a significant number of countries. The important diamond-producing countries heavily rely on peasant and small scale mining. Production of emeralds, semi-precious stones, talc and sulphur depend on a peasant labour force (Ackermann, 1998; ILO, 1999; Stear, 1990; World Gold Council, 2001).

Globalization of mining operations and markets is sustained by the inequality and vulnerability of women's labour in the economy of peasant mining. Consolidation of such markets has increasingly pushed women and children to peripheral mining frontiers. Women's labour directly or indirectly shapes the contested sites of mining frontiers. In the case of mining in the tropics, it is women and child labour that pushes the expansion of mining sites. Women are crucial in the formation of communities and communal life in peripheral mining frontiers, and this is the case for the Brazilian Amazon.

This chapter explores women's work in gold mining in the Brazilian Amazon. It describes the structural vulnerabilities organizing women's labour in mining communities, in particular the gender aspects of production and technology and their intersection with historical relations of class, colonialism, ethnicity and Westernization. Drawing upon debates of feminist political economy and geography, the chapter analyzes how female labour in the mining frontier of the Brazilian Amazonia is functionally and spatially connected to the contemporary process of capital accumulation in peripheral tropical regions.

Gendered labour in peripheral mining regions: Analytical propositions

The study of women's labour in mining is not prominent in the development field. On the one hand, studies on the political economy of mining are not part of the sexy, hot and trendy mainstream topics of development. This is due in part to the fact that development theories of structuralism and dependency assume that 'technological backwardness' and 'underdevelopment' arise from specialization of the periphery in raw materials (Cardoso and Faletto, 1969; Furtado, 1970). Neo-structuralism and world-system approaches also see in mining the perpetuation of the condition of peripherality. They define mining as an economy of *enclave* serving the logic of capital accumulation (Sunkel, 1993; Wallerstein, 1985). As such, there is a well-established anti-mining bias in mainstream and critical development fields. On the other hand, research on women's labour in extractive production heavily focuses on agriculture. These studies emphasize women's exploitation by lack of farm credit, unequal exchange relationships and land distribution policies. While they call attention to structural instabilities associated with poor agrarian policies, they remain silent with regards to mining issues affecting women (for women in agriculture, see Deere, 1995).

The few studies that exist on women in mining have focused on the description of women's work strictly at the local level; that is, their subordinated roles in sexual service sectors, household economy, and provision of security, reproductive work and income-generating work. Assuming 'Third World women' as a homogeneous category of study, this literature has not attempted to build strong analytical connections between women's labour in mining in specific geographic contexts and broader functional and spatial dynamics of capital accumulation (for a different perspective, see Barrios de Chungara, 1978; Harcourt, 1994).

In an attempt to overcome the theoretical shortages described above, this chapter defines three main propositions to build a social scientific understanding of women in mining. First, that 'women in mining' refers to a complex social and historical category; second, that it intersects with other historical relations (class, colonialism, ethnicity and Westernization); and third, that it is linked to the continuation and contestation of capital accumulation and its hegemonic practices and ideologies in peripheral regions of the global economy.

The first proposition acknowledges diversity in range and orientation of women's work, ideas, roles and beliefs in mining. It refers to the heterogeneous nature of the experiences of women in peripheral tropical regions, in particular their different ideas about their social positions and relations. The second proposition refers to contextually-specific forms of power that structurally connect historical social relations and locate them in terms of identities, material base and ideologies. It calls the attention to the manifestations at the micro-level of structural power relations running along and across multiple scales. These scales, or spatial networks, refer to a country's dependency on multinational mining investments, conflicts between foreign companies for land control and mineral rights, and national mineral legislation, among others. The third proposition refers to the linkages between the local and the global that defines subordination or super-ordination of gendered labour to the logic of capital accumulation. As such, it refers to the organization and functioning of a political economy of mining that enable and constrain global capital accumulation.

Women's labour in gold mining in post-development Amazonia[1]

The geography of mining in the Amazon region is characterized by the organization and functioning of hierarchies linking impoverished women and children to regional, national and global networks of mineral production. These hierarchies have deepened the structural unequal gender relations that historically characterize the insertion of women in mining economies. The penetration of corporatist mining capital organized the main hierarchy of mining production exploiting impoverished women to its own advantage. Mining corporations aggressively penetrated the Brazilian Amazon region in the 1970s, marking the beginning of Amazonia's development era. The main characteristic of this era was the impetus and optimism of the federal government to guide the development of the mineral capacity of the region. Two major development schemes promoted the expansion of corporatist and peasant mining and the creation of infrastructure for the integration of Amazonia into the national economy: the Second and Third National Development Plans (known in Brazil as PDN-II and PDN-III, respectively). Framed as an import-substitution strategy and to alleviate the problems caused by the first oil crisis, PDN-II concentrated on expanding infrastructure for exploiting the primary productive capacity of the Amazon. The government attacked the deep recession, fiscal crisis, inflation and unemployment of metropolitan economies by expanding mining and agricultural production. Mineral resources with a comparative advantage were the main focus of the

[1] Data presented here was collected during the period 1999–2001 in the mining communities of Serra Pelada, La Cutia, Miranda and La Cruz in Central Amazonia. A total number of 328 structured survey interviews and 36 in-depth interviews were conducted during that period.

program, although it included timber, beef, crops, and private colonization schemes in selected areas of the Amazon (Browder and Godfrey, 1997; Pompermayer, 1982).

By the end of the 1970s, the federal government established PDN-III, coinciding with the second oil shock of 1979. It was a comprehensive plan for mineral development in the Amazonian state of Pará, based on the financial assistance of the World Bank, Japanese Export-Import Bank, Japanese Commercial Bank Syndication, the US Export-Import Bank and the European Coal and Steel Community. At the time, the mining sector was responsible for more than 10 per cent of Brazil's exports (1.3 per cent of GNP). Besides the potential of industrial mining, the government saw in *garimpagem* (a Portuguese word for peasant mining) a short-term mechanism to alleviate the economic crisis and generate foreign exchange. Mathis and coworkers documented that in the early 1980s, the director of the Ministry of Mines and Energy, César Cals, openly encouraged the expansion of *garimpagem* in the Amazon (Mathis et al., 1997). The government rationale for promoting *garimpagem* was to colonize the Amazon region with a productive and docile population that was going to provide for the government and corporations cheap access to mineral inputs. The strong historical record of production in *garimpagem* supported this idea (Benedito Da Silva, 1993; Cleary, 1990)[2].

During the early years of these programs, men mass-migrated to the Amazon region while women attempted to sustain their households by entering the service economies. Having taken on temporarily the roles of chiefs of households, women started to migrate to gold mining sites (or *garimpos*) (Rodrigues, 1994). With intensification of migration to *garimpos*, a hierarchy of economic activities emerged that attracted female labour to the region (MacMillan, 1995). The immediate impact was the deepening of a structural gender division of labour whereby women were to occupy low- and non-remunerated jobs in the *garimpos*. Manual mining, elaboration of food on a permanent basis and sexual service sectors were at the bottom of the hierarchy of *garimpos* production and rested on women's labour. With migration of women to gold *garimpos*, these activities flourished based on an abundant supply of female labour.

Studies on the expansion of *garimpagem* conclude that women massively joined the *garimpos* labour force in response to the expansion of corporatist mining that occurred in the 1970s and 1980s. During this period, hierarchical organization of production of corporatist capital needed a massive influx of female migration to

---

[2] *Garimpagem* has a respectable record of production since the 1960s. It directly supports the industrial mineral sector of the Brazilian Amazon, by assuming the costs of exploration, discovery, prospecting, mining and processing of secondary deposits. Indeed, the most important phase of mining, exploration, bearing heavy investments and geological risks, is probably *garimpagem*'s most important contribution to mining. As a matter of fact, Brazilian geologists agree that most gold ore-bodies that today are being worked industrially were discovered by *garimpeiros* of the Brazilian Amazon.

fill its labour needs. In the early 1980s, coinciding with a neoliberal revolution that would open the Amazon mineral resource base to mining capital, women massively migrated to the Amazon. The need to generate income to sustain households, in the absence of men, coupled with the desire to venture into at the time a promising enterprise, explain female migration to *garimpos* (Lopes Da Silva et al., 1997; Rodrigues, 1993, 1994). Given that women were rapidly incorporated into gold-mining manual activities, mainly the techniques of *batear* and *requiar* (to wash and collect gold in conical pans), they contributed to the growth of peasant production.

After the government interventions through PDN-II and PDN-III to promote extractive economies, coalitions of global mining capital took over the development of Amazonia's mineral wealth. A new hierarchy consolidated, which displaced the State, marking the beginning of post-development Amazonia, characterized by deregulation and liberalization of the markets of production and trade for extractive economies. This post-development era has shaped a new division of labour not only in the Amazon region, but in other tropical frontiers. According to a report commissioned by the World Bank, governments of mining regions have restricted their intervention to promote the mining sector 'leaving the ownership, operation and management of enterprises to the private sector' (SAPRIN, 2002: 30). The effects of this post-development era in Amazonia have so far been the expansion of global mining capital operations, increasing conflicts and violence between peasant miners and corporations and expansion of small scale mining settlements in selected areas of the region[3]. Several multinational corporations have dominated the mining landscape, competing against each other for increasing shares of the markets for mineral production and trade (Anglo American, BHP Billiton, Glencore International, Rio Tinto, Barrick International, Placer Dome, Falconbridge, Phelps Dodge, Newmont Mining, Teck Cominco and Companhia Vale do Rio Doce). The expansion of operations of multinational mining corporations throughout Central Amazonia has reinforced the historical structural vulnerabilities of women's labour.

The first structural vulnerability is that women continue to provide the cheap labour force for the many diversified extractive operations of corporations. This is

---

[3] Mining in post-development Amazonia increased in part due to strong juridical and political incentives put in place by the federal government. A constitutional amendment approved in 1995 provided greater flexibility for investment in fuel and non-fuel mineral sectors. The Mining Code, revised in 1996 (Law no. 9314 of 11/14/96) granted the National Department of Mineral Production (DNPM) and the Development Concession of the Ministry of Mines and Energy (MME), the release of authorizations for exploitation of mineral deposits through the Exploration Authorization Permit. The National Institute for Mineral Research (CPRM) started to develop programs of basic geological mapping and prospecting in areas of potential mineral extraction. It became more involved with foreign investors, by assisting them with geological and economic databases, particularly for mining of coal, copper, diamond, gold, kaolin, nickel, peat and zinc.

particularly so for the impoverished women from the northeast region of the country. For example, CVRD encourages women's labour in its agro-industry production (Browder and Godfrey, 1997). Given the production cycles of agro-industry and its heavy reliance on capital-intensive technologies, women's labour is useful only on a temporary basis.

The value of women's labour stands at the bottom of the production hierarchy, aggravating historical class tensions and inequalities. Women in mining frontiers are part of the historically dispossessed populations that wandered into the areas in search of the possibility of a better economic life. They do not enjoy the diet, health and education that their counterparts can enjoy even in the most impoverished urban settlements of *favelas* (slums). They are worse off than the urban poor.

Women migrants from the Northeast, coming from the most impoverished region of the country, do not have access to formal education[4]. Most of these women do not have the economic resources to complete school years, or any other formal training that can allow them to compete in urban labour markets. According to the findings of my fieldwork, women come to mining regions because these represent a possibility of enhancing their economic livelihoods. They escape class inequalities in urban and peri-urban settings by migrating to the mining frontiers (for a complete discussion on peri-urban settlements in Brazil, see Browder and Godfrey, 1997).

Moreover, a significant number of women are affected by tropical diseases like malaria, yellow fever, parasites and pneumonia and have not had access to health services. As a case in point, shortages of medicines in the hospital of the community of Serra Pelada has obliged nurses to give priority to children, which has further affected the provision of health services to ill women. Women working in gold mining are particularly vulnerable to this situation, given that mercury contamination poses a permanent threat to their health.

A second structural vulnerability refers to the functioning of a closed spatial migratory circuit for women. Women inhabiting gold mining communities (Serra Pelada and Miranda) tend to come from Amazonian states with a long-standing tradition of peasant gold mining, such as Maranhão, Pará, Tocantis and Mato Grosso. The majority of women who live in stone-mining communities (La Cutia and La Cruz) come from the states of Bahia, Pernambuco, Piauí, Goiás, Ceará, Minas Gerais and Paraná, which have a long-standing tradition of stone mining. What all women have in common is that they come from peri-urban settlements, escaping poverty and violence from the cities.

These migratory circuits reproduce unequal exchange relationships between central and peripheral towns, whereby the most vulnerable populations – women and children – are spatially displaced, without any real possibility of breaking the circuit. They deepen the historical colonial relationships between urban population settlements of modern cities like Belém, São Paulo and Rio de Janeiro, and the

---

[4] Indeed, approximately 87 per cent of the sample had two or less school years.

Amazonian settlers, and the inequities of the economic colonization schemes promoted during global boom cycles of extractive economies.

As the number of corporatist operations increases, migratory circuits spatially expand to accommodate male migration and efficiently exploit this labour force. Meanwhile, the increase of corporatist operations seems to intensify the obstacles for female spatial mobility, reinforcing the limits of migratory circuits. The geography of the mining frontier, characterized by lack of paved roads, far from urban centres and markets, and lack of frequent transportation services, makes it difficult for women to move out of *garimpos*. Generally, men endure periods of declining productivity by moving out to explore side-line lands of corporatist projects, and join the seasonal labour force for agriculture and cattle ranching of nearby towns. Women stay in the *garimpos*, sustain their households with the scarce resources available to them, and, as the next section discusses, support male exploratory ventures.

Garimpagem and women's labour: Property, labour and production arrangements

A more complex dimension of the vulnerability of women's labour responds to the hierarchical agreements of property, labour and production of *garimpagem*. These function at the bottom of the broader hierarchy of mining production. As such, women's positions and roles in them function 'at the bottom of the bottom'. Property regimes encourage (and are encouraged by) a historical structural inequality between men's labour and women's labour. Men dominate the phase of land exploration (either through absentee investment in the venture, or through migration to the region). Through violent means, some men, self-proclaimed pioneers of land exploration, entitle themselves with ownership of *garimpos* sites. Since most *garimpos* are owned by men, women's labour occupies a subordinated position in the hierarchy of *garimpagem*. Men owners of land decide the schemes for income sharing and profits for their labour force.

Manual and mechanized production systematically exploits women's labour. Production starts with exploration and it involves the previous definition of a deal among an investor or speculator and usually a male miner (known as *garimpeiro*). The basic condition of the deal is that the investor pays for the supplies and equipment needed and *garimpeiros* venture into the forest in search of gold. Anthropologist David Cleary identifies many of the tools employed in this phase: metal shovels (*pá*) and picks (*picareta*); the equipment includes oil lamps, matches and basic food supplies (rice, beans, flour, coffee, oil lamps, matches, gas) for at least two months, depending on the remoteness and accessibility of the area to be explored (Cleary, 1987). *Garimpeiros* migrate in teams of three to eight people, depending on the deal previously arranged. By absorbing the costs of economic risks, women's labour reduces the costs of the phases of exploration (*exploração*) and prospecting (*prospeção*).

Exploration gives the erroneous idea that men share the economic risks of this phase with investors. It is women's labour that shares the economic risks with investors. As women support their households and husbands' exploratory venture, male miners shift their share of exploration risks to women's labour. Moreover, if *garimpeiros* receive a fixed fee irrespective of the outcome of the exploration, women's labour operates as a built-in incentive to encourage further exploration. It is also a built-in incentive for under-exploration, because it allows the *garimpeiro* to move to potential *garimpos* and wait, without having to work, for any outcome or discovery by another *garimpeiro*. *Garimpos* production creates, organizes and exploits the built-in incentive provided by women's labour.

Moreover, because women's labour bears no geological risk at the initial exploration phase, it also reduces the *garimpeiro*'s economic risk of failure in the second phase known as prospecting. Prospecting refers to the discovery and analysis of gold-bearing gravel deposits. Many Brazilian geologists consider it the most effective phase of *garimpos* production (Hünh, personal communication, 2000). Based on a sample of gold nuggets collected in a *cuia* (conical pan) and the depth in which it was found, the *garimpeiro* calculates the amount of time to be invested in mining. This is a very important phase, because most of the value added to deposits occurs at the moment of discovery (Godoy, 1985). It is also an inherently unstable phase, given the potential accidents (geological risks) associated with it.

Women's labour in this phase operates as a safety net for the *garimpeiro*. It allows the *garimpeiro* to undertake prospecting and expect some form of household or domestic compensation to bear the geological risks. The most common ways in which women's labour compensates for the hardships of prospecting are by the provision of food, caring in case of accidents and sexual services[5 6]. Since this safety net appears as 'outside' the market of *garimpos* production, *garimpagem* reproduces and exploits it as its invisible, non-economic sphere.

---

[5] In structured survey interviews, to the question, 'What would you do if your husband decides to go to northern Amazonia in search of gold?', 68.4 per cent of the sample in the *garimpos* responded they would encourage the husband. The remaining 31.6 per cent firmly stated that they would go with the husband, only if the new *garimpos* promised to be a lucrative enterprise. None of the women interviewed in the *garimpos* said they were against the husband's pursuit in exploration or prospecting. Of the women interviewed in the nearby towns, an outstanding 96.2 per cent said they would oppose the husbands' decision. Indeed, these women described *garimpos* as a dirty place to live, as backward communities, even when many of them had lived previously in *garimpos*.

[6] Obviously, this is not a risk-free pursuit for the male *garimpeiro*. As Cleary discusses, the *garimpeiro* has to take into account gold price, accessibility of the place where the steep bank will be located, infrastructural costs, investment returns and if the intention is to open several shafts, it has to consider extraction costs for the entire riverbank or the complete site (Cleary, 1990, 1987).

Vulnerability of women's labour to exploitation is more manifest in the activities of mineral extraction *per se* (a phase known as *lavra*). *Lavra* in manual mining, done today to a large extent by women, involves the collection of gold-bearing gravel in a *bateia* (a conical pan for the washing of gold). The woman moves the *bateia* in a circular motion to wash sand and water away and keep the gravel.

Interestingly, while many male miners assure that women are the 'masters' of the technique of *batear*, they do not see this activity as deserving equal remuneration to mechanized extraction. The value of the gold that women extract using a *bateia* is significantly less than that from mechanized extraction. This difference becomes legitimate as a compensation for operation costs of machines.

In mechanized mining, done predominantly by men, peasants burn the vegetation to clear the land. They break up the soil with a pressure pump using water, with a machine known as *bico jato*. Only male miners operate this machine. Water mixes with the sand and argil, and forms a pulp. Another pump, attached to the *bico jato*, called *chupadeira* sucks this pulp. This second pump has a rubber hose that the *garimpeiro* fixes inside a small plank opened near the shaft to receive the pulp. From there, the pulp goes to a machine located near the bank, for the first recuperation or concentration of gold.

Mechanized concentration of gold is done by men although some women participate in it. Two processes are used to recuperate gold: gravimetry and amalgamation. Once the pulp reaches the concentrating box the mixture goes through a chain of boxes for recuperation of gold. The amount of pulp to be processed determines the size and length of the cages. The main tool employed is the concentrating box called *cobra fumando* (which literally translates into 'smoking snake'). Made of wood from forest trees, it has an angle of inclination of approximately forty-five degrees and approximate length of fifty centimeters for each box. For an efficient recuperation of gold, *garimpeiros* put wooden hurdles inside each box to serve as barriers for the flux of the pulp (Cleary, 1987; Mathis, et al., 1997). Each hurdle allows for the passage of the mixture whilst holding back the gold nuggets. To facilitate gold recuperation, they cover each box with woven jute to trap the gold. The material collected in the hurdles and in the woven jute goes to a *bateia* for amalgamation. After the work with the *bateia* is done, *garimpeiros* squeeze the mixture in a cotton cloth. Like in manual mining, they burn the mixture of gold and mercury in a *cuia*; mercury vaporizes into the air and gold remains, ready for the market. The value of gold processed through mechanized mining tends to be higher, because higher amounts of the mineral can be extracted in a relatively short period of time. However, introduction of technology and its operation by men do not mean in all cases that it is more efficient in procuring a higher volume.

In amalgamation in manual mining, conducted by women, mercury is added to the *bateia* containing the gravel. Generally, the amount of mercury is kept to less than half a gram. The *bateia* is then moved in clockwise rotation to wash sand and water away and concentrate the heavier metal, mercury mixed with gold, in the centre. The material is then collected and burned in a *cuia*, after which mercury vaporizes into the atmosphere. Gold is then recuperated and is ready for the market[7].

Women's labour in the phase of concentration of gold (mechanized and manual) shares the environmental costs associated with the use of mercury. While it is true that water and air contamination affects the health of women and men alike, women are the care-takers of the family. They have to alter constantly their work and domestic routine to make time to take care of the ill. Diseases of children due to mercury poisoning seem to affect women harder than men: women in Serra Pelada repeatedly brought up this concern in the life-story interviews. In this sense, women have higher opportunity costs than male *garimpeiros*, given that their next best alternative will always include care-taking of the ill without economic remuneration.

Production schemes increase the vulnerability of women given that, with the introduction of technology, only men have the possibility of earning a sustained level of income in mining. Regardless of the amount of material processed in mechanized operations, men receive a salary paid by the owner of the machines or by the *garimpos* investor. Machine-owners rarely employ women. So women's income in mining comes from the amount of material they can work in the *bateia* and not by any contractual agreement with the owner of the *garimpos*.

Of the total number of women interviewed in each mining community at the time fieldwork was conducted, approximately one third or less generated income[8]. The most important income-generating activity for women is manual mining, although there are other income-generating activities such as preparation of food, street vending, and domestic work, deskwork, washing of clothing, rice cultivation and charcoal-making.

Income gaps between different groups of women reveal that generation of income requires reinforcing inequalities among different groups of women (mainly along the lines of ethnicity and place of origin). As a case in point, most women in

---

[7] In structured interviews, to the question, 'Can you explain how to separate mercury from gold?', 94.3 per cent of the sample in gold *garimpos* answered that the mixture had to be burned so that mercury could vaporize. 5.7 per cent of the sample said that the mixture was burned so that mercury could be eliminated easily from it. Virtually all the women of the sample knew about the process of *beneficiamento*. In Miranda, only 63 per cent of the sample said the mixture was burned in order to separate mercury from it. The rest (37 per cent) said they did not know about such a process.

[8] The monthly family income is defined in terms of the regional minimum salary, which was R$ 135.00 (approximately US$ 67.50) when fieldwork was conducted. Women in gold *garimpos* have the lowest incomes. In Serra Pelada, 93 per cent of the interviewed women receive 0-1 MRS (minimum regional salary) and 7 per cent receive 1-2 MRS. In La Cutía, 86 per cent receive 0-1 MRS, 14 per cent receive 1-2 MRS. In La Cruz, 100 per cent receive 1-2 MRS. Women in *garimpos* Miranda, however, seem to enjoy a higher economic status. 55.7 per cent receive 1-2 MRS; 25.7 per cent receive 2-3 MRS, and 18.6 per cent receive 3-4 MRS.

the communities of Serra Pelada, La Cutia and La Cruz that came from other gold mining regions generate a monthly income equal to or less than US$ 67.50 (at the time fieldwork was conducted). Of the women in the community of Miranda, 56 per cent receive one or two times this amount. Common experience, tradition of gold mining, and ethnic ties among women function as social determinants of income differences. Women are more willing to assist other women coming from the same mining towns, and sharing ethnic backgrounds. As such, some groups of women are more economically vulnerable to the structural inequalities of *garimpagem* than others.

As many researchers note, Serra Pelada, La Cruz and La Cutía today experience declining production trends, possibly in response to declining price levels in the international gold markets, extinction of secondary deposits in the region and land conflicts with mining companies and the federal government. As such, *garimpeiros* (men and women alike) engaged in gold mining are in fact worse off than *garimpeiros* of other minerals in the Amazon region, such as emeralds. Another related factor is that the region of Pará is one of the richest sources of minerals and raw materials of Brazil. Since it has the largest and most lucrative industrial mine of Brazil (that of Carajás), it is the target of incessant federal governmental and military intervention justified on the basis of security of national interests. This increases social and economic problems of miners settled in this region.

Vulnerability of women's labour in *garimpagem* is inscribed in broader processes of capital accumulation in the Amazon region. The ways in which women's labour facilitates and contests the logic of capital accumulation is discussed in the next section.

Gender and capital accumulation in the mining frontier

Women's labour has been crucial in the expansion of capitalism and the reproduction of its modes of production in the mining frontier. A rich and diverse literature of feminist political economy has shown the ways in which gender hierarchies develop to facilitate certain forms of capitalism, that is, capitalism intersected by patriarchy, patriarchal capitalism, independent patriarchy with a trans-capital identity, global capitalist patriarchy (Plumwood, 1993; Prügl, 1999; Vogel, 1995). One of the most influential scholars in the field, Maria Mies, has argued that women's work in peripheral areas allows the 'ongoing primitive accumulation and colonization' necessary for capitalist expansion and development (Mies, 1986). Her work has fueled an on-going theorization about the connections between contemporary international division of labour and the maintenance and recreation of asymmetrical gender division of labour at the local level.

Contrary to the claims of critical development theories that increasing modernization, industrialization and urbanization in the periphery will lead to disappearance of structural inequalities, Mies argues that these processes reproduce patterns of oppression and violence against women and expand the scope of the system of patriarchy. The gender (and sexual) division of labour at local levels, whereby women are manipulated as both 'producer-housewives' and 'consumer-housewives' is the manifestation of an international division of labour based upon unequal man-woman relationships (Mies, 1995).

Whether patriarchy is independent from, or intrinsic to capitalism, is not a fundamental inquiry of this chapter. What remains central to the study of women's labour in mining in the Brazilian Amazon is the discussion of the structural inequalities that maintain women's work in a condition of peripherality and undervalues it, for the continuation of the accumulation process of capitalism. Mies' words along these lines are worth noting:

> Within a limited world aims like 'unlimited growth' can be realized only at the expense of others. Or, there cannot be progress without regression of another part, there cannot be development to some without underdeveloping others. There cannot be wealth without impoverishing others. Concepts like 'unlimited growth' or capital accumulation, therefore, necessary imply that this growth (progress, development, wealth) is at the expense of some 'Others', given the limits of our world.
>
> Mies, 1995: 3

Labour arrangements of *garimpagem* are organized to facilitate surplus extraction. Since labour relations are based upon a structural gender division of labour, gender relations are continuously restructured to facilitate capital accumulation. Absorption of the value of women's labour by *garimpos* absentee investors, men miners and owners of machines has allowed these to generate profits and remit them upward through hierarchical networks of production. The informalization of the formal labour relations in which women participate, has created an abundant supply of cheap female labour subjected to the needs of *garimpagem*. Concentration of women's labour in agro-industry operations of mining companies has allowed capital to remit this surplus to global markets through networks of international trade.

For capital accumulation to progress in mining peripheral frontiers, the dynamics of gendered labour must become global. In the first place, gendered labour, and the structural vulnerabilities that organize it, is an effective conduit for surplus appropriation. Second, gendered labour helps to maintain peasant labour in reserve in the mining frontier. To the extent that women's labour sustains the household economy and men's ventures of gold exploration, it serves a crucial role in the reproduction, maintenance and captivity of peasant labour force. Third, gendered labour allows the mining frontier to create consumption markets for manufactures produced outside the periphery. It is the settlement of women in *garimpos* that fuels the creation of intraregional trade networks for household goods. This is what David Simon calls the 'captive markets' of the periphery; to this idea, one must add that these are facilitated by women's settlement in resource frontiers (see David Simon, 1995).

Two further questions need to be addressed. First, the extent to which globalization of gendered labour is a homogeneous process in peripheral mining frontiers. The most accurate response to this question is no. There is a selective (and somewhat uneven) incorporation of peasant mining communities into hierarchical networks of mineral production and trade. In a given time and space, peasant mining communities can occupy different positions in such hierarchies, depending on multiple factors, such as extinction of secondary deposits, land conflicts with mining companies and governments, male migration, etc.

Second, the extent to which this case study sheds light on potential arenas of contestation of the logic of capital accumulation. Although women's labour in *garimpagem* is vulnerable and subordinated to structural-historical inequalities, it is inscribed in complex and contradictory ideological constructs that need further research.

For example, women's ethnogeology[9] challenges the credo of mining capital with regards to the technologically-driven efficiency of corporatist operations. It legitimizes peasants' claims for land and mineral rights. In many developing countries, it exerts pressure on governments to include women and children in participatory development mineral policies (Quiroga, 2002). This ethnogeology contests the negative ideas that mining industries promote about *garimpagem*. These invariably depict *garimpagem* as destructive, violent, unstable and responsible for mercury contamination in the Amazon. They portray women in *garimpos* as prostitutes, ignorant, dirty and lazy peasants, which justifies police and military intervention and the expulsion of women from some *garimpos*.

Another example is women's ideas about entrepreneurship in *garimpagem*. As many women revealed in the life-stories interviews conducted during fieldwork, the decision to migrate to *garimpos* responds to the idea that extraction of gold, despite the hardships it entails, necessarily represents an upward economic mobility. Even when these ideas connect to the needs of the economy of *garimpagem*, they suggest complex and disarticulated relations of subordination of women's labour that deserve further research. Probably the strongest example of this is the role women have played in the struggles against mining companies over land issues, principally the CVRD, and against the government for its opposition to recognizing *garimpagem* as a small scale enterprise and *garimpeiras* as professional miners. Related to the above is the importance women ascribe to the tradition and history of *garimpagem*. Belonging to a *garimpos* community, albeit its poverty and violence, works as a powerful ideological construct to exert pressure upon governments to extend basic infrastructure services to the community.

Ideological constructs shed light on the power (and naturalization) of Western ideas that are needed for capital accumulation in the mining frontier. Women seem to perceive *garimpos* as the spaces that put them within the reach of a way of living

---

[9] The term 'ethnogeology' comes from the work of Cleary (1987), who defines it as a form of peasant knowledge based on their livelihoods and experience with the forest.

that could not be possible otherwise. Indeed, some women see *garimpos* as the opportunity to reach luxuries that the cities cannot offer for them. Possession of gold chains, bracelets, jewellery, and even gold-covered teeth, gives some women a higher status in the *garimpos*. This 'higher status' involves amassing a form of symbolic capital that confers legitimacy to women's labour in the *garimpos*.

Women's work in Brazilian Amazon thus reproduces (and is reproduced by) a gender division of labour needed for capital accumulation. Development policies put forward by international organizations, inspired by the rhetoric of *laissez faire*, reinforce directly or indirectly this gender division of labour. It is worthwhile to note that the first major intervention of the World Bank Group in the mineral and energy sector of the Brazilian Amazon coincides with expansion of *garimpagem* and women's insertion into this economy, reinforcing the consolidation of gendered labour in the region.

While international organizations have started recently to recognize the gender biases of their mining policies and loan instruments, they continue to hold to a patriarchal understanding of labour and production. It is imperative to redefine the loan instruments and mining policies put forward by international organizations, as well as the mineral policy reforms that developing countries have engaged in.

Also, investments of global mining capital in peripheral tropical regions create the incentives for women to migrate to these frontiers. Increasing foreign direct investment in gold mining operations in Latin America encourages the globalization of gendered labour and the deepening of its structural vulnerabilities. Literacy, health and development programs for women and children promoted by mining companies, do not aim at eliminating structural vulnerabilities affecting these populations. Quite the contrary: they attempt to solve the problems that directly impinge upon the stability and smooth progression of corporatist operations, and make invisible the historical-structural causes of vulnerability. If systematically women's labour continues to be exploited by hierarchies of mining production, then it is beyond question that capital accumulation continues to function by exploiting women.

The Brazilian case raises several theoretical and policy ruminations that are worth stating. The neostructuralist argument about 'high versus low productivity', based on 'high versus low capital coefficient per worker', cannot be sustained in the case of women's labour in *garimpagem*. Given the different nature of *garimpagem*, productivity of women's labour cannot be defined using the structuralist categories or 'pre-capitalist' or 'rudimentary' production. Historically, *garimpos* production has been sustained by women's labour. Its share of the economic costs of *garimpagem* and its direct mining production must be taken into account in the analysis of mining production.

In my previous work, I have defined several policy avenues to address the main economic problems of women's labour in mining (Graulau, 2001). The discussion of women's labour and capital accumulation has shown that policy debates must transcend the 'trickle-down versus bottom-up' recommendations. Both approaches have treated poverty of women in mining regions as the redeemable outcome of the unchallenged, unmistakable end of 'economic growth'. A policy framework is needed capable of delivering immediate reparation of the costs of women's labour in mining. Along these lines, governments must recognize women as workers of small-scale mining enterprises, and create the juridical conditions for their labour to be compensated. This must include granting mining licenses, improving access to credit and lending instruments, technical assistance in mining processes and commercialization, and, above all, land rights. A good example is the project known as TAWOMA (Tanzania's Women Miners Association), a non-governmental, non-profit organization that seeks to enhance women's economic livelihoods in mining in Tanzania (World Bank, 2001).

Any policy intervention, however, must aim at decolonizing women's knowledge and advance a rigorous understanding of their production in mineral economies. This includes collecting and analyzing, on a systematic basis, women's knowledge about natural resources and their interactions with their immediate environments. Above all, it must attempt to promote what geologist Héctor Lacrau calls 'geoscientific literacy', a form of knowledge able to extend citizens' horizons, the possibility to conquer the power to judge, reward or punish development policies and the fight to improve the quality of life.

References

Ackermann, R., 1998. 'Is Mining Compatible with Sustainable Development? A World Bank Perspective', *ICME Newsletter*, International Council on Mining and the Environment, vol. 6, no. 1.

Barrios de Chungara, D., 1978. *Let me Speak! Testimony of Domitila, a Woman from the Bolivian Mines*, New York: Monthly Review Press.

Benedito Da Silva, A. R., 1993. '*Garimpagem* como meta política do Estado', Paper delivered at the Conference *Conseqüencias da* Garimpagem *no Âmbito Social e Ambiental na Amazônia*, Belém: Fase, Katalyse and Buntstift, March.

Briskey, J. A., Schulz, K. J., Mosesso, J. P., Horwitz, L. R. and Cunningham, C. G., 2001. 'Environmental Planning Issues and a Conceptual Global Assessment of Undiscovered Nonfuel Mineral Resources', Paper delivered at 31st *International Geological Congress*, Rio de Janeiro, Brazil, 6-9 June.

Browder, J., and Godfrey, B., 1997. *Rainforest Cities: Urbanization, Development and Globalization of the Brazilian Amazon*, New York: Columbia University Press.

Bunker, S. G., 1985. *Underdeveloping the Amazon: Extraction, Unequal Exchange and the Failure of the Modern State*, Urbana: University of Illinois.

Cardoso, F. H. and Faletto, E., 1969. *Dependencia y Desarrollo en América Latina*, México: Editorial Siglo Veintiuno.

Cleary, D., 1990. *Anatomy of the Gold Rush*, Oxford & Basingstoke: Macmillan Press.

Cleary, D., 1991. *The Brazilian Rainforest: Politics, Finance, Mining and the Environment*, Special Report no. 2100, The Economist Intelligence Unit.

Cleary, D., 1993. 'After the Frontier: Problems with Political Economy in Modern Brazilian Amazon', *Journal of Latin American Studies*, no. 25, pp. 331-349.

Deere, C. D., 1995. 'Contribution of Gender to Peasant Studies', *Feminist Economics*, 1, no. 1.

Furtado, C., 1970. *Obstacles to Development in Latin America*, New York: Anchor Books.

Godoy, R., 1985. 'Mining: Anthropological Perspectives', *Annual Review of Anthropology*, no. 14, pp. 199-217.

Graulau, J., 2001. 'Peasant Mining as a Development Strategy: The Case of Women in Gold Mining in the Brazilian Amazon', *European Reviews of Latin American and Caribbean Studies*, no. 7, October, pp. 76-101.

Harcourt, W. A. (ed.), 1994. *Feminist Perspectives on Sustainable Development*, London: Lynne Rienner Publishers.

International Labour Office, 1999. *Social and Labour Issues in Small-Scale Mines*, Geneva, Switzerland: ILO.

MacMillan, G., 1995. *At the End of the Rainbow? Gold, Land and People in the Brazilian Amazon*, New York: Columbia University Press.

Mathis, A., De Brito, D. C. and Brüseke, F. J., 1997. *Riqueza Volátil: A Mineração de Ouro na Amazônia*, Belém: Editora CEJUP.

Mies, M., 1986. *Patriarchy and Accumulation on a World Scale: Women in the International Division of Labour*, New Jersey: Zed Books.

Mies, M., 1995. *Women and Work in a Sustainable Society*, Retrieved on 15 March, 2002, from http//www.jaysquare.com/resources/workdocs/wdoc10a.htm.

Plumwood, V., 1993. *Feminism and the Mastery of Nature*, New York: Routledge.

Pompermayer, M., 1982. *Estratégias do grande capital na fronteira amazônica brasileira*, Belo Horizonte: Universidade Federal de Minas Gerais.

Prügl, E., 1999. *The Global Construction of Gender: Home-Based Work and the Political Economy of the 20th Century*, New York: Columbia University Press.

Quiroga, E., 2002. 'The Case of Artisanal Mining in Bolivia: Local Participatory Development and Mining Investment Opportunities', *Natural Resources Forum*, no. 26, pp. 127-139.

SAPRIN, 2002. *The Policy Roots of Economic Crisis and Poverty: A Multi-Country Participatory Assessment of Structural Adjustment*, Prepared by the Structural Adjustment Participatory Review International Network, April.

Simon, David, 1995. 'The World City Hypothesis: Reflections from the Periphery', in P.L. Knox and P. J. Taylor (eds), *World Cities in a World-System*, Cambridge: Cambridge University Press, pp. 132-155.

Stear, W., 1990. 'Small-Mining – Strategic Opportunity in a Changing South Africa?', *Mining Survey*, no. 1, pp. 39-44.

Sunkel, O., 1993. *Development from Within: Towards a Neo-structuralist Approach for Latin America*, London: Lynne Rienner Publishers.

Vogel, L., 1995. *Woman Questions: Essays for a Materialist Feminism*, London: Pluto Press.

Wallerstein, I., 1981. *Historical Capitalism*, London: Verso.

World Bank, 2001. *Tanzania: Women in the Mining Sector*, Finding Reports no. 189, August.

World Gold Council, 2001. *The New El Dorado: The Importance of Gold Mining to Latin America*, Retrieved on 20 February, 2002, from http://www.gold.org/pub_archive/pdf/eldorado.pdf.

Chapter 15

Women Miners, Human Rights and Poverty

Ingrid Macdonald[1]

Introduction: Women's rights as human rights

All people – women, men, girls and boys – possess certain inalienable human rights that provide them with universal claims on society. These rights are expressed within international human rights instruments such as the *Universal Declaration of Human Rights 1948* (UDHR), the *International Covenant of Civil and Political Rights 1966* (ICCPR) and the *International Covenant of Economic, Social and Cultural Rights 1963* (ICESCR). There are also numerous other important human rights instruments[2]. For example, the International Labour Organization (ILO) *Declaration on Fundamental Principles and Rights at Work* represents the universal core labour standards which all ILO member nations are bound to promote and realize. These include the 'elimination of discrimination in respect of employment and occupation' which is a right enjoyed by both female and male workers. There are also eight 'core' ILO Conventions on labour rights, which seek to prohibit forced and child labour and protect the rights of freedom of association, collective bargaining, the right to organize, equal opportunity and equal treatment for men and women workers and workers with family responsibilities and the rights of indigenous and tribal peoples[3].

---

[1] I would like to thank Joanna Kyriakakis who assisted with research for and drafting of this chapter while undertaking an internship at Oxfam Community Aid Abroad. Joanna is undertaking a Doctor of Juridicial Science degree in International Law at Monash University, Australia.

[2] See Macdonald and Ross, 2003.

[3] There are eight 'core' ILO Conventions on labour rights, namely Convention 29 (forced labour), 87 (freedom of association and protection of the rights to organize), 98 (right to organize and collective bargaining), 100 (equal opportunities and equal treatment for men and women workers: workers with family responsibilities), together with ILO Conventions 182 and 190 (child labour) and 169 (the rights of indigenous and tribal people). Available at www.ilo.org.

The rights expressed under the international human rights system are universal, inalienable, interdependent, indivisible, and complementary. This means that it is necessary to protect and promote a person's civil, political, social, economic and cultural rights, and what are commonly called their collective rights, to enable them to enjoy full human dignity. It also means that these rights transcend national borders, economic paradigms and political structures, and some of them are more legally binding than others. For example, the prohibition on torture is a pre-eminent norm of international law whereas the right to social security is not (see Higgins, 1994: 19, for the validity of prohibition as a requirement of customary international law and Schacter, 1991: 85, for the binding nature of the UDHR).

Notwithstanding this on-going debate, rights are guaranteed to all human beings 'without distinction of any kind, such as race, colour, sex, language, religion, political or other opinion, national or social origin, property, birth or other status' (UDHR 1948: art. 2). Women are therefore entitled to the same protection and promotion of their human rights as men. The Vienna Declaration of 1993 provides, '[t]he human rights of women and of the girl-child are an inalienable, integral and indivisible part of universal human rights'[4], and Article 14 of the Fourth World Conference on Women 1995 ('the Beijing Declaration') declares that 'women's rights are human rights'[5]. Further, positive government action and policies are anticipated and obligatory to ensure women's rights are enjoyed without discrimination. For example, Article 2(2) of the ICCPR provides that State parties must 'take the necessary steps ... to adopt such legislative or other measures as may be necessary to give effect to the rights recognized' in the Covenant. Similar obligations toward the adoption of national legislation and measures to ensure the actualization of human rights standards for women are required by most human rights instruments[6].

The specific right of women to work and to work free of discrimination is also enshrined in a number of human rights instruments, such as Article 11 of the Convention On the Elimination of All Forms of Discrimination Against Women (CEDAW) (UNDFW, 1979), Article 26 of the Beijing Declaration, Article 23 of the Universal Declaration of Human Rights, and is implicit particularly in ILO

---

[4] Text available in *Human Rights Law Journal*, 1993, pp. 352-70.

[5] *The Beijing Declaration and Platform for Action*, op. cit, 'Mission Statement', para 1 Available at: http://www.un.org/womenwatch/daw/beiging/platform/plat.1.htm [Accessed November 4, 2002].

[6] See for example, the *Convention on the Elimination of All Forms of Discrimination Against Women*, which is framed by the obligations of State Parties to take positive measures to address the inequalities and discrimination faced by women; see also ILO Convention 111: *Discrimination (Employment and Occupation) Convention, 1958*, article 1, which obligates member states to take positive action and amend national legislation in order to promote and achieve equal opportunity and treatment in respect of employment and occupation with a view to eliminating any discrimination. This Convention has been ratified in Australia with the *Human Rights and Equal Opportunity Act* (Cth) 1986.

Convention 111 – *Discrimination (Employment and Occupation) Convention, 1958.*

However, in practice the duty to ensure women's human rights, including workers rights, is often far removed from the reality of implementation, application and enforcement, including in respect of the disadvantages suffered by women and children involved in small and large scale mining.

The barriers to realizing women's rights

Women are the most discriminated against, the most vulnerable and least empowered members of many societies, including those with rich mining, oil and gas resource endowments. It is therefore not surprising that women comprise 70 per cent of the world's poor and this proportion is growing. Throughout the world, women work longer hours for less pay than men employed in similar positions, and they are grossly under-represented in private sector management and political positions, leading the United Nations Development Group to comment that women represent the 'world's largest excluded group' (UNDP, 1993:25).

Women comprising 70 per cent of the world's poor is not a 'natural occurrence' or the result of biological differences between men and women. There are many different illegitimate social, political, economic, civil and cultural barriers that perpetuate systems of discrimination, oppression and the abuse of women's rights. As a result, the benefits of economic development are not distributed equally, and progress towards equality between men and women does not take place naturally. It can therefore not be assumed that women automatically benefit from economic and industrial development efforts, especially in respect of mining projects.

In India, for example, considering both market and non-market work (such as domestic care and subsistence farming), women work an average day 117 per cent that of men's, while in Indonesia it is 109 per cent and in the Philippines 121 per cent (UNDR, 2004: Table 28). In contrast, in both developed and developing countries women's wages are well below those enjoyed by their male counterparts, with the ratio in most east and south Asian countries near to or less than half that of men's (UNDR, 2004: Tables 24, 25). There is no country within the world where women earn the same as men (Seager, 2003).

Women also comprise the vast majority of the world's unpaid informal and subsistence agricultural workforce. In those countries where data is available, the United Nations Development Group reports that women invariably allocate the greatest portion of their time to non-market activities (such as domestic care and subsistence agriculture) in contrast to men whose time is predominately allocated to market, and hence income generating, activities. For example, in Indonesia, women allocated 65 per cent of their time to non-market activities and 35 per cent to market activities, in contrast to men who allocated 14 per cent and 86 per cent respectively. In the Philippines the figures show an allocation between non-market and market activities at 71 per cent and 29 per cent for women, and 16 per cent and

84 per cent for men, and in India at 65 per cent and 35 per cent for women, and 8 per cent and 92 per cent for men (UNDR, 2004: Table 28).

Yet much of the work of women, such as child-care, household responsibilities and daily food and water provision, has no value within the current neo-liberal system of economic development. This system, based on 'gender neutral' economic theories, places primary value on paid labour (Hunt, 1996: 25-6). As women are often not active or only partially active within the paid labour force, especially in relation to male dominated activities such as mining, their non-paid labour is not valued. In addition, underemployment is often not acknowledged in statistical data and hence considered in policy. As a result, economic development, has often served to marginalize the roles and responsibilities of women and exclude them from any benefits of such development. The Director-General of the International Labour Organization in his 2003 report on work and poverty acknowledges that 'far too much of women's work is still uncounted and undervalued' (ILO, 2003: art. 6). Further:

> The distinction between the public and private spheres operates to make the work and needs of women invisible. Economic visibility depends on working in the public sphere and unpaid work in the home or community is categorized as 'unproductive, unoccupied and economically inactive.
>
> Charseworth et al., 1991: 614

Throughout the world women also continue to suffer persistent and systematic human rights abuses for no other reason than they are women[7]. In the workplace, home, health system, the public domain and in conflict situations, women are subjected to violence, abuse and discrimination that is often sanctioned or ignored by judicial and political institutions. Examples include sexual assault, domestic violence, forced prostitution and lack of access and control over reproductive and employment choices.

Given the considerable barriers to the actualization of women's rights, there are key international human rights instruments dealings exclusively with removing the barriers to women's rights and empowerment. CEDAW (UNDFW, 1979) has been declared and adopted so as to address directly the disparities between men and women as regards their access and enjoyment of human rights, and the persistence of practices and attitudes that hamper the actualization of women's rights. The Beijing Platform for Action of 1995 is also specifically concerned with women's rights and empowerment. Such instruments are built upon the recognition that despite the existence of human rights instruments extensive discrimination against women continues and inequality persists (UNDFW, 1979: preamble). The protection and promotion of women's rights is fundamental to poverty alleviation and development.

---

[7] Human Rights Watch, *Women's Human Rights*. Available at: http://www.hrw.org/women/index.php [Accessed November 4, 2002].

Kofi Annan in the United Nations 1998 Annual Report states:

> The rights-based approach to development describes situations not simply in terms of human needs, or of developmental requirements, but in terms of society's obligation to respond to the inalienable rights of individuals. It empowers people to demand justice as a right, not as charity...[8]

The rights based approach to development recognizes that there can be no real solution to poverty, without respect for, understanding and promotion of human rights, particularly the rights of traditionally marginalized groups, such as women and girls, ethnic minorities and indigenous peoples. As such, it recognizes that economic poverty is not a natural state or phenomenon, but results from the direct denial, violation and abuse of the human rights of men, women, girls and boys, by entities that have more access to power, or through systems that are based on injustice, inequality and discrimination.

Human rights and human development operate in a symbiotic relationship, with the achievement of one based upon the other. This idea was encompassed by the United Nations Development Programme (UNDP) in their *Human Development Report 2000,* where it stated that 'human development is essential for realizing human rights, and human rights are essential for full human development' (UNDP, 2000). The UNDP also asserts that '...[h]uman rights and sustainable human development are interdependent and mutually reinforcing'[9].

Given the majority of the of the world's poor are women, the promotion of women's rights, gender equality and the empowerment of women are essential prerequisites to achieving sustainable people-centred development and combating poverty. Recognized under the 1986 Declaration on the Rights to Development[10], the active, free and meaningful participation of all people, including women in development, including their active, free and meaningful participation in the workforce, is fundamental to addressing the well-being of the society as a whole. Article 16 of the Beijing Declaration provides:

> Eradication of poverty based upon sustained economic growth, social development, environmental protection and social justice requires the involvement of women in economic and social development, equal opportunities and the full and equal participation of women and men as agents and beneficiaries of people-centred sustainable development. As such, promoting the right of women to work is fundamental for ensuring that development benefits flow to all members of the community equitably, and actualize goals to alleviate poverty.

---

[8] United Nations, *Striking a Better Balance*, vol 3, Annex 6: 143.

[9] United Nations Development Programme, 1998, *Integrating Human Rights with Sustainable Human Development: A UNDP Policy Document*, January.

[10] Adopted by the General Assembly of the United Nations, GA Res 41/128 of 4 December 1986, preamble.

The large scale mining companies and feminization of poverty

Globally, the push towards a free-market system has resulted in the increasing impact of mining companies' activities on the world's poorest and most vulnerable people, including women and children. Recent figures show that as a result of the increasingly rapid and unfettered movement of international capital, the revenues of five of the world's largest transnational companies add up to more than double the combined Gross Domestic Profit of the poorest 100 countries (Utting, 2002).

This process of economic globalization is being driven by transnational companies, wealthy governments and multilateral financial institutions including the International Monetary Fund, World Trade Organization, World Bank Group and Asian Development Bank. The resulting economic liberalization, market deregulation, privatization of industries and services has meant that, as stated in a recent Oxfam America briefing paper (Langman, 2003: 6):

> Foreign direct investment (FDI) … has become such an important part of global development strategies that it has replaced foreign aid as the main source of external capital for many developing countries. Today, FDI amounts to about 60 per cent of the international capital flowing into developing countries each year and is nearly ten times larger than official development assistance. In contrast, in the late 1980s, the amounts of annual aid and FDI in developing countries were roughly the same.
>
> UNCTAD, 2002

It has also enabled a rapidly growing mining industry, dominated by transnational corporations, to expand into some of the world's most remote communities, including those within the Asia Pacific (Langman, 2003: 5-16).

Governments, multilateral financial institutions and companies justify the promotion of large scale mining in poor countries in the name of economic development and reducing poverty. Yet despite decades of economic liberalization, deregulation and privatization easing the way for such activity, inequality between and within countries has increased[11]. The *Human Development Report 2004* reports that the previous decade saw an unprecedented backward slide in development outcomes for many nations with the result that in 46 countries people are poorer and in 26 countries more people go hungry than a decade ago (UNHDR, 2004: 128).

This backward development slide appears to be even more pronounced in resource rich countries. The World Bank's recent Extractive Industries Review found that:

> Data on real per capita gross domestic product (GDP) reveal that developing countries with few natural resources grew two or three times faster than resource-rich countries

---

[11] UNDP (2001) states that: 'The income gap between the richest one-fifth of the world's people and the poorest one-fifth measured on average national income per head, increased from 30 to 1 in 1960 to 74 to 1 in 1997'.

> over the period 1960–2000. Of 45 countries that did not manage to sustain economic growth during this time, all but six were heavily dependent on extractive industries, and a majority of them also experienced violent conflict and civil strife in the 1990s.
>
> World Bank, 2004

Otherwise known as the 'resource curse', numerous studies have shown that many of the world's most resource-rich countries are also the world's poorest in economic terms (Ross, 2001: 7). Oil, gas and mining industries are important to more than 50 developing countries, which are home to 3.5 billion people. Yet more than 1.5 billion of these people live on less than $2 US per day. Moreover, 12 of the world's 25 most mineral-dependent states, and six of the world's most oil-dependent states, are classified by the World Bank as 'highly indebted poor countries' with the world's worst human development statistics (Ross, 2001: 7).

Despite enormous funds being generated through resource extraction, these funds have largely not been used to combat poverty. Instead, the revenues generated by extraction have often been embezzled by corrupt elites, spent on military armaments by authoritarian regimes, or have even fuelled regional instability through groups warring over control of the revenue streams from resource extraction (Ross, 2001; Global Witness, 1999; Hafild, 2003). Many of the people within these countries remain impoverished and their rights undermined due to corruption and economic mismanagement, which has severe impacts on the majority of the poor: women and children.

Human rights duties of mining companies

The increasing global power and influence of transnational companies, and their impact on development and poverty means that they, alongside governments, must be responsible for upholding human rights, including the rights of women workers. As UN Secretary General Kofi Annan has noted:

> The fragility of globalization … poses a direct challenge to the self-interest of the corporate sector, and a central part of the solution is the need … to accept the obligations – and not merely the opportunities – of global citizenship[12].

Traditionally, governments have had the primary responsibility for ensuring that mining companies do not violate people's rights, including the protection of the rights of women mine workers. This reflects the power historically wielded by the state, both individually and collectively through institutions like the United Nations. Yet, rights and duties under international law are slowly being extended to non-state actors and individuals. Thus far, individuals have been found legally

[12] Address of Kofi Annan, Secretary General of the United Nations, to the World Economic Forum, Davos, Switzerland, 28 January 2001, available at www.weforum.org.

responsible for war crimes, crimes against humanity and other gross human rights abuses[13]. The UN *Universal Declaration of Human Rights* also codifies not only the moral responsibility of companies to uphold rights of those affected by their activities, but is increasingly seen implying their *legal* liability as 'organs of society' to respect, promote and secure human rights[14].

The UN Sub-Commission on the Promotion and Protection of Human Rights also recently adopted the *Norms of the Responsibilities of Transnational Corporations and Other Business Enterprises with Regards to Human Rights* ('UN Norms')[15]. The UN Norms codify company duties regarding human rights, and were drafted by independent human rights experts elected from different regions by the UN Commission on Human Rights. The UN Norms include the rights of women workers and would be applicable to the mining industry, including protections covering freedom of association and collective bargaining, a safe and healthy working environment, prohibitions on child and forced labour and the right to a living wage.

Oxfam Community Aid Abroad's focus on women and mining

Oxfam Community Aid Abroad[16] focuses on women and mining issues based on decades of work with local community women and men impacted by mining in the Asia Pacific. In the past few decades, the Australian mining industry has increased its activities in economically developing countries in the Asia-Pacific. Australian mining operations are therefore increasingly impacting on poor and vulnerable

---

[13] See cases concerning: *Trial of Major War Criminals Before the International Military Tribunal Nuremberg*, 1947; *Statute of the International Criminal Tribunal for the Former Yugoslavia*, UN Doc S/Res/827, 1993; *Statute of the International Tribunal for Rwanda*, SC Res 955, 49 UN SCOR (3452nd meeting), UN Doc S/Res/955, 1994; and the requirements of the *Rome Statute of the International Criminal Court*, 37 ILM 999, opened for signature 17 July 1998.

[14] *Universal Declaration of Human Rights*, (1948), op cit, preamble: Renowned international legal scholar Professor Louis Henkin states that: '…[e]very individual and every organ of society excludes no one, no company, no cyberspace. The *Universal Declaration* applies to them all'.

[15] United Nations Norms on Responsibilities of Transnational Corporations and other Business Enterprises with regard to Human Rights (with commentary); E/CN.4/Sub.2/2003/38/Rev.2 (2203); UNESC CHR, Sub Commission on the Promotion and Protection of Human Rights, 55th session agenda item 4.

[16] Oxfam Community Aid Abroad is an independent, secular, non-government development agency and the Australian member of the Oxfam International confederation. The agency undertakes long-term development projects, provides humanitarian responses during disaster and conflict, and advocates for policy and practice changes to promote human rights, justice, gender equality and women's empowerment in 30 countries across Asia, Africa, the Pacific, Central America and Indigenous Australia.

community women and men – the same women and men that Oxfam Community Aid Abroad has worked with for over 50 years. Many of these local women and men complained, and continue to complain, of human rights abuses and environmental degradation caused by, or on behalf of, Australian mining companies. These people often have no institution that they can access for fair and equitable redress, so companies have been able to disregard their concerns. As a result, Oxfam Community Aid Abroad established the position of the Mining Ombudsman in 2000. The primary objective of the project is to demonstrate the need and feasibility of the Australian Federal government formally establishing such a broad based industry mechanism within Australia.

The Mining Ombudsman is invited by local community women and men impacted by the activities of Australian listed mining companies to investigate their grievances and attempt to secure resolution or redress from the company concerned. There are cases currently in Peru, Papua New Guinea, the Philippines, Indonesia and Fiji[17].

Through the Ombudsman, we have found that the grievances of community women and men affected by such mining activities often constitute denial of their basic human rights – especially to free, prior and informed consent; to self-determination, land, and a livelihood. These grievances are often industry-wide, arising at every stage of the project cycle. Indigenous peoples are particularly vulnerable in part because they are usually the poorest, most marginalized groups in society, and because of the particular damage that can result from loss or damage to the lands, waterways, flora, fauna and sacred sites with which they have a longstanding spiritual, often custodial relationship.

In June 2002, we convened the *Tunnel Vision: Women, Mining and Communities* workshop in Melbourne. The first of its kind in Australia, the workshop brought together speakers from Indigenous Australia and the Asia Pacific speaking on the impacts of mining on women living in local communities. It found that women had overwhelmingly been excluded from the benefits of mining, and borne the brunt of its negative social and environmental impacts[18]. Issues raised echoed the grievances voiced by women from communities who have brought their complaints to the Mining Ombudsman. These grievances include:

1. Companies entering into negotiations only with men, making women neither party to the negotiations, nor beneficiaries of royalties or compensation payments, even in matrilineal societies where women are the culturally recognized landowners. As a result, women are often stripped of their traditional means of acquiring status and wealth.

---

[17] See Oxfam Community Aid Abroad Mining Campaign at www.oxfam.org.au/campaigns/mining and the *Mining Ombudsman Annual Reports 2001, 2002, 2003* and *2004* for further information.

[18] A copy of the report containing the papers presented by the speakers is available from the Oxfam Australia website www.oxfam.org.au/campaigns/mining.

2. Companies not recognising the religious and spiritual connections of indigenous women to their environments and land, especially when they are displaced by mining activities.
3. Women generally have little or no control over and access to any of the benefits of mining developments, especially money and employment. They therefore become more dependent on men who are more likely to be able to access and control these benefits through being recognized by the mining companies and being the ones who are paid the benefits or receiving work at the mine.
4. The traditional roles and responsibilities of women are marginalized as the community becomes more dependent on the cash based economy created by mine development.
5. The workload of women increases as men work in a cash economy created by mining operations, and women bear increased responsibility for the household and food provision through traditional means. This workload can be further exacerbated if there is environmental pollution or contamination of land or water sources. It is the women who bear the responsibility of finding alternative food sources, walking further distances to access clean drinking water and caring for the sick children or family members.
6. Women become more at risk of impoverishment, particularly in women-headed households. This is particularly the case where women are abandoned by their husbands.
7. Women bear both the physical and mental strain of mine development, especially when it involves resettlement. This is especially evident in increased levels of domestic violence.
8. Women suffer from an increased risk of vulnerability to HIV/AIDS and other STD infections, family violence, and sexual abuse – often fuelled by alcohol abuse and/or a transient male workforce or the need to find money to feed their families.
9. Women suffer active and often brutal discrimination in the workplace.

These gender impacts represent the denial of the basic human rights of women from local communities affected by mining and women engaged in the mining industry.

The exclusion of women from large scale mining

Large scale mining has high levels of productivity with low levels of employment, due to it being based largely on mechanized processes requiring specialized technical skills. It is also increasingly associated with foreign corporate management by large transnational companies facilitated through privatization, deregulation and liberalization as already discussed. Women are represented in very small numbers within the large scale mining industry. For example, women comprise only 8-14 per cent of the Australian mining industry workforce, of which only 2-20 per cent are within the professional technical divisions of the industry[19].

We can speculate on any number of factors that lead to the exclusion of women from large scale mining which include not least the male dominated nature of the extractive industries. In a report to the Australian Institute of Mining and Metallurgy 'Women and Mining' taskforce in November 1998, it was found that some of the main reasons why women were discouraged from becoming involved in the mining industry were the remoteness of mine locations, the male dominated 'old boys' culture and inadequate site infrastructure (Pattenden, 1998). Further, first hand accounts from women and women's representative bodies in communities impacted by large scale mining projects are available which identify a number of recurrent themes in women's exclusion from the large scale mining sector, or brutal discrimination when they are employed. For example, see the papers presented in the *Tunnel Vision* report, direct testimonials collected through the Mining Ombudsman work and the reports of the International Women and Mining Network (RIMM, 2000).

The exclusion of women from the large scale industry is contrary to the rights of women and reduces what, if any, development benefits may flow to communities from mining and is also contrary to the rights of women.

Exclusion from decision-making

The exclusion of women from engagement with large scale mining projects begins at the first interaction between the mining company and the local people. In the *Tunnel Vision* workshop speakers discussed the experiences of women from Indigenous Australian, Papua New Guinean, Indonesian, Philippines and Indian communities who were systematically excluded from dialogue pertaining to mining project developments. It was pointed out that predominantly male company heads and politicians tend not to recognize women and issues pertaining to women, and too readily accept the demands by community men that women need not be consulted, often despite national laws requiring that women participate in negotiations.

Yet, under the UN Convention on the Elimination of All Forms of Discrimination Against Women (CEDAW) women have a right to engage equally in decision-making affecting them. In the event of a discrepancy between traditional law and the convention regarding this entitlement, the latter must prevail (UNDFW, 1979, articles 2f and 5). This right of participation also arises pursuant to the 1986 *Declaration on the Right to Development* which provides that every person is entitled to 'participate in, contribute to, and enjoy economic, social, cultural and political development, in which all human rights and fundamental

---

[19] Jessica Horsley, *Women in Mining – The Statics* on behalf of the Women In Mining Working Group for AusIMM available at www.aussimm.com/women/women.asp.

freedoms can be fully realized,' (UNDFW, 1979, article 1) and as discussed, is essential for achieving sustainable development and combating poverty.

Women's exclusion is often compounded by cultural constraints and stereotypes regarding the appropriate roles of men and women. Speakers within the *Tunnel Vision* workshop reflected an array of experiences as to women's exclusion from the large scale sector. As a result the women missed out on the economic benefits of mining whilst bearing the burden of many of the negative social and environmental impacts, thus intensifying their impoverishment and disadvantage. In the Indian example, high levels of illiteracy amongst women, lack of technical skills and the social stigma associated with women's participation in the paid work force were cited as among the key factors operating to exclude women from employment in large scale mining, which is increasingly technology dependent. The commencement of large scale mining in areas such as Indonesia and Papua New Guinea, operated to supplant the traditional power base of women and their centrality in land ownership and economic production, largely through the exclusion of women in the dialogue process and involvement in the subsequent cash economy. Men, who often dominate the pre-project negotiations with mining companies, if such negotiations occur, largely assume the primary entitlement to the work that flows from those projects.

Hostile work hours

The work environment within the extractive industry is hostile to women. In its advocacy, the Construction, Forestry, Mining and Energy Union in Australia have attempted to draw attention to concerns over long working hours in the mining industry and the impact of such hours on families. One claim regarding the average hours of work in the Australian gold mining industry is 3,000 hours per person per year, which represents 60 hours per week. It was proposed that this represents world's best practice (Colley, 2003).

A recent Western Australian review into extended working hours across different industries, in the context of Australia wide data, found that the Mining Industry has the highest average weekly hours across all sectors considered (Hartley et al., 2004). In 2003 the average hours per week in the mining sector in Western Australia was 48.4. Results from research in the United Kingdom similarly show the mining industry to be characterized by pervasive long working hours, ranking the mining industry second after agriculture in terms of numbers of employees working above 48 hours and 60 hours per week (Kodz et al., 2003: Figures 5.15 and 5.17).

An industry wide culture of long working hours impacts in particularly onerous ways upon women and serves to marginalize and exclude women from full participation. As already discussed, women are largely responsible for child care and other domestic responsibilities, as a result, women's participation in jobs that require long work hours is hampered by their responsibilities to family and home, or they may be excluded automatically on the basis of a perceived inability to participate to the full extent demanded by the culture of the workplace (Hartley et al., 2004: 32-33). In this Western Australian study, the view emerged that:

> [E]xtended working hours is part of a male work culture ... with particular consequences for women since women's capacity to work extended hours is more constrained given that women continue to be the primary care giver. While working extended hours may be viewed as mandatory to career advancement by both men and women, it remains an expectation that women are less able to meet due to their family responsibilities. Thus, women are disadvantaged when working within a pervasive work culture wherein it is perceived that there is a correlation between long working hours and commitment and professionalism (*footnotes omitted*).

A recent United Kingdom study (Kodz et al., 2003) into long working hours suggests that women are marginalized in certain industries by both systemic long working hours and by the active promotion of that culture by male colleagues for the purpose of excluding women, for example by 'working long hours, holding late meetings, or using after work socialising for discussing work issues, and criticising female staff for leaving "early"' (ibid: 207). Women working in male dominated industries reported a heightened experience of pressure to work long hours, and a strong disinclination to engage in what is perceived as 'presenteeism', a male work characteristic. Women also suffer greater negative impacts upon their health and general wellbeing as a result of employment involving longer work hours than men (ibid: 241-2).

Yet the obligations upon ILO member states under Convention 156: *Workers with Family Responsibilities* demands that governments address women's exclusion from the formal mining sector as a result of hostile work cultures. For example, Convention 156 requires each Member state to, so far as possible, enable persons with family responsibilities to exercise their right to engage equitably in employment, and to minimize conflict between employment obligations and family commitments (ILO, 1981: article 3.1). More broadly, ILO convention 111: *Discrimination (Employment and Occupation) Convention, 1958* requires member states to take positive measures to address discrimination affecting access to employment and occupation, with discrimination encompassing 'any distinction, exclusion or preference made on the basis of ... sex ...which has the effect of nullifying or impairing equality of opportunity or treatment in employment or occupation' (ILO, 1981: article 1.1a).

Fly in/fly out operations

Another element identified as being hostile to women is the predominance of 'fly-in' and 'fly-out' operations ('FIFO') in the mining industry. As mining often occurs in remote locations, employees are literally flown in for intense work shifts and then flown out for rest. For much the same reasons as apply to long working hours generally, this mode of operation and employment is especially hostile to women employees, especially those with family and child care responsibilities.

In it's joint submission to the Western Australian Review Panel on extended working hours, the Chamber of Minerals and Energy of Western Australia (CME) and the Australian Mines and Metals Association (AMMA) reported that amongst their members two thirds of employees worked 12 hour shifts, the most common work roster arrangement was 2 weeks on/1 week off, and half of the sites surveyed operated on a FIFO basis (Hartley et al., 2004: 10). The overall roster pattern within the mining industry in Western Australia shows that up to 76 per cent of workers work on this basis.

Negative impacts of FIFO employment arrangements are disproportionately suffered by women. Women are subjected to increased risks of sexual and other violence, marital stress and breakdown, and infections such as HIV, due to the separation of families and the health risks associated with make shift mining communities that are more isolated, attract an influx of male workers, and often do not include easy access to health and educational resources.

The ILO (2003) reports an above-average risk of HIV infection for workers in traveling and itinerant occupations, due to the worker's separation from their families and the availability of alcohol and sex along travel routes. Women as a category are themselves more vulnerable to infection due to their sexual and economic subordination to men which can prevent women from having the power to negotiate safe sex, and can expose them to greater risks of sexual violence and harassment at work.

Equal opportunity policies

Some mining companies have equal employment opportunity or formal diversity policies that purport to protect the rights of women workers (Hall, 2003). However, in discussions with women workers engaged in the mining industry, both professional and non-professional, they state that where these policies exist they do not always significantly improve their work environment. Further, the 'Women in Mining' Taskforce report contained various quotes including: 'EEO is a term coined by Human Resources Departments who may wish it to exist for technical departments who do not *(female, 26-35)*' (Pattenden, 1998: 25). Also:

> I've had a few situations where I have seriously considered taking some (companies) to the Equal Opportunity Tribunal, but I know that as soon as I do that I won't get another job in the mining industry. The mining industry is so small in Australia – in fact, the whole worldwide mining industry – is so small that you do something like that and everybody knows about it and everybody remembers it. And you're finished. You can never expect to work in the industry again *(female, 40s)*.
>
> Pattenden, 1998: 29

Depression and frustration are common complaints of women mine workers who can become professionally and socially isolated in the mining industry. The 'Women in Mining' Taskforce report found, in respect of Australia, 'that harassment and discrimination remain a significant problem within the minerals industry across professions' (Pattenden, 1998: 26, 31). As stated in the Preamble of CEDAW, the negative effect of discrimination upon the actualization of women's rights will undermine development efforts and the well-being of society as a whole:

> Discrimination against women violates the principles of equality of rights and respect for human dignity, is an obstacle to the participation of women, on equal terms with men, in the political, social, economic and cultural life of their countries, hampers the growth of the prosperity of society and the family and makes more difficult the full development of the potentialities of women in the service of their country and of humanity...
>
> UNDFW, 1979, preamble

Supply chain management and promoting 'flexibility'

What is also evident is that women generally have higher levels of employment in areas such as catering, cleaning, food supply and other types of low skilled work (RIMM, 2000). The statistics emerging from the most recent United Nations *Human Development Report* bears this out, showing that across both developed and developing nations, the percentage of women in the services sector continues to be substantially higher, and in the industries sector substantially lower, relative to men. For example in the Philippines women make up 172 per cent of the services industry comparative to men. In Thailand, it is 119 per cent (UNDP, 2004, table 27, pp. 229-232). Women employed in these types of roles are more vulnerable given that companies can claim that it is not them that employs them but the contractor, and so they are not therefore responsible for the conditions under which these women work. These types of contract roles are characterized by low pay, high pressure, long overtime at short notice and shift work, casual and short-term contracts based on supply of the good or service to the company with limited to no employment protection (RIMM, 2000). In the Mining Ombudsman Case Study on the Vatukoula Gold Mine (Macdonald, 2004), the problems encountered by women workers at the Vatukoula Gold mine in Fiji are described as:

> Currently, women who work at the mine are engaged in tasks such as administration, cleaning and supervising. A good majority of women are also engaged as hand pickers – a task which requires them to sort by hand the good from the bad ore before it is taken through the gold processing machinery...Where a private contractor employs women as hand-pickers, their situation is even more vulnerable. Although contrary to the employment legislation in Fiji, women claim to work for a period of three or four years

> as casual workers without a written contract. Should they fall pregnant, they are automatically dismissed from their work.
>
> Bale-Tuinamola, 2003

The promotion of the idea of a flexible workforce within large scale mining risks what limited rights women do enjoy in the workplace being slowly eroded and inequalities exacerbated. Increased levels of contract labour and resistance to fundamental core labour standards such as the right of collective bargaining and the guarantee of a living wage[20] has severe impacts on vulnerable workers such as unskilled women working in supply chains (Oxfam International, 2004). Not only does this undermine the human rights of women workers, it reduces any development benefits that they and their families may receive.

The problems for women engaged in small scale mining

In a 1999 report, the ILO found that female participation in the mining sector becomes more and more pronounced the smaller the scale of the activity. It also predicts that this will be an increasing trend into the future.

Women's higher representation in smaller scale mining has important implications when we consider the problems particular to small scale and traditional mining. It also begs the question, why are women working in the small scale informal mining sector but largely excluded from the large scale formalized mining industry?

Some of the problems inherent in the small scale mining sector that are briefly detailed below are more fully discussed in the ILO (1999) report, the World Bank Extractive Industry Review *Striking a Better Balance* report (World Bank, 2004) and the Conference Report from the II International Conference on Women and Mining 2000 (RIMM, 2000). These problems involve basic labour rights violations such as the denial of the right to freedom of association and collective bargaining, adequate health and safety protections and the right to a fair wage, they also involve other human rights violations such as the denial of the right to fresh water, a healthy environment and to live free of violence. Such problems include:

1. The small scale and traditional mining sectors are largely unregulated. This means that women workers in this sector are more susceptible to significant risks to health and safety, with the likelihood of work site accidents largely increased and the employment of dangerous practices more widespread. The ILO reported that the dangerous conditions particular to the small scale

---

[20] A living wage is intended to provide female and male employees with a minimum wage that not only provides themselves and their families with adequate shelter, food, clothing, education, healthcare and transport, but also includes a reasonable amount of discretionary income.

sector results in a workplace fatality rate of up to 90 times that experienced in mining operations in industrialized countries. As women are often engaged in the processing of raw materials, including sieving and washing materials, unsafe practices occur such as undertaking those tasks from home or polluting watercourses that are used for domestic purposes, which places families at greater risks from silicosis and mercury poisoning. With women principally responsible for the care of families and the sick, this places the greatest burden on women. Women are further impacted by a lack of regulation, as they are more susceptible to sexual violence and exploitation, which are also a problem in the small scale mining sector.

2. Small scale and traditional mining is often subordinated to large scale mining, and miners in those sectors are placed at risk of dispossession of their land and mining rights by the commencement of large scale mining operations. This has led in cases to the subsequent illegality of small scale and traditional mining, the failure by governments to recognize traditional mining rights and the resultant risk of forcible eviction and violence against traditional and small scale miners by security forces. Examples abound throughout Indonesia, the Philippines, Africa and Latin America and some are documented within the Mining Ombudsman work, including the Indo Muro Kencana case and Kelian case from Indonesia (Macdonald and Ross, 2003). Further, when large scale mining comes to regions with established small scale and traditional mining, it is often the women who are the first to lose their jobs and who do not obtain employment within the large scale mining industry and therefore lose their source of independence and livelihood – this situation is particularly difficult for female headed households.
3. Small scale and traditional miners are also often stuck in a cycle of poverty, where they are only able to access less productive lands, receive less than the real value of their product and are largely unrepresented or serviced by unions to address these problems.
4. The ILO have also identified women's need to bring children with them to a mining site as a factor contributing to the problem of child labour in mining, as often the only alternative, leaving the child unattended at home, is considered a worse outcome. This problem is not just confined to small scale mining but also large-scale mining.
5. The health issues for makeshift communities surrounding small scale mining are also more acute than in other areas, which are not suffering influxes of migrant workers, shortages of basic resources and the like. Again, this burden is felt most heavily by women, who have responsibility for the domestic and care activities.

The way forward: Addressing the gender inequity of mining, what can we all do?

In order to address the disadvantage of women engaged in and impacted by the small and large scale mining sectors, a prerequisite is the recognition of women as key stakeholders in the management and development of natural resources, and economic actors in their own right. It should also be remembered that workers in small and large scale mining are also often women living in communities impacted by mining. As such, they are detrimentally impacted when there is environmental pollution, social problems, human rights abuses and illnesses that affect local communities – all of which are human rights infringements.

Basic minimums that should be committed to and actualized by governments and companies, include the right to collective bargaining, abolition of child labour, non-discrimination and equal opportunities and the right to a basic living wage. These are important protections for women that should apply to all types of mining, traditional, small scale, large scale and service industries. Protecting and promoting the rights of women is fundamental to achieving sustainable development and combating poverty.

In order to promote gender equality and women's empowerment, gender must be main-streamed into all stages of the project cycle from planning and design to implementation, monitoring and evaluation and in the workplace. This means that women's perspectives, needs and interests must be considered in all planning and programming, and women must participate in all activities as decision-makers and beneficiaries.

As a result, companies, small and large, must provide not only policy commitments, but a work environment where women can work free of discrimination and harassment. Governments are legally required to ensure this occurs. These policies must be monitored and enforced to ensure compliance and they must be applied down supply chains in order to protect female contract workers. The education of all workers, male and female, is also fundamental for securing change. If women have complaints concerning male colleagues, they must have the ability to access a safe, confidential and independent mechanism to seek redress. Retaliation against complainants by work colleagues can not be allowed under any circumstances.

Overall, women's rights must be respected and promoted by all of us in all of our activities. The promotion of women's rights and gender equality is fundamental to combating economic poverty and sustainable development. This is something that each and every one of us can do something about in our own sphere of influence, with the first starting point being recognition that oppression and discrimination of any kind – even that which is argued for under the guise of culture, including a workplace culture – is unacceptable.

Unless we actively ensure that the rights of women are promoted, including the rights of women workers, economic poverty will only get worse – for in the end, what is the benefit of mining if it does not benefit women, who comprise half of the world's population and the majority of the poor?

References

Bale-Tuinamoala, Laisa, 2003. 'The Impact of Mining on Women in Fiji', *Voices for Change 2003*, Fiji Women's Rights Movement, Madang, Papua New Guinea, August 2003.

Charseworth, H., Chinkin, C. and Wright, S., 1991. 'Feminist Approaches to International Law', *American Journal of International Law*, vol. 85, pp. 613-614.

Colley, Peter, 2003. *CFMEU Mineworkers: Partner or Production Factor*, Speech Presented at the Sustainable Development Conference, Brisbane, Australia, November 2003, available at http://www.cfmeu.asn.au/mining-energy/policy/Colley_MCA2003a.pdf [28 June 2004] :10.

Global Witness, 1999. *A Crude Awakening*, at http://www.globalwitness.org, The Conflict Diamond Campaign.

Hall, Brigette, 2003. *The WIMNet Diversity Questionnaire – Summary of Responses to Date*, available at www.ausimm.com.au/women/women/asp.

Hartley, L., Creed, H., Gilroy, P. and Todd, D., 2004. *Extended Working Hours Review*, Final report to the West Australian Government, Perth.

Higgins, Rosalyn, 1994. *Problems and Processes: International Law and How We Use It*, Clarendon Press, Oxford, p. 19.

Horsley, Jessica, 2004. *Women in Mining – The Statistics*, on behalf of the Women In Mining Working Group for AusIMM available at www.aussimm.com/women/women.asp.

Human Rights Watch, 2002. *Women's Human Rights*, Available at: http://www.hrw.org/woman/index.php [accessed November 4, 2002].

Hunt, J., 1996. 'Situating Women's Development Needs Within the Human Rights Framework', in G. Moon (ed), *Making Her Rights a Reality: Women's Human Rights and Development*, Community Aid Abroad, Melbourne, pp. 25-26.

International Covenant on Civil and Political Rights (*ICCPR*) 1967, adopted 16 December 1966, entered into force 23 March 1976, G.A. Res. 2200A (XXI), UN Doc. A/6316, 999 UNTS 171, reprinted in 6 ILM 368. article 2.

International Covenant on Economic, Social and Cultural Rights (ICESCR) 1967, adopted 16 December 1966, entered into force 3 January 1976, G.A. Res. 2200A (XXI), UN Doc. A/6316 (1966), 993 UNTS 3, reprinted in 6 ILM 360.

International Labour Organization (ILO), 1958. *Convention 111: Discrimination (Employment and Occupation) Convention*, article 1.

International Labour Organization (ILO), 1981. *Convention 156: Workers with Family Responsibilities*, article 3.1.

International Labour Organization (ILO), 1998. *Declaration on Fundamental Principles and Rights at Work*, article 2.

International Labour Organization (ILO), 1998. *Declaration on Fundamental Principles and Rights at Work*, article 2 (d).

International Labour Organization (ILO), 1999. *Social and Labour Issues in Small Scale Mines*, International Labour Office, Geneva, available at http://www.ilo.org/public/english/dialogue/sector/techmeet/tmssm99/tmssmr.htm [accessed 28 June 2004].

International Labour Organization (ILO), 2003. *Report of the Director-General: Working Out of Poverty*, ILO, Geneva, pp. 95-96.

Kodz, J., Davis, S., Lain, D., Strebler, M., Rick, J., Bates, P., Cummings, J., Meager, N., 2003. *Working Long Hours: A Review of the Literature*, United Kingdom Department of Trade and Industry, Employment Relations Series no. 16, available at http://www.dti.gov.uk/er/emar/errs16vol1.pdf.

Langman, J., 2003. *Investing in Destruction: The Impacts of a WTO Investment Agreement on Extractive Industries in Developing Countries*, Oxfam America and Make Trade Fair, June, p. 6.

Macdonald, I., 2004. *Mining Ombudsman Case Report: Vatukoula Gold Mine*, Oxfam Community Aid Abroad, available at www.oxfam.org.au/campaigns/mining.

Macdonald, I. and Ross, B., 2003. *Oxfam Community Aid Abroad Mining Ombudsman Annual Report*, Oxfam Community Aid Abroad, available at www.oxfam.org.au/campaigns/mining.

Oxfam International, 2004. *Trading Away Our Rights: Women Working in Global Supply Chains*, Make Trade Fair and Oxfam International.

Pattenden, Catherine, 1998. *Women and Mining*, A report to the 'Women and Mining' Taskforce: The Australasian Institute of Mining and Metallurgy, November, available at www.aussimm.com.au/women/women.asp.

RIMM, 2000. *II International Conference: International Network 'Women and Mining'*, 16 September 2000, Bolivia, available at www.iwmn.net.

Ross, M., 2001. *Extractive Sectors and the Poor*, Oxfam America, p. 7.

Schachter, Oscar, 1991. *International Law in Theory and Practice*, Martinus Nijhoff, Dordrecht, Netherlands, p. 85.

Seager, Joni, 2003. *The Atlas of Women: An Economic, Social and Political Survey*, The Women's Press, London.

UNCTAD, 2002. *Least Developed Countries Report 2002: Escaping the Poverty Trap*, United Nations, New York, p. 12.

UNDAW, 1995. *The Beijing Declaration and Platform for Action*, Fourth World Conference on Women, Beijing,

UNDAW, 1995. *The Beijing Declaration and Platform for Action Mission Statement*, Available at: http://www.un.org/womenwatch/daw/beijing/platform/plat.1.htm [accessed November 4, 2002].

UNDFW, 1979. *The International Convention on the Elimination of All Forms of Discrimination Against Women*, United National Development Fund for Women, available at: http://www.unifem.org [accessed November 4, 2002].

UNDP, 1993. *Human Development Report*, Oxford University Press, New York.

UNDP, 2000. *Human Development Report*, New York, Oxford: University Press.

UNDP, 2001. *Human Development Report*, New York, Oxford: University Press.

UNDP, 2004. *Human Development Report*, New York, Oxford: University Press.

UNESC CHR, Sub Commission on the Promotion and Protection of Human Rights, 2003. *55th Session Agenda*, item 4.

UNHCHR, 1993. 'Vienna Declaration and Programme of Action', United Nations World Conference on Human Rights, Vienna, *Human Rights Law Journal*, 352: 18.

United Nations Development Programme, 1998. *Integrating Human Rights with Sustainable Human Development. A UNDP Policy Document*, January.

United Nations, 2003. *Norms on Responsibilities of Transnational Corporations and other Business Enterprises with regard to Human Rights (with commentary)*, E/CN.4/Sub.2/2003/38/Rev.2 (2203).

United Nations, *Striking a Better Balance*, vol. 3, Annex 6: 143.

Universal Declaration of Human Rights, 1948. Adopted 10 December 1948, G.A. Res. 217A (III), UN Doc. A/810, at 71, article 2.

Utting, P., 2002. *Regulating Business via Multi-stakeholder Initiatives: A Preliminary Assessment*, UN Research Institute for Sustainable Development, available at www.unrisd.org.

World Bank, 2004. *Striking a Better Balance: The World Bank Group and Extractive Industries*, vol. 1: Ch. 2, available at www.eireview.org.

Chapter 16

Roti do, ya goli do! (Give us bread, or give us bullets!): Stories of Struggles of Women Workers in Bhowra Colliery, India

Lindsay Barnes

Introduction

I first visited the mines of the Jharia coalfield in 1983. A few months before that, I had come from the UK to Delhi to study, and had been invited by some trade union activists to visit Dhanbad and the surrounding mines. I was naïve, young and ready for adventure. Dhanbad was well known throughout India for its mafia, its violence and its coalmines. The Jharia coalfield, which lies mostly within the district of Dhanbad, is the biggest source of coking coal in India. When I first visited the area, it was part of Bihar (now in the newly created state of Jharkhand), and is one of the oldest coalfields of India. Coal has been mined here since the late 19th century.

And it shows. The chaos, devastation – total environmental destruction – greets you as soon you move out of Dhanbad town. If there is hell on earth this must surely be it. Open cast craters, mountains of debris, stagnant ponds of filthy water, abandoned gear heads from decades past, all sitting atop a fire burning away below ground. The underground coal seams have been burning away since the 1920s.

I was taken to visit the coalmines by male activists of the Bihar Colliery *Kamgar*[1] Union (BCKU), a left wing militant trade union of coal miners. On reaching each colliery town I would be quickly introduced to a group of women workers and left to fend for myself. With broken Hindi, and no one around to translate, we managed rather well. They showed me around the coalmines, explained their work, and introduced me to their bosses and comrades.

The greatest impression and impact was, in retrospect, neither the filth and chaos, nor the violence of the mafia, but the presence of women miners in the colliery settlements. They were everywhere. In the bazaar drinking tea whilst

[1] *Kamgar* in Hindi means employee.

chatting with their friends, buying provisions, returning from work with baskets on top of their heads. The bazaar area was not the domain of men alone. These women had carved a space for themselves. They moved around in the bazaar with their heads held high, with a confidence I had not witnessed elsewhere in India till then. I had experienced life in Delhi, and had visited a few cities of northern India. Travelling around on buses in the big metropolis had been pretty harrowing. Never once did I feel unsafe whilst moving around the mining towns.

It also struck me that these mining women also made their own decisions, without consulting their men folk. Men were there, but in the background. And they were usually the men of the union that I came into contact with through them, not their own husbands. They seemed to be remarkably free from household work, and domestic concerns. They also had money to spend. Many of them wore good quality *sari*s, with matching blouses, bangles and *bindi*s. Since the nationalization of the mines in the early 1970s the miners have earned relatively good wages compared to what they used to earn during the time the mines were privately owned, and compared to what manual workers in the area earn.

Many of the women I met were the most militant, and loyal, members of the union, the BCKU. They would narrate tale upon tale, with much relish, of their fights with musclemen, colliery managers and bosses. These were not women content to be on the sidelines of struggle, but were very much on the front line. Union leaders and activists also told me that women were the most militant section of the colliery workers. Colliery women had clearly struggled to carve out a place for themselves, smashing the stereotype of the submissive, oppressed Asian woman. In spite of the violent, dangerous and corrupt world they lived in, colliery women were far from cowed down. Or perhaps the atmosphere had forced the women to become tough.

It was with much sadness that I soon learnt that the coal industry was vigorously pursuing a policy to remove women from the workforce. During the days of private ownership women workers had largely been extremely badly paid, employed on a casual basis, and denied most legislative benefits. Ironically soon after women's employment was regularized – and wages and benefits increased – management began to seek ways to get rid of them.

The Voluntary Retirement Scheme (VRS) was introduced initially in 1976. Under this scheme the woman worker is 'encouraged' to retire and donate her job to her husband, son or son-in-law. Employment of women in the Jharia coalfield fell from 14,000 in 1975 to 9,000 in 1984 (DGMS, various dates). New employment opportunities for women in the coalmines were practically closed.

So, although mining women were the most militant and loyal union activists, they were losing the very basis of their strength – their jobs. Their union, the BCKU, though they never supported the VRS, never much opposed it either. Since the women were not being actually 'retrenched', and since the sons of the women were given a job, opposition was subdued. In an era of shrinking job opportunities, and rampant unemployment in the coalfield, even a militant union like BCKU did not oppose this scheme. Other unions in the coalmines vigorously supported the scheme, and demanded its renewal whenever it was temporarily withdrawn.

This apparent contradiction led me to begin asking many questions. These were women who had made history, but why could they not preserve their important role in the industry? So it was that part of the history I explored. How had these women made history – what sort of struggles did they initiate or take part in? What were the organizing strategies of the coal miners, and how did women shape, or how were they shaped, by them?

Methodology

To unearth the struggles of the women colliery workers the most useful source was the women themselves. The use of oral testimonies is of particular use in bringing to light women's experience of struggle. In India, *Stree Shakti Sanghatana* recorded the oral testimonies of women in the Telengana struggle. Since the struggle had been the subject of earlier research, they effectively challenged the view that women 'contributed' to the struggle by playing a 'supportive' role, and emphasized that women actually shaped the movement and were involved in the making of history (*Stree Shakti Sanghatana*, 1989).

In the case of women in the coalminers' struggles, however, there is no such documentation to critique. There have been no attempts to document the struggles of the coalminers, so there is no published, documented history of struggle to place women in, or to challenge. There are few written documents about the colliery workers of India, in spite of the number of workers employed and the importance of the industry. The organized struggles of the coalminers have received scant attention, and none at all has been paid to women's struggles within it.

The politically sensitive area in which the research was located also determined the methods used for gathering information. For in spite of the presence of tough and protective women workers, the coalfield is a violent and corrupt place. I was identified closely with the union, the BCKU, for practical and ideological considerations. In the highly charged atmosphere of the coalfield, there is no fence sitting. If you are not on the 'side' of the workers, then you are pro-management.

It would not have been possible to gain an insight into the lives of women colliery workers had I not spent many months living amongst them. These women are adept at telling lies, and acting. One group of women admitted to me that they told lies to another research scholar: 'Why should we tell all our tales of woe, what would she have done? Did she ever come here again? She probably wrote a book about us, and made a lot of money'.

In putting together a history of the coal miners' struggles in Bhowra colliery based on oral testimonies, I am aware of the weaknesses of the method. Women, particularly the elderly and retired women, were prone to lapses of memory, and were more preoccupied with the contemporary problems. They often devalued or marginalized their own importance in the struggle, a characteristic found by other researchers. 'How do we evaluate women's work, when not just the official records, but the women themselves continually make distinctions that devalue and marginalize their own importance?' (*Stree Shakti Sangatana*, 1989: 31).

I had met colliery women through the BCKU, and my subsequent fieldwork and research was facilitated by this union. The union arranged accommodation with a trade union activist's family in Bhowra colliery. I stayed there for the winter months from 1985 to 1988, collecting the testimonies of around a hundred women – mostly workers, some retired, and a few who were wives of coal mining men. It is from these testimonies that I attempt to put together the pieces to build up a coherent past of women's involvement in Bhowra's history.

Bhowra colliery

Bhowra is a colliery settlement 12 kilometres from Dhanbad, at the southern end of the coalbelt, bordered by the River Damodar. It is one of the earliest coalmines in Jharia coalfield. Coal began to be extracted here in 1901, when it was owned by the British company, Eastern Coal Company, and managed by Mackinnon and Mackenzie. Ownership passed into Indian hands in 1955, when Karam Chand Thapar and Company took over. In 1971 it was nationalized along with all the other coking coalmines, becoming part of the Bharat Coking Coal Limited (BCCL), a subsidiary of the public sector undertaking, Coal India Limited.

In the mid 1980s there were around 8,000 miners employed in Bhowra – which included the inclines, opencast mines and the coke plant. Of these, 800 were women. Since it is located on the fringe of the coalfield, it is less crowded than other mines. Up-country miners – from the poor districts of Gaya, Monghyr and Bilaspur of Uttar Pradesh and Bihar – live in arched-roofed, one room, and company-built quarters. Local miners live in the hamlets that the mining settlement has now swallowed up, or in the crowded bazaar areas.

The good old days

Although coal mining had started at the turn of the nineteenth century, there are few references to any organized struggles of the miners until 1946. During the first few decades of the industry the conditions prevailing in the mines made organization difficult. Each mine was different from the next, workers were separated by caste, regional background (language), and the workers were largely migratory.

By the mid 1940s the coalfield witnessed widespread discontent, with many spontaneous, militant struggles erupting in the different coalmines. During the war years the coalmines had seen a labour shortage. The ban on women working below ground, which had been introduced in 1929, was lifted in 1943, and then reintroduced in 1946. In 1947, 6,000 women lost their jobs in the coalmines of

Jharia. The rationing of essential goods, which the government had set up during the war to attract labour, was allowed to lapse now that labour shortage was a thing of the past. In 1947 the Chief Inspector of Mines noted 77 strikes in the collieries (CIM, various dates).

In Bhowra colliery one of the biggest strikes of its history was organized in 1948. It was from a retired woman miner, Girija Devi that I came to learn of this strike after staying in Bhowra for four months in the winter of 1985–1986. As was usual with conversations with retired workers, she was lamenting the 'good old days' when people were honest, the place was clean, and rice was cheap. Wages were low, she admitted, but the 'company' provided them with rice, pulses and oil. Then she broke into a smile and added, 'Once it was so rotten that we went on strike for three months and thirteen days':

> We went on so many demonstrations, with our red flag. We'd shout, 'We won't take your rotten gram!' or 'Roti do, ya goli do' (Give us bread or give us bullets). Then the *goonda*s (musclemen) of the 'company' would come to beat us up. The men would run away, but the women stayed put.
>
> Girija Devi, retired colliery worker

Women were not spared by the *goonda*s[2] because of their gender. Many old time trade union activists and leaders informed us that the women on the picket line – and women were always at the front – were beaten up and jailed during the strike:

> There was one woman, Dassin we'd call her, the wife of Tetar Das. She was the most militant of all the miners. She organized men as well as women. Women were experts at picketing, not like the men. Women and children would lie down in front of the pit – allowing no one in or out. The management couldn't bring in blacklegs, though they tried. There was even one woman with a two day old baby on the picket line.
>
> Chinmoy Mukherkee, Communist Party leader

Most of the miners who permanently resided at the colliery were poor, landless, lower caste migrants from the nearby districts of Gaya and Monghyr. Local villagers would come and go, combining coal mining with agriculture. Still they also remember and supported the strike in their own way. They usually left the mines for their village, but donated rice – the most essential of food items – to the strike fund. Budni Rajwarin, a villager who still comes to the colliery bazaar to sell vegetables, used to work in the colliery. She also recalls that the strike was about bad quality rations:

> I remember the strike, we villagers left for home, otherwise how could we survive so long? The leaders also had to hide from the police and the 'goondas', so we had to hide them. We told so many lies, I became an expert liar in those days.
>
> Budni Rajwarin, retired colliery worker

---

[2] Local hooligans.

Several retired women miners informed us that the main issue was rotten gram, which the 'company' tried to foist upon the workers. The male workers that I met, however, had a different slant. They invariably informed us that the issues were 'bonus' and wages.

In Bhowra, the 'Three month, Thirteen day' strike was crushed by co-opting a section of the workforce, with the help of labour contractors. Leaders were put into jail, and musclemen were brought in to terrorize the workers. Frequently miners remained 'loyal' to the contractor who had brought them to the mine in the first place. These contractors set up a rival union, with management support. One such labour contractor was Kedar Nath Singh, a Congress leader from Patna, who had the full support of the management. In fact he identified more with management than with workers. He claimed credit for breaking the strike of 1948:

> I captured the union in my own way, as I had 50 per cent of the workers under my direct control. For three years before the strike I had inducted a lot of people… I had made a place for myself amongst management.
>
> Kedar Nath Singh, Congress party leader, Bhowra

Women workers, however, revealed that it was not so easy to get rid of the 'Red flag' union of the Communist party, and replace it with the management-supported 'Tricolour flag' of the Colliery Mazdoor Sangh affiliated to the Congress party:

> The *tiranga jhanda*[3] *wallahs* (the supporters of the tricolour flag of the Colliery Mazdoor Sangh) would come with the police and the bosses, then there would be a fight with the supporters of the *lal jhanda* (the red flag of the Communist Party union). They came like this many times until finally the good leaders had to leave.
>
> Saro Bhuini, retired colliery worker

As a last gesture of frustration, an angry mob of workers beat up one of the contractors who also claimed credit for breaking the strike:

> I was on my way home, from *bustee* Number Ten to work, when an angry mob of around 150 workers pounced on me, they broke all my arms and legs… look, the scars are still there. I had to spend a month in hospital in Dhanbad. The manager and Agent of the colliery looked after me though. They sent food and milk and a car for my family to visit me every day.
>
> Budhu Mia, retired labour contractor, Bhowra

The colour of the party's flag, however, did not always matter to management. As far as possible, leaders would be bought off and co-opted, whatever the colour. In the coke plant of Bhowra, though there was a Communist party union, there was no labour trouble. The union leader was on the payroll of the company, and was

---

[3] *Tiranga* is tri-colour, conventionally the Nationalist Congress Party of India, *jhanda* is the flag.

supported by the owner and management (information gathered from the Crime Directories at the Jharia Police Station).

The *tiranga jhanda* in the coalfields

The 1950s and 1960s were mostly labour trouble-free years for the private mine owners to expand production and maximize their profits. The Congress Party affiliated union, the Colliery Mazdoor Sangh, consolidated its monopoly in the coalfield. It 'served mining interests well, in that it has helped to maintain industrial peace' (CIM, 1950). Wherever rival unions existed, the Colliery Mazdoor Sangh would seek to crush them in any way it could. In their endeavour they had the full support of the administration, the police and coalmine owners. Many of the union leaders were contractors, or became contractors, and were powerful in their own right. The hired musclemen – the *goonda*s – of the collieries were also involved in money-lending, lending to miners at extremely high rates of interest. Sporadic violent outbreaks took place, but these were mainly due to inter- and intra-gang warfare. The use of gangsters by colliery owners and trade union leaders became increasingly common in the 1950s and 1960s (Rothermund, 1983: 242-43).

In Bhowra colliery this pocket union did not call a single strike during these years. Bhowra's management, though paying the legal minimum wages and more or less complying with labour legislation, crushed any attempt at labour militancy. As in most collieries they had their musclemen to keep the miners subdued and compliant.

These were dark days for miners – men and women. The ownership of the mine passed from British to Indian owners, Karam Chand Thapar (KCT) in 1955. No new miners' houses were built, sanitation and even water supply to the settlements was neglected. Many villagers were displaced during the ownership of KCT and received scant compensation. Organized protest against displacement of the villagers, or to demand adequate compensation, did not take place. The recognized and dominant union, the CMS, was pro-management. Besides it was dominated by up-country, non-local employees and contractors. By the late 1960s only around 7 per cent of Bhowra's workforce were members of the union (Bannerjee, 1981: 124). The CMS leader in Bhowra, KN Singh, told me that they set about organising 'social activities' for miners, such as religious meetings and repairing temples and mosques.

For women these years were grim. Although women had been active and militant during the strikes and trade union struggles of the 1940s, the CMS had no need of them. When machines were introduced, the union accepted the retrenchment of women workers without any voice of protest. Women's employment in the Jharia coalfield fell from 26,000 in 1950 to less than 10,000 by the late 1960s (DGMS, 1950). The overall employment of women in the coalmines probably did not fall as much as these figures indicate, since management often reduced the 'permanent' workforce to avoid paying minimum wages and other benefits.

Instead the CMS involved itself in the setting up of 'welfare centres' for women and children. These centres were supposed to make better housewives for the coal miners. They were to be taught sewing and knitting, advized on family planning, hygiene and food preservation and other necessities for the 'ideal' mining family household. They were, however, largely unsuccessful (Sen Gupta, 1960: 149).

In Bhowra the permanent workforce was decreasing, though production was growing. Total employment fell from 5,000 to less than 4,000, and women's employment from 1,000 to less than 500 between 1968 and 1970. One explanation for this contradiction was that colliery owners resorted to retrenchment instead of paying the minimum wages as directed by the Wage Board in 1967. Many of these retrenched workers continued to work as contract or casual labourers. In 1971 in Bhowra North Colliery, which employed around half the total workforce of Bhowra, only 187 women workers had permanent jobs (DGMS 1972). These women were mostly older women employed on time rated jobs like track cleaning, shale picking, sweepers and other lighter jobs. Piece rated women workers, such as wagon loaders, were usually younger women, and employed by the various contractors.

These contract labourers, men and women, formed the bulk of the colliery workforce by the end of the 1960s and they were totally alienated from the union activities of the CMS. In fact the union leaders were often labour contractors anyway, obstructing the possibility of organized resistance. However the severity of oppression during these years resulted in the most violent phase of colliery workers' struggle from the late 1960s.

Militancy in the Jharia coalfield

1967 marked the beginning of the most militant decade of colliery workers' struggles in the history of the coalmines in India. It is useful to brief the reader about this movement, in order to understand the struggles of women in Bhowra during this time. The year 1967 marked a new era in the history of the coal miners' struggles in the Jharia coalfield. Many industries witnessed labour unrest at this time. The coalfields of West Bengal initially saw more protest, perhaps influenced by the changing political scenario and the (extreme left) radical Naxalbari movement in north Bengal. Some of this new radicalism and upheaval 'spread' from West Bengal to Bihar.

On the other hand the economic crisis had deepened. Shortage of food grains in the coalfield's hinterland became acute. Five blocks of the Dhanbad district were declared famine areas (Coalfield Times, various years).

The Wage Board had recommended a significant increase in wages and other monetary benefits for coal miners in 1967. However, in order to evade increased payment many of the mine owners started to retrench old workers, and replace them with new ones through contractors (Ghosh, 1992: 280-281).

Meanwhile Mr A K Roy, a militant communist trade union leader from Sindri, was elected as a Member of the Legislative Assembly in 1967. This had a remarkable effect on the labour movement in Dhanbad. He had defeated a famous, influential, leader-cum-contractor, much to the delight of local villagers and colliery workers. This boosted the morale of the much-exploited, and hitherto subdued, villagers and miners and led to the formation of the Bihar Colliery Kamgar Union in 1970.

By the late 1960s the exploitation and oppression of coalminers and the neighbouring villagers had reached its peak. Moneylenders became influential and rich, with most miners indebted to them. One contemporary journal estimated 40 per cent of the total wage bill went into the hands of the moneylenders (Coalfield Times, Dhanbad, 1973). Collieries hired their own army of *lathi*[4]-wielding strongmen from the west – Gangetic Bihar or further afield. A strong caucus of contractors, pocket trade union leaders and moneylenders had emerged. Violent clashes between the musclemen of the colliery owners and the miners became widespread, often under the leadership of the BCKU.

In the midst of violence and labour unrest the coking coalmines in the Jharia coalfield were nationalized in 1971. The mines under private ownership were ridden with corruption and mismanagement, and unscientific mining methods – known as 'slaughter mining' – were common. With the nationalization of the collieries, with more money around, corruption increased. Huge numbers of outsiders were inducted into the workforce, mostly higher caste men from the Gangetic plains of Bihar and Uttar Pradesh. Around 50,000 miners lost their jobs within one week of nationalization due to the influx of these outsiders (Sengupta, 1982: 15).

The so-called 'mafia' of the coalfield came into its own during nationalization. The 'mafia' started as union leaders/contractors/moneylenders (or all three), and had the support of the administration and police. They called their own relatives to get the now lucrative mining jobs, sold jobs to outsiders, and upgraded existing workers – for a price. Even the Minister for Steel and Mines at the time, Mohan Kumarmangalam, estimated not less than 30,000 people were inducted at the time of nationalization (Kumarmangalam, 1973). Meanwhile local miners, contract workers, and even many who had been permanent, found themselves jobless.

With the total sell out of the established, recognized trade unions, the frustration and exploitation of the colliery workers was growing. Following nationalization coal workers' expectations also increased. Violent struggles in many collieries of the Jharia coalfield took place, leading to the observation that mass movements by day and warfare at night is fairly common in Dhanbad (Pradeep and Das, 1979).

---

[4] Stick, commonly used for beating and to scare people with.

The movement involved both miners and villagers for the first time in the history of the coalmines. The increasingly conscious villagers turned to the Jharkhand Mukti Morcha (a political party) from the early 1970s. The nationalization of the coalmines in 1971 greatly boosted the JMM, for the massive influx of outsiders, grabbing jobs in the mines, added fuel to the simmering discontent and resentment.

The miners and villagers also had a common enemy in the *lathi*-wielding nexus of the 'contractor-cum-moneylender' combine. This caucus of exploiters oppressed local villagers and coalminers alike. During this decade, a 'worker-peasant' alliance became a reality in the Jharia coalfield with the JMM and the BCKU joint struggles.

This decade of militant struggle also saw the involvement of women colliery workers. After two decades of being marginalized in the workforce, exploited by the company musclemen and ignored by the established trade unions, they were frustrated, angry and threw themselves into the struggle. The fight to rid the collieries of the *goondas* and the moneylenders was an issue women workers felt particularly strong about.

Other issues fought by the coalminers and the BCKU during this decade were often central for women workers, too. Colliery women participated in struggles, alongside their men, insofar as the issues touched their lives.

The form of struggle was influenced, as well as facilitated, by the involvement of women workers. Women often remembered the fights, the clashes, the demonstrations and the confrontations as much as the issue at stake. Compromise and negotiation was not their domain. The *goonda*s, or the mafia for which Dhanbad was well known, had changed their 'modus operandi'. They no longer needed to directly squeeze the workers, but made a handsome living for themselves by cheating the government's coffers through the nationalized mining industry. After the militant movement of the coalminers died down in the 1980s and 1990s, women workers no longer played a leading role. The following words of one militant woman worker sums up the general feeling of many women I met through the BCKU:

> After the struggle life is much better. We got jobs in BCCL, so now we can feed our children properly. The red flag has liberated us from the trap of the moneylenders, and the rule of the *goonda*s is over. Now these men are scared of us! They are still here, but they have rectified themselves. I roam around until 11 in the night, nothing will happen. I have nothing to fear. Before the struggle I wouldn't have dared to do such a thing because of the *goonda*s… We had to fight for everything… Now there is no struggle, comrade. About what should we fight? There are no moneylenders or goondas.
>
> Sumitra Bhuini, Colliery worker, Bokaro-Jharia colliery, interviewed in 1994

Meanwhile in Bhowra...

In Bhowra the nationalization of the mines did not immediately benefit the women workers. The *chaprasi*s (guards) or *goonda*s appointed by the company during private ownership to control the workforce should now have found themselves jobless, but instead the influx of money boosted their position. They became 'leaders' of the pocket union, and charged workers rent for living in the company-built houses and extracted money from the workers on some pretext or other (based on interviews with many BCKU activists in Bhowra):

> Those leaders were corrupt to the core. They would try and extract money from us. And women! What to say? They would pick on village women mostly, those coming from across the river to sell vegetables. They would molest them, harass them – and much worse.
>
> Shanti Mallick, woman worker

Corruption in the colliery office was rampant too. More money meant more corruption – the stakes were now higher:

> The company increased our wages and then took it back from us. We have to pay for sick leave, for light duty, we have to pay the loading *babu* and the office *babu*...
>
> Parvati Devi, woman worker

It was at this time that one woman worker – a *kamin*, as she is known throughout the coalfield – narrated her story to a young employee of the mine, who wrote a poem on the basis of it (see Appendix 1), It reveals the frustration she felt after nationalization, for not only was she harassed by her bosses, the *hazri babu* (attendance clerk), but the leader *babu*, too.

Trouble surfaces at 4B quarry

The workers at 4B quarry in Bhowra in the early 1970s were mostly women. They had been made permanent two years after nationalization in 1973. There were around 1,000 workers there, and the contractor – who now had no official role in the company – charged two to three Rupees per month from each worker for his services.

The women also had to pay for the repair of their tools, as well as the services of the colliery staff. The staff would under-measure the amount of coal the women had cut and loaded, and the surplus would be made into another wage in the name of an absent worker and drawn by them.

The contractors and the colliery staff were on good terms with the CMS union leader, Ramlal Bhat. Though this nexus was established during the days of private ownership, it was much boosted by nationalization now that there was much more money around. But whilst this nexus tried to extract more and more from the workers, the workers too were changing. With nationalization their expectations also grew. A small group of workers in Bhowra began to try and establish a rival union, to challenge the corrupt, pocket-union of the CMS. This rival union was to be a branch of the Bihar Colliery Kamgar Union (BCKU), which had by then established itself in many collieries through violent and militant struggles against a powerful caucus of union leaders, moneylenders, contractors and police.

The CMS, however, considered Bhowra its own territory, and violently opposed any encroachment. The fights took place in both the villages in and around the colliery – the *bustee*s, and the one-roomed, company built lines of quarters for miners – the *dhowrahs*:

> The *tiranga jhanda* (tricolour flag of the CMS) violently opposed the *lal jhanda* (the red flag of the BCKU). If we hoisted the *lal jhanda* in any *bustee* or *dhowrah* it would lead to a fight. When we called some BCKU sympathizers to 4B, to help us oppose the tactics of the contractor, staff and union leaders, they weren't allowed in. The mining supervisor and his assistant threatened them with guns, and told them only workers were allowed.
>
> Shanti Mallick, woman worker

In order to deal with this violent opposition, the BCKU decided to organize a mass demonstration – outside the 4B quarry. Shanti Mallick, one of the most militant woman workers in the quarry at the time, organized all the workers from the quarry, and the wife of a colliery worker, Mrs Indira Ghosh – or *boudi* (sister-in-law, brother's wife, a common way to address married women in Bengal) as she is still known – organized the women of the *bustee*s and *dhowrah*s. The huge demonstration of 'thousands' of mostly women and children presented their list of grievances to the managers in the area office in Bhowra. Although the demonstration had been peaceful, 11 workers, including Shanti Mallick, were suspended. Ten days later she was told to pick up her identity card or I-Card from the colliery office, and that her services had been terminated:

> So we went along to see Roy *babu* (A. K. Roy, the leader of the BCKU) in Dhanbad. We asked what to do next. He told us to fight, and stay united in struggle. He taught us that if we are united, we will win the fight. So we fought, and 22 days later we all got our jobs back.
>
> Shanti Mallick, woman worker

The CMS tried to force workers on demonstrations organized by them, and tried to intimidate them, but the days of the CMS were numbered. The struggles in 4B helped oust the CMS from Bhowra. And women like Shanti Mallick and Indira Ghosh were crucial in the struggle.

Village women fight back

Women workers told us that village women were subjected to much worse harassment than they were by these outsiders or *rangdars* (bullies/musclemen) as they were known:

> The Biharis used to harass women who came from the villages to sell vegetables in the bazaar. They would pick on them because they were not organized, and didn't come in big groups. They didn't know to fight like we did.
>
> Shakuntala Mahatain, woman worker and local villager

Sometimes these women would be stopped and molested on their way to the bazaar, or in the evening on their way home from work. In the bazaar the Biharis or *rangdars* would snatch their wares and not pay for them. Soon after the victory in 4B quarry, these village women also started to raise their voices. The women began to protest and everyday arguments and small fights broke out in the bazaar.

This was around 1974–1975, when the 'Lalkhand-Jharkhand' alliance was in full swing. 'Lalkhand' was the 'red land' of the miners' communist-led BCKU, and the 'Jharkhand' was the movement for separate statehood of the oppressed and exploited tribals and poor villagers, under the banner of the Jharkhand Mukti Morcha. This was the height of the alliance between coal miners and villagers. So it was little wonder that the daily fights in the bazaar in Bhowra attracted the attention of the movement. BCKU's leader A. K. Roy and the JMM's leader B. B. Mahato both visited Bhowra at this time and urged village-worker unity to fight to common enemy – in this case the Bihari (outsider) *rangdar*s.

The issue of women's harassment brought many local men into the BCKU, which had hitherto been dominated by outsiders. One day these men took on the higher caste outsiders in the bazaar and thrashed them. Following this incident, harassment reduced.

Women's protests had been to establish the right to come and go unhindered by men, but these local men had not fought, however, for women's right to sell vegetables in the market, or the right to move around, but had fought because 'outsiders' were molesting 'their' women. It was a question of 'honour', not of women's rights. One of the BCKU union leaders even suggested that women should stay at home to stop such harassment, and men should instead come and sell the wares in the market. Till today, however, most of the villagers coming to Bhowra bazaar to sell vegetables are still women.

The right to work

Soon after the protests in 4B Quarry and Bhowra bazaar, the Emergency in India was declared. After the lifting of the Emergency in 1977, though, women's struggles were again launched. Although many workers had gained permanent employment in 1973, many – especially women labourers – were still working under contractors. Bharat Coking Coal Limited (BCCL), however, had no more need of women workers. Now they were well paid, and at long last began to receive other benefits such as maternity leave, the industry tried to remove them. The fight for the right to work by the women contract labourers was narrated by one of the women involved:

> We *kamin*s (women colliery workers) had gone to Koyla Bhawan (the head office of BCCL) with two of our union leaders. The bosses were sat in a big AC room, and tried to give us tea and biscuits. Tea and biscuits! Did we go for tea and biscuits! I told them, 'We want jobs, not tea and biscuits!' Then I smashed my fist on the table like this… and smashed all my bangles!
>
> Durgi Bourin, woman worker

The women ultimately did secure permanent jobs in the colliery, Durgi firmly believing that her convincing arguments and the strength of their union were the main reasons. What this incident brings out, though, is that now the bosses had to listen to women workers like Durgi, and even had to offer them tea and biscuits – something unthinkable during private ownership of the mines. And, moreover, women like Durgi Bourin were sufficiently confident and articulate to argue their own case. The success of their action has greatly boosted the confidence of the women workers that were involved.

The frequent transfer of women workers was a strategy used by management to harass them, thereby 'encouraging' them to opt for early retirement. They had to fight to get transferred back to their 'home' colliery. The following narrative again shows us how the confidence of women workers was built:

> We had gone to the GM's (General Manager's) office in that colliery – most of us were *kamin*s and a few men. We hadn't thought about what to say, or who would lead. We *kamin*s expected the men to speak, but they didn't. We stood there in silence for a few minutes, and then I realized that it was useless to depend upon the men, and so I argued the case for all of us. I thought to myself, 'This GM is a human is he not, and so am I'. So I explained how we had to set off at 4 in the morning, and get back home after dark. I told him how it was dangerous for us to be roaming around the coalfield at night. Anyway we got transferred back after a few days. After that, workers – both men and women – started coming to me for help, since I had learnt how to talk.
>
> Dhanpatia Devi, woman worker

The most militant section of women colliery workers has been the wagon loaders. This is the job of loading coal, in cane baskets on their heads, into railway wagons in the sidings at the colliery. This was, traditionally, a 'female' job. Their militancy was for several reasons – there are usually many women working at one place, they are all working under the same conditions, and receive the same sort of wages. By 1985, when I first visited Bhowra, there were 814 women workers, and 625 were wagon loaders.

This concentration of women in this activity facilitated their militancy, but it also left them vulnerable to mechanization. When management tried to introduce mechanized loading in the colliery in 1980, it was the women wagon loaders that fought to oppose it:

> We were around one hundred *kamin*s in the siding when they brought the pay loader. We *gheraoed* (surrounded) the machine and the operator – we wouldn't let them use it. The police with their *lathi*s were brought in, and they tried to threaten us *kamins*. I told the police, 'You make your living by wielding the *lathi*, and we make ours by loading wagons. Take your *lathi*s and leave us our wagon!'
>
> Basanti Roy, woman worker

The wagon loading *kamins* also fought for many other issues that were important to them as workers, and as women. Corruption, for example, is a major cause of the problems facing the coal mining industry. Few unions have managed to mobilize workers against this, and the administration is incapable. Corrupt leaders, managers and colliery staff have political clout, rendering them immune from legal action.

Most workers accept corruption as part of life. But when it gets too overwhelming, they retaliate. In Bhowra it was frequently the wagon loading *kamin*s that fought on this issue. No incident came to my knowledge of male workers taking up this cause. In the following narrative it was the growing extortion of the much-hated loading *babu* that caused the women to retaliate.

Three wagon loading women had come to see Durgi Bourin earlier that day. The loading *babu* had tried to extract two hundred rupees from each wagon loader at the siding:

> It's time we taught the 'loading' *babu* a lesson. He has made enough money from us – he has two jeeps now. He's a notorious chap. He burnt his wife for dowry! But he got away with it because he has contacts, you know, with big people. I'll organize all the *kamin*s, come in the morning to see the *tamasha* (showdown).
>
> Durgi Bourin, woman worker

The *kamins* also fought for the right to be treated with respect. They narrated one incident when a woman worker was admitted into the colliery hospital. Since most *kamin*s belong to the lowest caste in the Hindu hierarchy, they were often treated disrespectfully by the doctors, who were usually upper caste. One such doctor made crude comments to one woman worker and within hours more than a hundred angry *kamins gheraoed* (surrounded) him, garlanded him with sandals and demanded his transfer. The women succeeded in getting him sent away from their hospital.

In most of these small, day-to-day struggles and fights, the *kamins* organized spontaneously, without leadership from the union. If the union was needed, the leaders would usually be called in after the struggle had been launched. Often the managers would call them in to pacify the angry, 'unreasonable' women.

Kamins* and *rangdars

Women workers not only had to struggle against their bosses, contractors and musclemen, but some of their own union leaders too. The *kamins* may have been the most militant colliery workers in their union, but they often had to fight within it. By the late 1970s opportunistic elements had joined the now-strong union, and they were no longer the idealistic, revolutionary working class leaders of the earlier years. Money – too much of it – the *kamin*s informed me, was the cause of the corruption.

It was the behaviour and attitude of some of their union's leaders towards the women workers that alienated the *kamins* from the BCKU. It also signalled the beginning of the decline of the union, and the end of militancy:

> I had to move my house from the village to the *bazaar bustee* due to him (one of the union leaders). He would come in the night… When we fought against the Biharis, when they were harassing village women, he fought alongside us. Then he was a hero! Then we had the union's support, and all the village men. But when he did the same, did the union help? No, none of them! Men… they are all the same!
>
> Shanti Mallick, woman worker

The issue was highly problematic for the union. It did not fit into the 'worker versus boss' or 'local versus outsider' struggles – for both the exploiter and exploited in this case were workers and both were local.

The women also had a role to play in the making of such *rangdars* (musclemen). At times they admired their courage and strength, the dividing line between heroes and *rangdars* is fine. Durgi Bourin recounts the following story of one such (erstwhile) hero, who had been condemned by most *kamins* for sexually harassing many women:

> Look at the leaders now! All they do is sit around chit chatting. Now if Anando was alive, would he have wasted his time like this? No way! He was a man of action. He would shout the filthiest abuses at the manager, once he even threw a chair at him! The bosses were shit scared of him! That's what's needed nowadays.
>
> Durgi Bourin, woman worker

When a fellow worker commented that Anando only inspired fear and not respect, Durgi replied that the bosses will only concede demands if they are scared. Basanti Roy, another militant *kamin* informed us:

> See, we learnt about communism from A. K. Roy, but we learnt how to fight from Amitabh Bachan!

Even Indira Ghosh, who had organized the women in the struggle in 4B quarry, felt that the present union leader was lacking in masculinity, 'What sort of a man is he?'

Women workers had made their own history in the coalmines. They also helped to create the heroes or *rangdars* as they became. In this we also have a glimpse of why women are getting marginalized in the workforce, and are almost part of history. Union leaders were not all heroes or *rangdars*, far from it. Some of the most decent human beings are to be found amongst the union leadership. But the leaders are all men, and most of them are not from the same castes and communities that the working women are from. There is a distance between the leaders and the women they claim to represent. Some of the local union male activists looked down upon the crude, loud-mouthed and independent *kamins*. The women know it well. The following was told to us by Durgi Bourin during a program to commemorate the Bengali freedom fighter Subhash Chandra Bose. The *kamins* were all sat on the floor, whilst the Bengali middle class women, along with the union leaders, were seated on chairs at the front:

> See our leader, over there… looks like a real Bengali *babu*, doesn't he? But look at the women. With their starched saris, will they fight the police and the goondas? Of course not! Then he will come to me and ask us to mobilize the *kamin*s.
>
> Durgi Bourin, woman worker

Although women complained loudly and frequently about their union leaders, they felt that there were no alternatives available:

> The BCKU is like a family … and like every family there are black sheep, bad boys that go astray. Now should we disown our family because of them? If we have problems in our union here, we are to blame also, and we should try to sort them out. Even if we leave the union, those bad elements will not leave Bhowra. If we leave the BCKU, which union can we join? The Congress union is spineless; they are the ruling party so they can't fight for us. And the Janata Party? I once went on a demonstration organized by them, just to see how they behave. All of the men were Biharis; no women came along at all. They keep their own wives well hidden at home, how can they expect us to join their party, and stand shoulder-to-shoulder with them in their struggles?
>
> Champamoni Majhain, woman worker

Conclusion

These testimonies shed some light of the sort of struggles women workers had initiated, or had taken part in. It was these struggles, and the ongoing everyday struggles, that gave colliery women the relative freedom of movement they displayed. The struggles that women initiated were not only about wages and bonus, but encompassed a wide spectrum of issues, which were important to them as women, and as workers. In the 1940s they fought for better quality food rations, for the right to organize in the early 1970s, for the right to move about unhindered, for the right to be treated with respect. They often launched their struggles without male leaders and were more likely to be involved in direct action than negotiations.

Often the women were on the front line of struggle, on the picket lines or against the *lathi*-wielding police. But when negotiations with the management began, the male leaders stepped in.

The nature of colliery workers' struggles has changed. No longer do the workers fight pitched battles with the *goonda*s, the hired thugs of the coal mine owners. Union leaders negotiate settlements with management over the table, with both sides being overwhelmingly male. As far as women are concerned, they talk in the same language. The future in the coalmines of Jharia is, unfortunately, male.

But if the unions accept the demise of women workers, they also lose their most loyal and militant supporters, their cutting edge. If the nature of the workforce of the future looks gloomy, the trade unions that represent them are reflecting this change. The coal industry of Jharia is in a big mess, with the mines making loss upon loss. Massive World Bank loans are bailing the industry out. The trade unions, of all colours, are merely treading water. For how long will a loss-making, mismanaged and inefficient industry, as crucial to the nation as coal mining, be tolerated in these times of economic liberalization?

At colliery level the mines are starved of investment, with idle and broken machinery lying around. Corruption has grown in leaps and bounds. The world that the *kamin*s have to negotiate now is far more chaotic, complex and corrupt than ever before. The forces that women have to confront now no longer end at the GM's (General Manager's) bungalow, the *goonda* or the moneylender next door. Those forces are now faceless bureaucrats in Koyla Bhawan, or in Kolkata (Calcutta) or New Delhi or even in far away Washington, DC. These are not the fighting domains of Durgi Bourin or Shanti Mallick any more.

References

Bannerjee, S., 1981. *Impact of Industrialization on the Tribal Population of Jharia-Ranigunge Coalfield Areas*, Calcutta, p. 124.

Chief Inspector of Mines, various dates. *Annual Reports*, Chief Inspector of Mines, Dhanbad.

CIM, 1950. *Annual Report*, Chief Inspector of Mines, Dhanbad.

Coalfield Times, various years. Dhanbad.

DGMS, 1972. *Annual Report*, Director General Mines Safety, Dhanbad.

DGMS, various dates. *Annual Reports*, Director General Mines Safety, Dhanbad.

Ghosh, Ranjan, 1992. *A Study of the Labour Movement in the Jharia Coalfield, 1900–1977*, Unpublished PhD Thesis, Department of History, Calcutta University, pp. 280-281.

Kumarmangalam, S. M., 1973. *Coal Industry in India: Nationalisation and Tasks Ahead*, Oxford and IBH Publishing Co., New Delhi.

Pradeep, P. and Das, A. N., 1979. 'Organization of the Future? A Case Study of the Bihar Colliery Kamgar Union', *Human Futures*, Autumn.

Rothermund, Indira, 1983. 'Trade Unions and Trade Union Leadership in the Coalfield', in J. P. Neelsen (ed.), *Social Inequality and Political Structures*, Manohar Publications, New Delhi, pp. 242-243.

Sengupta, N. (ed.), 1982. *Fourth World Dynamics: Jharkhand*, Authors' Guild Publications, Delhi.

Sengupta, N., 1982. 'Background to the Jharkhand Question', in N.Sengupta (ed.) *Fourth World Dynamics: Jharkhand*, Authors' Guild Publications, Delhi.

Sen Gupta, Padmini, 1960. *Women Workers in India*, Asia Publishing House, Bombay, p. 149.

Stree Shakti Sangatana, 1989. *We were Making History*, Kali for Women, New Delhi.

Appendix 1

The 'Kamin's Lament'

In midwinter we rise at four in the morning,
To cross the river, to reach the quarry,
Seven miles we trek each day,
Returning exhausted before we sit down to eat,
Sleeping without a proper bed,
But relieved to have reached home at last.

In midwinter we rise at four in the morning,
In the darkness we leave our children sleeping,
We have to hurry, to reach the quarry,
If we are late then we will not get work.
Just a few minutes late and the 'hazri babu' shouts,
'Go away!' he orders. Or 'See me later,' he grins.
If we miss work for one day, 'Go back home!' he orders.
Or 'Bring me a chicken tomorrow!' he demands.
Sometimes our digging sticks get broken,
Cutting the earth or breaking the rocks,
The 'lohar babu' of 'Four Number Colony' repairs them,
For two rupees a week.
The 'sahibs', the 'sirdar', the 'munshi', the 'hazri babu' –
All of them make us weep.

At four in the evening our shift is over.
The 'leader babu' calls us to his house.
We carry coal for his stove,
We carry bricks and lime to build his house,
He gives us food, but we have to wash his dishes,

Bring his shopping from the market,
And so it gets late.
'Stay here overnight,' he tells.
'Where will you go at this late hour?'
'How will you go home now?
Why to take so much trouble?
You will have to return in the early morning anyway,
And if you are late you will lose your job.
How will you get your job back then?
What will you eat?
There are no more contractors.
Now you have a government job.
Soon you will get a permanent job.
You workers will run this colliery!
There are no more private companies.
So, come, come. Come and sit by me.'

If I refuse, his eyes turn red, his face turns black.
We give our bodies,
Our honour is robbed,
Our life is shattered.
But when the photos are taken, and the I cards are made,
Where are our names?
Nowhere to be found!
Will anyone listen to the lament of a *kamin*?

S. P. Thakur

Chapter 17

Globalization and Women's Work in the Mine Pits in East Kalimantan, Indonesia[1]

Kuntala Lahiri-Dutt

Introduction: Global mines, local lives

A 'new geopolitics of resources' through the imposition of a 'New World Order' (Howitt et al., 1996) is emerging as the post-industrial nations increasingly exhaust their mineral extraction and mining breaks 'new grounds' in developing countries[2]. Large scale mining is now a globalized industry spreading its tentacles into the remote parts of the developing countries. Mining has increasingly spread across international boundaries as many huge multinational organizations expand their operations into the developing countries where, instead of turning into enclaves as during the colonial era, they are setting off processes that have important gendered consequences. This large scale mining industry has come under severe criticism for its poor social and environmental performance in these countries, with several international institutions taking up reviews that examine questions of compliance and transparency. At the same time, some of the international bodies have called for gender equity and a greater empowerment of women in the mining communities and workforce[3]. This call is related to the increased focus on gender equity concerns in other natural resource management areas such as water, drawing attention to their potential for participatory and sustainable development of local

[1] I would like to acknowledge the support of all concerned in this project: the senior managers of the company, particularly Mr Evan Ball, the Managing Director, Mr Barry Schafferius, Manager Pit Bintang, Mr Harry Miarsono, General Manager, Community Empowerment and Sustainable Development, Mrs Retno Nartani, Manager, Mining Services, and employees such as Mr Piara Barus and of course the operators, trainers and supervisors for speaking with me.

[2] See Mines, Minerals and Sustainable Development, 2002 report with the same title.

[3] In Papua New Guinea, for example, the World Bank has run two international conferences on 'Women and Mining: Voices of Change' in 2003 and 2005.

communities in regions of mining operation[4]. Thus, gender is emerging as a major area of concern for the large scale extractive industry.

It has been emphasized that the introduction of large-scale mining affects women in more complex ways than is commonly understood. The gendered impacts of mining have been put into three broad categories, health and wellbeing, women's work and traditional roles, and gender inequalities in the economic benefits from mining activities (Hipwell, 2002: 11). Reporting from Andhra Pradesh, India, Ghosh (2004: 409) writes:

> The production plant and the housing colony were built on land acquired by the company from the local community. Extensive mining has ushered in myriad problems such as alienation of lands, loss of economic and livelihood opportunities, loss of forest cover accompanied with its diverse impacts, social and cultural changes due to a migrant population coming into the region, degradation in the physical environment due to pollution and contamination of air, dust, and water by the company's extraction and processing activities. However, even more important are the problems faced by *adivasi* women who are the original inhabitants of the village.

The context may be different but similar issues ring through in other studies. In her early ethnographic work on the political economy of development in a mining town in Indonesia, Robinson (1986: 12) says:

> The fundamental change in Soroako has been the loss of the village's most productive agricultural land to make way for the mining project. As a consequence, wage labour for the company has become the principal stable form of livelihood. However, a large proportion of village residents do not enjoy regular employment, and they have been reduced to a semi-proletariat, living by occasional waged work and a variety of activities in the informal sector… For all village residents, there have been critical changes arising from the general effects of capitalist penetration, through the development of new forms of social relations mediated by the capitalist marketplace.

How the multinational mining operations negotiate the local, culturally-grounded gender roles in the mine pits in least developed areas has, however, remained a more complex area. In her later (1988: 64) work, Robinson notes the gender aspects of these changes, that 'in the change from peasant agriculture to wage labour the women have been subject to a decline in their economic independence', but at the same time 'women have become more economically dependent on men, changes in cultural forms of the expression of gender have resulted in a decline in some of the restrictions on women's personal freedom which hitherto prevailed in the community'. The fluidity of gender interacts with global economic imperatives and changes continuously in the local mine pits. This chapter examines such a

---

[4] See International Council of Minerals and Metals (ICMM) 'toolkit' for Sustainable Community Development published in 2005.

context: a large coal mining operation[5] located in the midst of equatorial forests in Sangatta, East Kalimantan (formerly East Borneo), Indonesia. However, deviating from the two approaches of environmental justice and political economy of development, this chapter sees the globalized mine pits as gendered workplaces where the local concepts of gender roles conflict with imported values and conform or change in myriad ways. My approach in this paper has more to do with looking at women as actors in the globalized mine pits, examining whether at all or not, or how far such global processes that use women's labour have led to their empowerment.

This approach derives from the recent literature on women and their work under economic globalization. Initially, there was a polarization of positions with regard to changes in women's work with globalization: it was either 'benefits increased wellbeing' or 'increased poverty/victims of changes' for women workers. It is now generally agreed that the context is more complex. Whereas many new post-Fordist production systems[6] are known to have included women as cheap labour in the informal sector, they have also led to what is known as the simultaneous casualization and feminization of work (see Roy Chowdhury, 2004; Jhabvala and Sinha, 2002; Neetha, 2002). Discrimination against women can take place outside of the labour market such as in unequal access to formal education and skills training, restricting their employment opportunities (see Chakravarty's 2004 work in several manufacturing units in India). This limitation in employment can result in 'the lack of power to choose' (see Kabeer, 2001). However, Jhabvala and Sinha (2002: 2037) noted, 'In a way, the changes in economy have brought about a visibility for the woman worker which did not exist before. The term "feminization of labour" is now widely used and women are becoming more visible in areas of work which was traditionally barred to them.' The term means two changes in women's labour. Firstly, it refers to the substantial increase in the proportion of women in paid work over the last two decades[7]. Secondly it refers to the 'flexibilization of labour' – a fallout of the changing nature of employment itself (Kanji and Menon-Sen, 2001: 1).

In mining, a growing body of academic as well as activist literature has been looking at the negative impacts of neo-liberal global expansion of mining on women (Hinton, 2003; Bhanumathi, 2003; Ahmad, 2002; Tauli-Corpuz, 1997). My contention is that the gendered impacts of globalized mining capital do not entail only a simplistic black-white effect, but more complex than seeing women as either 'victims' or 'agents of change'. Whilst State-owned mining in India is excluding women, and informalizing women's work in the ASM almost throughout the entire Third World, it is also incorporating women, albeit at the

---

[5] PT Kaltim Prima Coal (KPC or the company henceforth).

[6] The nature of work itself has changed through deregulation of labour markets, fragmentation of production processes, de-industrialization and emergence of new areas of export specialization.

[7] Globally, about 70 per cent of women in the 20-54 age group are in the workforce.

lowest level, in other mine pits in other locations. The hiring of women for work in the mines is indeed new in Indonesia, and the number of women employees is rather small and has slightly declined in recent years. Yet, of the small numbers of women employees, the largest job group is that of 'operators' in the company hierarchy – women who are employed to work in the 'field', in the open cut mine pit to drive huge trucks. 'Mining operations' is a large employee category in the various divisions within the company, but the number of women in this category is much lower than those in other/office job areas. As against the office staff, the operators enter the company with lower education attainments and from lower social classes. The hiring of women in the mines is barred under a protective law in the country. Why, then, does the company hire women? Is it opportunistic? Is it part of any company/Government policy? Is hiring women in the mines part of the Company's commitment to the local community? Who are these women and what are their attitudes to jobs that are not yet quite usual? Several other mining companies in Indonesia have been hiring female operators to drive heavy vehicles for the past 12 years or so, although women are grossly under-represented in the mining workforce. These companies try to maintain international profiles in their operations and wage standards, although complaints against their functioning styles, especially the lack of attention to environmental care, are not uncommon. Some are now owned by Indonesian interests, their capital flows are primarily global in nature, and in most cases the markets are also located abroad. For Indonesian women, the large majority of whom are employed in uncertain and low-paid informal sector jobs or are self-employed in petty trades, women's employment in mining operations is thus problematic. Examining one company to seek answers to the above questions may, in my view, provide us with an indication as to how the processes of globalization affect women and their work in the formal mines. The answers have great relevance in determining whether at all, and how much, the employment of women in the globalized mining industry empowers women, and directing us towards finding ways towards creating the conditions for a greater empowerment of women in the mining industry.

Of the global trends discerned in the mining industry (Lahiri-Dutt, 2004), the ones relevant to our subject here are the rapid expansion of mining in the resource-rich, poor countries of Asia, Africa and Latin America where the imperatives of providing electric power, adding revenue to the governments exchequer or providing a better standard of living are driving the nations to invite global capital to exploit their mineral resources. Indonesia is now the fifth largest coal producer and the revenues from mineral resources provide additional impetus for economic growth for this country. Even in the developing countries, mining is becoming more mechanised, large scale and more high-tech, with a much larger ecological footprint, but certainly with better-paid jobs. International attention, especially on the environmental concerns, is bringing these multinational mining companies under the strict compliance laws and various review processes. How gender is interacting with the operation of global capital in mining is the question that I wish to explore. This chapter looks at the mine pit as a place of gendered work and it is not unusual that cultural and social attitudes towards what constitutes 'male' or 'female' jobs would result in occupational segregation, the extent of the problem depending on the specific nature of the job, and according to the country where the operation is located (Hunt, 2001). In most mining companies women tend to be concentrated in the feminized job areas such as office administration and teaching (horizontal occupational segregation), which are lower job categories than men (vertical occupational segregation). However, women continue to enter, in a small way, the non-traditional fields such as mining operations, and related engineering and technology based areas such as geology, computer science, and information communication technology (Hinton, 2003).

The previous chapters in this book have adequately discussed how organized mining continues to be seen as a risky and hard job, and excludes women in the name of protection. I would like to begin by noting that in the formal sector such as the one being discussed, the mining jobs in the fields are no longer as manually demanding as they used to be in the early days of industrial mining. In the mine studied, operations are by enormously large, state of the art, shovels and dumper trucks that are fitted with radio communication, global positioning systems, and also air-conditioned driver's compartments. The wages, even at the lowest – operators – scale, are much higher than those offered by local companies in areas other than mining. There cannot be any possible reason why women should not undertake these jobs that require training and some mechanical skill, rather than education.

The research process

KPC has employed women in its workforce since 1987, when it went operational. Originally a joint venture by the two mining giants BP and Rio Tinto, it was purchased by Bumi Resources, a fully Indonesian-owned company, in 2003. The Company remains international in the nature of its capital and its products are intended for the global market. Located in Sangatta, the capital of East Kutai Regency about 50 km north of the equatorial line, the lease covers 90,000 hectares of coastal land, the concession for which expires in 2021. Production has risen steadily since 1989, and the company has turned into one of the largest export collieries in the world, playing a catalytic role regionally through changes in the physical, social and cultural landscapes. The changes have been so significant as to have been the subject of two extensive research projects by Australian researchers. Kunanayagam in her 1995 work described the mining company, KPC, as the quintessential patron, building up a patron-client relationship with the local community, as it preceded the arrival of local government in the area. The more recent work was the one by Garza, looking at the changes in the local community brought by the presence of the mining company in the classical ethnographic model. Garza noted (2003:5):

> Sangatta is itself often the attractor, pulling in friends or relatives of those already settled there to work either for the mining company or one of its many contractors. Furthermore, Sangatta has drawn in an international element required for support of the mining company operations at KPC. Although residents may lack international travel experience, they can generally identify the headquarter countries of the various big name corporations providing local support to the mining company… Despite its relative remoteness from more 'central' locations in East Borneo, Sangatta is now linked not only locally with principal cities such as Balikpapan and Jakarta in Indonesia, but also … globally with such countries as Australia, Japan, Taiwan, Germany, and the Netherlands. These new connections bring an ironic international element to a formerly unknown – and only relatively recently named – locality in an inconspicuous corner of East Kalimantan.

KPC truly has brought the globalized world to the doorsteps of Sangatta. It has been the harbinger of many changes – through its impacts on the environment and through the creation of a dependent community (Kunanayagam and Young, 1998). Massive social changes are inevitable in such a situation, but how do gender roles change? Has the presence of the company or women's employment in the mine been empowering for them? Gender has been a vexed issue as the company seems to be caught in a conflict of values with regard to hiring women to work in the pits. KPC follows a masculine corporate culture as do many other mining companies, and the tokenism in 'celebrating' femininity is apparent in its following of a 'special day in the year' as 'Kartini Day'[8]. The small proportion of women in its workforce is indicative of this prevalent attitude as are the culturally and socially rooted perceptions that I came across among male managers.

This study is based upon a series of interviews and conversations with a vertical cross-section of 54 women (and men), combined with official data of past and present employment statistics. In selecting the participants, there was no conscious effort on my part to undertake 'scientific or statistically sound sampling'; however, the resultant interviewees can indeed be described as a 'sample' of women related to KPC who were interested in my research and were willing to participate in it. The interviews were conducted on location over two weeks in 2004. Besides trying to explore the attitudes of the employees on the issues in the workplace, I also spoke with the company managers to assess their attitudes to hiring women and men, hoping that these attitudes would informally provide clues with regard to the justifications of hiring of women in the mines. Whilst some phenomena (such as the percentage of females in various job categories) are more measurable than others[9], different research themes (such as the attitudes and perceptions of/towards the female operators) require qualitative research techniques (Whitfield and Strauss, 1998). Such qualitative methods for examining workplace issues have

[8] Kartini was a legendary woman of Indonesia, a famous scholar-poet.

[9] Although not directly used, Malhotra et al.'s 2002 work on measuring women's empowerment and Kabeer's influential papers on empowerment as such have been useful in the study.

focused on describing, decoding, translating and coming to terms with meanings of the employment relationship as seen by those involved (Whipp, 1998), and have had a profound impact on diverse enquiries in gender issues in workplace research. A valuable source book was Jacobs' (1995) *Gender Inequality at Work* with a collection of papers on sex segregation and gender inequality in the workplace.

Globalization and women's labour in the mines

The process of global connectivity in the contemporary world is leading to far-reaching changes in both mining and women's work. Globalization is about 'making things global'; the phenomenon of 'creating languages, services and products that apply not just to an individual neighbourhood or city or country, but to the whole world' (UNPAC, 2002). The process is also creating a 'global culture' – most important in mining, especially in developing countries such as Indonesia, where many new mines are being established with international capital. Women, usually at the bottom of the social hierarchy in Third World countries, are generally having to bear the worst brunt of the negative effects of the globalization process – increased workloads, lower pay, more insecure jobs, and more stress. In 1981, Elson and Pearson argued that women's 'nimble fingers' were the reason for their recruitment. This refers to that factory situation, 'whereby women are thought to be more dexterous than men and capable of performing manual assembly and finishing work at high speed' (Chant and McIlwaine, 1995: 149). The 'nimble fingers' argument of seeing certain jobs as 'suitable for women' turned them into cheap fodder for globalization in the new factories located in Asian countries. According to Elson and Pearson (1981) women were believed to have a patient disposition, and a capacity for high levels of concentration which equip them for carrying out monotonous, repetitive tasks. Women were also trained at home by their parents to do the work that the companies expected them to do, so they were the preferred work force in the new global factories. Clearly, these arguments would be at a loss to explain the hiring of women in the mining jobs. The theoretical position is elaborated by Saptari (2000: 149), who opposes these simplistic assumptions and argues that:

> Such an argument assumes too much of a fit between labour supply and labour demand. Although it is true that the division of labour by gender operates in every aspect of society, the way in which it operates varies by industry and is historically and geographically specific. The stereotype of 'nimble fingers', although prevalent, has failed to address various issues such as the differentiation among women themselves, the changing nature of job definitions and skills in the workplace, and the uneven development of industrialization in different Asian countries.

I use this as my starting point in examining the gender issues in the workplace of mining. There is no doubt that hiring the cheaper labour of women anywhere has to deal with enhancing the profits of an organization. This happens when female labour force is hired at a low wage. There are many variations of this argument found all over the developing world, especially in the current situation of globalization. By hiring women in the mine pits, KPC is overriding the country's laws[10], a situation that needs adequate justification in terms of economic profits. It must be noted that the company does not discriminate between women and men in terms of wages: women work and earn at par with men, despite not being well-represented in the higher levels of management. Within the company, women operators are considered less likely to take risks than their male counterparts. Could their safe behaviour be the only reason for hiring women in the mine pits, particularly in view of the importance of safety and the liability responsibility that globalized operation standards demand? Or does the company impose its own corporate gender values on Indonesian culture by providing an equal opportunity workplace?

The conventional thinking in Indonesia identifies two spheres – the public and the private – and idealizes a woman's domestic role within the family. Besides reproductive labour, women are not recognized as contributing any other forms of labour especially outside of their homes (Robinson and Raharjo, 2000). Therefore, if women work to support families on farms, in fisheries or in forestry, that work tends to be hidden. In fact, women form the mainstay of the informal sector working in petty trade and food processing, and being self-employed in unstable businesses[11]. At the same time, however, urban middle-class families have taken advantage of education and training opportunities for their daughters, who have entered the mainstream 'public' spheres of work. Increasingly, the typical working woman, portrayed as an essential part of the nation-building process, is not the working class woman but a professional, middle-class woman working in schools, offices and elsewhere in tertiary sector jobs (Sen and Stivens, 1998). A growing number of urban, middle-class professional women now refuse to be seen as the archetypal mother or *Ibu*, thus raising important questions of gender codes. As a result, the workplace has become more diverse and gender equality has been accepted as a formal state policy. Although agriculture remains the backbone of the economy, extractive industries are becoming important, contributing about 4 per cent to the Indonesian GDP in 2001.

The rules/laws guiding women's employment in Indonesia in general, and in the mines in particular, are 'protectionist' but conditions can be sidestepped. For example, Paragraph 3 on Employment (of women) Law or Act, Republic of Indonesia, Number 13 of 2003, states that women younger than 18 years are not allowed to work between 11 p.m. and 7 a.m., and employers are not allowed to employ pregnant women to work between these hours if the jobs are medically harmful for their condition. However, an employer can employ women workers

---

[10] Much of the protective legislation was initiated in the late 1940s, and included restrictions on women's night work, working in the mines, lifting heavy weights over 25 kg, 3 months' paid maternity leave and 2 days a month 'menstrual leave'.

[11] CIDA, Gender Fact Book, State Ministry of Women Empowerment, 2000.

between 11 p.m. and 7 a.m. if they provide nutritional food and drink, keep the workplace secured and protected from sexual misbehaviour, and provide transport to and from work. These provisions, it is noted, cannot be replaced with cash. These restrictions reveal the state's concern about women working during the night. However, the local government has the power to give written exemptions[12] to these restrictions and KPC has obtained permission to maintain uninterrupted shift work schedules on the basis that it complies with certain conditions for women workers. Currently, besides having to get special permission to employ women on night shifts from the *Bupati*[13] and from the parents/guardians of the women, KPC has to provide to both women and men operators separate toilets, separate restrooms, separate locker rooms, and separate prayer rooms in recognition of their gender needs. However, it appears that in the mine pits, toilets are not separate for women and men, neither are rest rooms adequate in number.

Women employees in KPC

Since it began operations in 1987, 309 women have entered the KPC workforce, of whom 161 have since left. The trend since 1992 shows that there has been a slight decrease in the proportion of women in the total workforce. This is not apparent to many top level managers, probably due to an invisibility of women as a whole to the company. For example, female operators were already working in KPC when the current Human Resources Manager joined the company in 1997. He remembered his early days and commented that there are 'more women operators now', although the data indicate that the numbers of women employees as a whole have been falling steadily. Women employees are generally concentrated in the lower ranks of the organizational hierarchy. Currently just under 5 per cent of the KPC workforce is female, all being Indonesian. By contrast, women make up nearly one third (31 per cent) of the Indonesian workforce. Just over half (56 per cent) of the women work in the administration section of the company, with the remainder in various operations related areas. Office work is commonly preferred by the more educated women, for reasons of comfort, status/inclination and perceived abilities, and also reflects the availability of better education to girls from middle class households.

Only 3 per cent of employees being female in 'mining operations' seems small but is significant in absolute terms because of the large number of all employees that this category comprises. Indeed it represents 34 per cent of the total female workforce. Of the total female workers, 56 per cent work in white collar jobs, but the figure of 34 per cent in mining operations division is significant because this group of women employees of KPC has the least education and seem to be from the lowest classes. Women in mining operations are probably greater in number

---

[12] This instrument is called stblt (*staad blat* – Dutch colonial law).

[13] District Chief.

than the average participation of women in Australian mining (7 per cent, as per MCA, 2005) or in Indian collieries.

The average age of women workers in KPC is normally distributed about a median age of 33 years and ranges from 22 years to 49 years. Clearly, the average woman employee is young (about 33 years), and married, drawing lower average salaries than males because they are concentrated in lower paid jobs. This is probably due in part to the fact that about one-third of them are operators in production-related work. Mining operations work requires that women are fit and young, and consequently women wishing to work as operators join the company at 18-20 years, whereas women undertaking clerical level work start at 22-24 years. The age of retirement is 50 years in the company whereas it is 54 in the country as a whole. Their average length of employment (LOE) with the Company is around 8-9 years but after about 4 years, there seems to be a crisis period which demands the resignations of many women, the reasons in many cases being a marriage following job, and childbirth following the marriage, leading to either resignation or 'unfit termination'.

Table 17.1 Female workers in different divisions, July 2004

Division	Total staff	No of females in category	Percentage females in category	Percentage of total females
Chief Financial Officer	28	9	32.1	6.4
Commercial	98	13	13.3	9.2
Development	139	16	11.5	11.3
External Affairs & Sustainable Dev	37	11	29.7	7.8
Health Safety and Environment	97	14	14.4	9.9
Human Resources	38	10	26.3	7.1
Managing Director	3	1	33.3	0.7
Marketing & Development	12	5	41.7	3.5
Mining Operations	1609	48	3	34
Mining Support	526	8	1.5	5.7
Processing & Infrastructure	275	6	2.2	4.3
TOTALS	2862	141	4.9	100

Of the 148 current female employees, 42 (28 per cent) are single with ages ranging from 22-44 years and with an average of 30 years. The married women (106) have a similar range (22-49 years) but on average are 5 years older. Of the total married women, 34 (32 per cent) of marriages took place after they joined the workforce and 59 (56 per cent) have husbands who are employed by the same company. I cannot, with certainty, comment on the degree of overlap between the two. However, in my view, the fact that 56 per cent of women are married within the company might be seen as a possible benefit in terms of savings towards providing perks such as accommodation. Of the 31 'operators – heavy equipment' including the two trainers, 26 (84 per cent) are married, and 12 (46 per cent) are married to KPC employees. This means that a large number of female operators are married but not significantly within the company.

All but two of the 148 women in KPC's employment were hired at one of five locations – Balikpapan, Jakarta, Surabaya (another east Kalimantan town), Sangatta (the original village that housed the mine) and Samarinda (the largest town cum port in east Kalimantan). Originally this was done because of the lack of a skilled workforce being locally available. Still, the biggest supplier of female workers was the most local, namely Sangatta. The negative correlation between the point of hire and length of employment – locally hired women tending to have longer periods of service – could possibly be because of post-marriage lack of child-support especially for those with maternal families being located at far away places.

Quite a few issues come to the fore from a review of the numbers: the fact that women operators constitute a large proportion of overall female employment in the company, that many of the women are leaving the company after about 4 years of service, the attitudes of the operators towards their work, and above all attitudes of others towards the women truck operators and their work in the mine pits. The major gender issues for women include the conflicts of the notions of motherhood and family with work or *Peran ganda* (dual role or double shift); the lack of child care support at home forcing early resignation; the lack of access to informal networks of men; the need for appropriate mentoring; and an inability to navigate the organizational political maze. For female operators some unmet gender needs included future options such as promotions; and practical requirements such as the provision of an adequate number of Indonesian style toilets with better lighting and more water supply in the pits; two days per month of period leave without affecting bonus; the need to accept culturally rooted attitudes such as fear of the dark while on nightshifts; and some issues related to ergonomics in handling of the large machines. Overall, women tend to be excluded from the informal workplace networks, vital for career development but commonly dominated by men. The problem is compounded by many employers' assumption that women, unlike men, are not able to devote their full time and energy to paid work because of their family responsibilities or the assumption that women are emotional and not meant to be leaders. Consequently, women in KPC often do not have the opportunity to do the more prized jobs which would advance their careers, lack promotion opportunities and find it better to either change tack or leave the company.

Negotiating gender identities in the pits

I will now turn my attention to those women who are working in the field, as operators. Working in the mine pits is a new experience for both women and men in Indonesia: the newness of Sangatta, the strange landscape of the mine, and the intimidating machinery. However, for those women from remote villages from distant islands, it offers a completely new experiential world. When she first moved there, Martha felt 'very different'. The overall ambience of Sangatta was initially more urban and difficult for her to get used to: 'In Sulawesi, houses are along the road, but here one has to walk more, or use *Angkot* (public vehicle)'. Again, when she saw the big wheels of the dumper trucks, she was amazed. She noted how scared she felt initially about her job in the mine pit:

> When I drove the big equipment, I felt small by the open spaces all around. The size of the machines, especially the height of the trucks, made me feel scared. During night shifts, I was scared of ghosts. I could not avoid feeling uncanny about dangerous places in the pit.

Sometimes, sitting in the trucks, she felt lonely and bored and played music to avoid falling asleep. She also used the radio communication to chat with other girls working on the field – they talked about their family, especially at the time of loading, when they did not have to operate the trucks, she would chat a little with other girls. When it rains, the operations stop because the roads become slippery. 'Then I would chat with other girls.'

Many operators are also proud of their job and feel economically empowered. Hafsah said: 'When I compared this work with the jobs I would have had to do back there in my village, it made me feel proud'. This was because the job itself was 'so different' but also because it made her proud of her income: 'I can now make choices of clothes to buy, of sending money to my parents back home in the village, of sending my daughter to English school'. She could not hope to receive that sort of money in any village job. She said, 'But, again to be honest the salary should've been higher than what we actually get in view of the risks the operators face'. These risks do not only include the possibility of accidents and physical discomfort but also the risks of feeling depressed and lonely sitting in the operator truck alone. About halfway into the eight-hour shift, the job begins to get boring. This is also the time when the supervisor or the 'ledihan'[14] would push the operators to work harder. Although it is commonly believed that there has been not

---

[14] 'Ledihan' sounds rather apt and nice, but it is actually a corrupted form of 'leading hand' meaning the foremen in supervisory positions.

a single accident involving women, this lady narrated a story of one of her female friends hitting the electric pole on the roadside and toppling over. She could also vaguely recollect one more case during the nightshift when a female operator's truck toppled over, fracturing her arm.

An interesting question is how women's work in the mines as operators is seen by the staff wives. This is particularly so as identities of a 'traditional' and homebound woman whose main job is to look after children and the household, and that of a woman working in the pits alongside men are at two extremes of gender roles. This conflict became quite apparent in my conversations with the members of Periska, the Ladies' Club. The attitudes of Ladies' Club members to female operators reflect their own ambivalent gender role perceptions. For example, when asked about what they feel about women driving heavy trucks and other machines, many of them responded that 'there is no problem in that'. However, when I asked if there should be many more women doing such work, they were unsure. Similarly, when we talked about combining work and home, and working night shifts, a member noted that 'doing such double shift is very hard, nearly impossible'.

This critical question of conflicting gender identities remains unresolved for all women, but is more acute among the operators than the office workers. As one woman operator pointed out, a woman working in the mine does not want to stand out by being too different. In view of their small numbers and employment uncertainties, they feel that if they demanded too many facilities as women, the decision-makers would say, 'If you cannot work like us, then do not try to work here in the mines'. An operator said that she does not 'feel special'. When she goes back home, she just behaves like any other housewife, so that the neighbours see that she is just 'like them', nothing different. Another operator remarked that her girlfriends laud her as 'working in the mines is a man's job', giving her the proud feeling that she might indeed be different and special.

Although a female operator said that the rights and obligations set by the company do make life better for her, it is evident from other women's comments that women are not assured of career prospects as the company offers only a limited opportunity for their promotion into supervisory positions such as trainer. Several other operators too hoped for promotions into office jobs which are perceived as more comfortable and more feminine. This career stagnation in the pit jobs seems to be a problem for all women in the operations division. There is only one woman manager in a senior position in the mining operations area in KPC, and according to her, it took her much longer to reach this position than her male colleagues. Other women qualified with mining engineering degrees have switched over to purchase, marketing or environmental divisions because of the lack of promotions.

'Unfitness' again is a contested area as many women, unable to negotiate the conflicting gender identities, choose to leave the job as unfit. K is 31, married, and claims that she still suffers from lower back pain after 11 years of working as an operator. She joined the company with a school degree. At 20, she claims that she was fit enough for the job of an operator in pit Bintang. Nuraini feels that the job in the mine became difficult for her once she had a child. When she got pregnant, she changed her job assignment, but when the child was born, she had to get one of her sisters from back home to provide her with support in looking after the baby. This went on for two years, after which the sister went back to look after the mother. For her, doing both jobs, at home and in the mine, was fairly difficult, tiring and boring. Sometimes after returning from the night shift, all she wanted was to fall asleep, but the child would cry and wouldn't let her sleep. This increased her frustration and did not allow her any peace of mind. She would feel tired all the time and hoped to catch up with sleep all the time. Hence she was already exhausted with the job when she was found to be unfit for work and terminated. She is all praise for her husband, because unlike many others, her husband does not drink or beat her, and also helps with cooking and the household chores. Clearly, the job did not lead to dramatic changes in gender roles within the family, although the long term effects on empowerment cannot be denied. Nuraini told me that earning the money had brought about many changes in her life. For example, it has taught her how to handle and take care of money. She took time to describe in detail how she now looks after her family's finances including the savings, investments and expenditures for the long run, not just the day to day running of the household. She feels that working outside of home has indeed changed her a lot, especially because her work was in a male domain such as mining. This, she emphasized, was possible because of her earning a reasonably good wage. She is no longer the little Sulawesi girl who came to Sangatta to live with her relatives, but is a person aware of the world outside her home. She is Muslim, but took the decision not to cover her head. One way of finding out about one's perception of a type of work is her possibility of sending the children into it. When I asked her whether she would send her daughter to work in the mine, she was not certain. She still thinks that this is a hard and risky job, and would prefer to give education to her daughter so that she may work in an air-conditioned office. At the same time, she added that if the girl really wishes to work in the mine, although it is a very hard decision, it will be her choice in life.

Why have women in the mine pits?

Let me now deal with the questions that I began with. It is still a novelty to have women to work in the mine pits, although the reasons are not yet quite clear. To seek these reasons, we might try and get a flavour of the interviews expressing different opinions. Joe, one of the expatriate managers I met in Balikpapan commented after hearing that I was researching women operators: 'the presence of women settles the men down – reduces their bravado'. When I asked him to elaborate on his meaning of the term 'bravado', he was unclear. According to him, operators in general tend to take more risks whereas women do not. The presence of women in the workplace tends to reduce the risk-taking propensity of men. Joe also noted that the presence of women brings a politeness in the behaviour of men – something that according to him is rooted in Indonesian culture – swearing less in front of women. This is not exclusively the view of the expatriate managers. The senior Indonesian managers also believe that having women in the mines neutralizes the wild behaviour of some men. Whether this is specific to Indonesian culture or an expression of chivalry on the part of all men is not clear. The senior woman manager of operations feels that this is not particular to Indonesian men. According to her, 'expatriate men swear a lot, but they don't swear at women operators. When I am in a meeting, they stop automatically'. A senior expatriate manager observed: 'women are more careful in their jobs and as a result not one of them have had any accidents. They can also cope better with repetitive and tedious jobs, are easier to deal with, and tend to have a steadying impact on men. Above all, women do not take time off for prayers interrupting work schedules in the field, and cope well with colleagues at the workplace.' A senior Indonesian manager added that 'women create less trouble than men' clarifying that his impression has been built up during industrial disputes most of which came from men. A (male) supervisor noted that 'women are easier to give instructions to'. Another Indonesian manager commented that it is commonly perceived in the company that women operators are more careful – a woman driver had not had a collision so far. However, on the same breath, she added that the number of the operators is so low that it is hard to tell. 'But, from what we hear from the supervisors and superintendents, we get the impression that women operators are more careful and more efficient than men.'

An issue that needed to be explored was whether women operators are more productive then men or not. In certain areas of work, the unanimous answer is in the positive, although this is hard to prove with statistics. This is an important question because of the complications over 'menstruation leave'[15]. Women are more punctual than men, are more regular at work, more devoted to the work and absent for fewer days than their male colleagues. One of the Managers pointed out that the efficiency is noted only till the women operators get married. This is corroborated by the data on length of employment. As long as women operators are single, they are attentive to their jobs, but once they get married, their attendance drops, and, especially after they have children, they often have to quit.

The general consensus among KPC managers was that 'women operators were easier to handle'. I tried to probe these comments in more detail. According to the senior woman manager in the operations area, women operators are more meticulous and diligent. She invokes cultural background rather than nature as an

---

[15] The Indonesian government gives two days monthly paid leave to each working woman. As the 2003 changes in this provision makes clinical inspection mandatory before giving the days off to women, office workers generally do not take advantage of it. Operators feel that they are working in a field job and have the right to stay at home these two days. However, in KPC regular attendance is related to productivity which is rewarded by a huge amount of bonus that can be equivalent to up to six months' wages.

explanation: 'a woman in our tradition is thought to be careful. She learns early to be careful as parents are harder on the girl and this is etched on her mind'. This is, however, not reinforced by the comments of the one female trainer. Yosepina has disciplined both men and women in her job, teaching them how to operate trucks and introducing them to safety procedures. According to her, whilst it is not different to teach women and men, often women are in fact more embarrassed initially to reveal their quick learning capacities as they feel that they must follow the cultural norms of modesty. Again, a male operator, after a few years at work, tries to be 'brave' and does things that he was not instructed to do.

With regard to safe behaviour, three reasons were offered by various people I interviewed. An Indonesian manager reasoned that 'women operators being safer' has more to do with lifestyle than anything else. To work night shifts, one needs to have sufficient rest and food. Men are more mobile. On a night shift, they get the day off and spend time in the nearby town, seeking fun, coming to work tired. Women take sufficient rest at home and come to work fresh. This view was not supported by my discussions with many women who felt that combining night shift work with child care at home is extremely exhausting. The second reason for women being more careful, offered by several people such as the mine manager, was their inherent nature: 'Women are careful by nature and men are careless. Men have many things on their minds – they try to get over the present job quickly and go on to the next one. This is how they get into accidents'. The third reason that many women operators offered had to do with their different attitudes to work, a fact that women feel more proud of, more confident and hence take extra care in their job. This is evident from the comments of Nengseh, a local Kutai (a local ethnic group) girl with little education:

> When I came to work in the field, it was better than the idea I had about it. In my mind, I had imagined it to be a dirty place. In my imagination, I had also visualized the truck to be very hot, but I found that it was quite comfortable… I was not sure of my ability to drive such huge trucks and kept thinking, 'Will I be able to drive such gigantic trucks? I cannot even drive a car, how will I drive this huge truck?' It was bigger than anything I had ever seen before. When I was all by myself, I was delighted and very proud too to be sitting in the cabin of the truck by myself. All my fear had vanished, and a confidence overtook me. Holding the steering wheel of the truck, I felt a surge of power coming on to me.

Several Australian expatriates are in high managerial positions in KPC[16]. They are rather ambivalent towards women operators working for a mining company. The ambivalence arises because of the fact that expatriates are weary of imposing western values of gender equality on Indonesian traditional culture, but some of them, at the same time, prefer to see more women coming out of homes to take up

[16] The expatriates live in an isolated bubble of their own, and the lack of inter-racial understanding is not uncommon. The relationship of the expatriates with locals is described in the chapter 9 of Robinson's 1986 book.

non-traditional roles within an organization. Yet, one manager I met was critical of my enquiry itself, thinking I had come to impose my western views on the women of Indonesia: 'what is the need for women to work in the mines? It is not a fit place for women to work'. Also, there is a tendency for them to club all Indonesian women into a single category, thus avoiding acknowledgement of the complexities of class, social status, ethnicity, education, and personal backgrounds. The emphasis these expatriate managers tend to put on women operator's efficiency, creating a 'business case' for hiring women as operators, is unclear. Some male managers I interviewed hinged their observations on the operational benefits of hiring women. The overall attitudes to expatriates working in the mining companies in the country are mixed; they themselves are notorious for setting examples of loose morals. Difficult questions such as gender equity are not raised as a rule. The female operators, in their turn, are pleased with the expatriates they work with, noting that they are more patient and well-mannered than some local managers.

The reasons for hiring women as workers in the pits apparent from these interviews are several: making men behave/improving workplace conditions; easygoing and amenable to handling; more careful or safe behaviour and less risk-taking propensities; lack of aggressiveness in industrial disputes; punctuality at work and lack of absenteeism; being more docile and easier to handle; and being more productive. They reveal a combination of misogynic ideologies with sound business sense of retaining control over the labour force. Thus, what is women's work is constantly being redefined within the mine, women being identified for certain jobs exclusively on an *ad hoc* basis. The important issues that are arising out of women's work in the mine pits are not being adequately addressed and women too are choosing to keep silent rather than demand their rights for fear of being labelled as 'unsuitable' for what is being portrayed as men's work.

Conclusion

Working in the mine pits does not necessarily carry a negative connotation for the operators in KPC; rather, this work provides new identities and new senses of the self that they acquire in these masculine places. In KPC, women operators are at the bottom of the company's ladder. The attitudes to them and their non-traditional work, too, reveal conflicting ideologies. These gender ideologies reflect the social relations of gender and perpetuate the differential positions of women and men in mining work. Whereas economic differences among women result from the inequalities of class and ethnicity, the gendered social relations in mining are structured by the mode of production as in many other workplaces (according to an ILO report published in 2004). Within a mining company like KPC, women's share of managerial positions is extremely small or non-existent and their rates of career progress slow and uneven. Men in general are in the majority among managers, top executives and higher level professionals, whilst women are concentrated in lower level administrative positions. In offices, they are in 'soft' jobs such as receptionists, secretaries and office clerks, making women operators seem unusual. This ideology often affects the operators themselves who are inclined to feel that they are in a masculine job area. In KPC, both visible and invisible rules have been constructed around the 'male' norm which women sometimes find difficult to accommodate: supervisors and trainers do not see women as equal with men, women tend to have to work much harder than men to prove themselves, and sometimes they have to adopt 'male' working styles and attitudes[17]. It is always an issue that women are expected to behave 'like men' – being at the bottom of the pile, they cannot afford to be 'different'. Yet, when they go home, they are expected to fulfil their roles of a mother and wife. The perceptions that continue to prevail in the heads of women and men, in the absence of any definitive support to women from the company, have the potential to be entrenched as norms if not dealt with immediately.

Finally, I would like to offer a few thoughts regarding gender equity in the mining organizations. What does the hiring of women as operators in the globalized mine pits of a developing country like Indonesia mean for other countries in similar situations? If gender equity means cultivating the freedom of the individual woman as a person, then clearly the hiring of women has no developmental significance. I say this because the perception that women operators are more docile and easier to control has been the prime reason for hiring women. The fact that women are more productive tends to be ignored. If the globalized mine pits are to bring development to the remote corners of Borneo or elsewhere it would need to respect women's capacities and empower them. Whilst hiring women the companies would need to think of a clear gender strategy as part of their mine design and planning. Mining must be an empowering process for women in developing countries, and there is no reason why women cannot be offered the benefits as actors on their own, rather than as part of the community, as wives and *Ibu*s. There are many job areas within a mine operation besides truck operators – electrical goods rewiring is one, for example – in which some women are already working. These areas need to be explored for hiring women. If the company's mandate is to work in harmony with local communities, then it must make a priority of hiring local women. These are some simple and effective measures which have the potential to make mining development empowering for women.

[17] Many examples can be cited from KPC: although many women still wear the headscarf covering their heads, all operators are made to wear jeans and boots as a kind of uniform, the only female Mining Services Manager is not seen in the drinking bars haunted by her colleagues, and rumours about her 'abilities', 'emotional nature' and 'poor relations with husband' are rife.

References

Bose, Sharmistha, 2004. 'Positioning Women within the Environmental Justice Framework: A Case from the Mining Sector', *Gender, Technology and Development*, vol. 8, no. 3, pp. 407-412.

Canadian International Development Agency, 2001. *Gender Profile: Indonesia*, available at http://www.acdi-cida_ind.nsf/0/d05d652cd5fca2af85256.

Chakravarty, Deepita, 2004. 'Expansion of Markets and Women Workers: Case Study of Garment Manufacturing in India', *Economic and Political Weekly*, November 6, pp. 4910-4916.

Chant, S. and McIlwaine, C., 1995. *Women of a Lesser Cost: Female Labour, Foreign Exchange and Philippine Development*, Pluto Press, London.

Elson, D. and Pearson, R., 1981. 'The Subordination of Women and the Internationalization of Factory Production', in K. Young (ed.), *Of Marriage and the Market: Women's Subordination in International Perspective*, Zed Books, London.

Garza, James, 2004. *Yearning for the Community: Ethnography of a Settlement Adjacent to a Multinational Mining Company in East Borneo*, Doctoral Thesis submitted to Monash University, Clayton.

GenderLink, 2003. *Best Practice Guidelines for Creating a Culture of Gender Equality in the Private Sector*, Commission for Gender Equality, South Africa.

Hayes, Adrian C., 2000. *Poverty and Deprivation in Indonesia: An Assessment of Data Sources and their Use in Managing and Monitoring Development Assistance Projects and Programs*, AusAID, Canberra.

Hipwell, W., Mamen, K., Weitzner, V. and Whiteman, G., 2002. *Aboriginal People and Mining in Canada: Consultation, Participation and Prospects for Change*, Working Discussion Paper Prepared for the North-South Institute.

Hinton, Jennifer, 2003. *Women in Mining: Voices for Change*, Summary Report of Conference in Madang, Papua New Guinea, August 6-10.

Howitt, R., Connell, J. and Hirsch, P., 1996. *Resources, Nations and Indigenous Peoples: Case Studies from Australasia, Melanesia and Southeast Asia*, Oxford University Press, Melbourne.

Hull, Terence H., 2001. *Counting for Democracy: Challenges in the Development of National Statistical Systems in a Decentralised Indonesian State*, AusAID, Canberra.

Hunt, Juliet, 2001. *Institutionalizing Gender Equality Commitments in Development Organisations and Programs*, Gender at Work website, Practical Tools section.

International Labour Office, 2004. *Breaking through the Glass Ceiling: Women in Management*, International Labour Office, Geneva.

Jacobs, Jerry A., 1995. *Gender Inequality at Work*, Sage Publications, Thousand Oaks.

Jhabvala, R. and Sinha, S., 2002. 'Liberalisation and the Women Workers', *Economic and Political Weekly*, vol. 37, no. 21, pp. 2037-44.

Kabeer, Naila, 2001. *Bangladesh Women Workers and Labour Market Decisions: The Power to Choose*, Vistaar Publications, New Delhi.

Kanji, Nazneen and Menon-Sen, Kalyani, 2001. *What does the Feminisation of Labour Mean for Sustainable Livelihoods?*, International Institute of Environment and Development and Regional and International Networking Group, London.

Kunanayagam, R. and Young, K., 1998. 'Mining, Environmental Impact and Dependent Communities: The View from Below in East Kalimantan', in P. Hirsch and C. Warren (eds), *The Politics of Environment in Southeast Asia: Resources and Resistance*, Routledge, London.

Kunanayagam, R., 1995. *In the Shadow of the Company: Patronage and Status in a Mining Settlement*, Unpublished Masters Thesis, Monash University, Clayton.

Lenz, Ilse, 2003. 'Globalization, Gender and Work: Perspectives on Global Regulation', *Review of Policy Research*, vol. 20, issue 1, pp. 21-43.

Malhotra, Anju, Schuler, S. R. and Boender, Carol, 2002. *Measuring Women's Empowerment as a Variable in International Development*, World Bank, Washington DC.

McMahon, Gary, 1998. *Mining and the Community: Results of the Quito Conference*, The World Bank, Washington DC.

Minerals Council of Australia, 2005. *Gender Participation Data for the Australian Mining Industries*, Canberra.

Neetha, N., 2002. 'Flexible Production, Feminisation and Disorganisation: Evidence from Tirupur Knitwear Industry', *Economic and Political Weekly*, vol. 37, no. 21, pp. 2045-57.

Robinson, Kathryn M., 1986. *Stepchildren of Progress: The Political Economy of Development in an Indonesian Mining Town*, State University of New York Press, New York.

Robinson, Kathryn M., 1988. 'What Kind of Freedom is Cutting your Hair?', in Glen Chandler, Norma Sullivan and Jan Branson (eds), *Development and Displacement: Women in Southeast Asia*, Centre of Southeast Asian Studies, Monash University, Melbourne, pp. 63-78.

Robinson, Kathryn M. and Bessell, Sharon (eds), 2002. *Women in Indonesia: Gender, Equity and Development*, Institute of Southeast Asian Studies, Singapore.

Robinson, Kathryn M. and Raharjo, Yulfita, 2000. *Population and Development in Eastern Indonesia: Women and Development*, AusAID, Canberra.

Roy Chowdhury, Supriya, 2004. 'Globalisation and Labour', *Economic and Political Weekly*, vol. 39, no. 1, pp. 105-109.

Saptari, Ratna, 2000. 'Production Processes and the Gendering of Industrial Work in Asia', *Asian Studies Review*, vol. 24, no. 2, pp. 147-59.

Sen, Krishna and Stivens, Maila (eds), 1998. *Gender and Power in Affluent Asia*, Routledge, London and New York.

Tandjung, I. R. and Sutomo, Haryoto, 1998. *Updates Tables for People, Land and Sea: Development Challenges in Eastern Indonesia*, AusAID, Canberra.

Tauli-Corpuz, Victoria, 1997. 'The Globalisation of Mining and its Impact and Challenges for Women', Proceedings of the *International Conference on Women and Mining*, Third World Network, Baguio City.

United Nations Platform for Action Committee (UNPAC), 2002. *Women and the Economy*, Manotoba.

Whitfield, Keith and Strauss, George, 1998. *Researching the World of Work: Strategies and Methods in Studying Industrial Relations*, Cornell University Press, Ithaca and London.

Index

For Product Safety Concerns and Information please contact our EU
representative GPSR@taylorandfrancis.com
Taylor & Francis Verlag GmbH, Kaufingerstraße 24, 80331 München, Germany

www.ingramcontent.com/pod-product-compliance
Ingram Content Group UK Ltd.
Pitfield, Milton Keynes, MK11 3LW, UK
UKHW020023200926
475702UK00014B/2026
* 9 7 8 1 1 3 8 2 6 4 4 0 3 *